Statistics in Civil Engineering

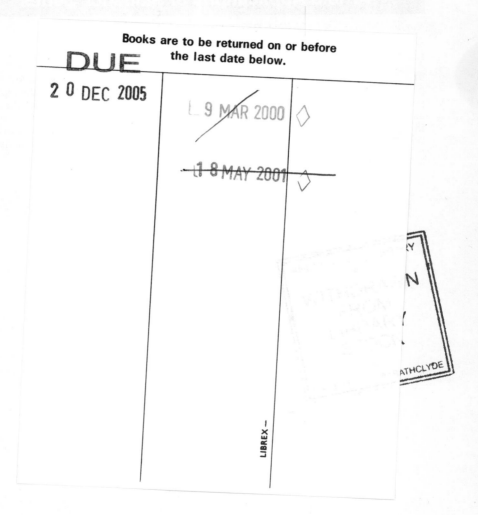

Arnold Applications of Statistics Series

Series Editor: **BRIAN EVERITT**
Department of Biostatistics and Computing, Institute of Psychiatry,
London, UK

This series offers titles which cover the statistical methodology most relevant to particular subject matters. Readers will be assumed to have a basic grasp of the topics covered in most general introductory statistics courses and texts, thus enabling the authors of the books in the series to concentrate on those techniques of most importance in the discipline under discussion. Although not introductory, most publications in the series are applied rather than highly technical, and all contain many detailed examples.

Statistics in Civil Engineering

Andrew V. Metcalfe

Department of Engineering Mathematics
University of Newcastle upon Tyne, UK

A member of the Hodder Headline Group
LONDON • NEW YORK • SYDNEY • AUCKLAND

Copublished in North, Central and South America
by John Wiley & Sons Inc.
New York • Toronto

First published in Great Britain 1997 by Arnold,
a member of the Hodder Headline Group,
338 Euston Road, London, NW1 3BH
http://www.arnoldpublishers.com

Copublished in North, Central and South America by
John Wiley & Sons, Inc., 605 Third Avenue,
New York, NY 10158–0012

© 1997 Andrew V. Metcalfe

British Library Cataloguing in Publication Data
A catalogue record for this book is available from the British Library

Library of Congress Cataloging-in-Publication Data
A catalog record for this book is available from the Library of Congress

ISBN: 0 340 67660 4
ISBN: 0 471 19484 0 (Wiley)

Publisher: Nicki Dennis
Production Editor: Julie Delf
Production Controller: Sarah Kett
Cover design: M2

Typeset in 10/11 Times by MCS Limited, Salisbury, Wiltshire
Printed and Bound in Great Britain by J W Arrowsmith Ltd., Bristol

Contents

Preface   viii

Notation   x

1. Statistics in Civil Engineering   **1**
    1.1   Introduction   1
    1.2   Computer simulation   1
    1.3   Quality assurance and reliability   2
    1.4   Extreme value distributions   2
    1.5   Relationships between variables   3
    1.6   Bayesian analysis   4
    1.7   Time series   4
    1.8   Spectral analysis and dynamic systems   5
    1.9   Design of experiments   5
    1.10   Computing   6
    1.11   Using this book   6

2. Probability Distributions and Monte Carlo Simulation   **7**
    2.1   Introduction   7
    2.2   Stochastic (Monte Carlo) simulation   7
    2.3   Applications of discrete distributions   14
    2.4   Applications of continuous distributions   19
    2.5   Summary   34
    Exercises   36

3. Quality Assurance and Reliability   **39**
    3.1   Components of variance: reducing variability in the strength of concrete   39
    3.2   Case study: polished stone value inter-laboratory trial   43
    3.3   Control and CUSUM charts   50
    3.4   Reliability   55
    3.4   Summary   75
    Exercises   76

4. Extreme Value and Related Distributions   **81**
    4.1   Introduction   81
    4.2   Extreme value type 1 distribution (Gumbel distribution)   81
    4.3   Order statistics and their uses   85
    4.4   Generalized extreme value distribution   92
    4.5   Generalized Pareto distribution and peaks over threshold (POT) analysis   94

4.6	Seasonality, trends and other explanatory variables	99
4.7	Wakeby distribution	102
4.8	Improving the precision of flood frequency analysis	103
4.9	Multivariate extreme value distributions	105
4.10	Summary	112
	Exercises	112
5.	**Relationships Between Variables**	**117**
5.1	Introduction	117
5.2	The standard multiple regression analysis	118
5.3	Multicollinearity	134
5.4	Recursive calculations	137
5.5	Generalized least squares	138
5.6	Intrinsically linear models	140
5.7	Non-linear regression	143
5.8	Measurement errors in the predictor variables	145
5.9	GLIM	148
5.10	Summary	156
	Exercises	157
6.	**Bayesian Methods**	**163**
6.1	An introduction to Bayesian methods	163
6.2	Asset management plans and the Bayes' linear estimator	166
6.3	Bayesian overview	172
6.4	Dynamic linear models	174
6.5	Summary	176
	Exercises	176
7.	**Modelling Variability in Time and Space**	**179**
7.1	Introduction	179
7.2	Markov chains	179
7.3	Time series	188
7.4	Stochastic models based on point processes	212
7.5	Spatial processes	217
7.6	Summary	231
	Exercises	231
8.	**Spectral Analysis and More on Dynamic Systems**	**237**
8.1	Introduction	237
8.2	The spectrum	237
8.3	Linear dynamic systems	247
8.4	Wave loading on offshore structures	260
8.5	Wavelets	262
8.6	Stochastic dynamic programming	264
8.7	Summary	266
	Exercises	267
9.	**Design of Experiments**	**269**
9.1	Introduction	269

9.2	Completely randomized design (CRD) with one factor	269
9.3	Randomized block design	274
9.4	Completely randomized design with two factors	276
9.5	Difference between a RBD design and a CRD design with two factors	281
9.6	2^k factorial designs	282
9.7	D-optimum designs	283
9.8	Summary	286
	Exercises	286

Appendices **289**

1.	Introduction to statistical methods	289
2.	Random number generation	319
3.	Maximum likelihood	321
4.	Bootstrap methods for confidence intervals	325
5.	Optimization algorithms	327
6.	Theoretical background to CUSUM charts	330
7.	Derivation of EV1 distribution	335
8.	Fitting the Wakeby distribution	337
9.	Expected value of the residual sum of squares	339
10.	Bayes' linear estimator	341
11.	Simulating from ARIMA processes	343
12.	Neyman-Scott streamflow model	345
13.	Fourier series and transforms	349
14.	Wavelet transforms	353
15.	Statistical tables	358
16.	Selected data tables	366

Glossary **385**

References **399**

Author Index **411**

Subject Index **415**

Preface

A common feature of civil engineering projects is that they are expected to operate under a wide range of environmental conditions, and that these conditions appear to include substantial random components. For example, subsea pipelines are at risk from iceberg keels, bridges have to withstand hurricanes, the possibility of earthquakes has to be allowed for in many areas of the world, and meteorite showers are a hazard for space stations. The physical environment is not the only source of uncertainty. Transport engineers, for example, have to design systems that can manage variable demand, now and in the future. Whatever their speciality may be, civil engineers are expected to make allowance for random environments in an accountable manner, using up-to-date statistical methods.

Nicki Dennis, of Arnold, suggested that there would be a market for a statistics text at an intermediate level within the community of civil engineering students and professional civil engineers. This book is a consequence of her suggestion, and I am grateful for her encouragement. I hope the text will help students progress from the material covered in a typical introductory statistics course to applications described in civil engineering, and related, journals. It has been my intention to motivate readers by emphasizing the applications of the methods described. I have tried to explain the mathematical basis of most of the techniques discussed, in an informal manner. A proper understanding of the basic mathematical principles is valuable for several reasons. The assumptions behind mathematical models are never satisfied exactly. We need to be able to decide whether or not a particular model is appropriate for a particular application, and a knowledge of the mathematics involved enables us to reach sound decisions. A second reason is that familiarity with the underlying mathematical concepts is a prerequisite for understanding more advanced texts and the research literature, as well as for modifying techniques or developing new ones.

I am grateful to students, friends and colleagues, from the Department of Civil Engineering at the University of Newcastle upon Tyne, for encouraging my interest in applications of statistics to civil engineering. There are too many such people to name individually, but some will be apparent from the references. However, I would like to thank the following people associated with Newcastle University for specific help with this book: James Bathhurst,

John Bull, Aidan Burton, Evelyn Kelly, Thomas King, Christine Lachecki, Dale Mellor, Benoit Parmentier and Tony Rhodes. Many individuals from the civil engineering industry, and research laboratories, have also contributed. They include: Peter Adamson (Sir William Halcrow and Partners), G. Boon (Sheffield City Museums), D. Cowdery (Environment Agency), Ken Day (Concrete Advice Pty Ltd), K.M. Gordon (Proudman Oceanographic Laboratory), Tony Greenfield (Greenfield Consultants), J. Kemp (ARC Northern), J.R. Meigh (Institute of Hydrology), David Peel (Yorkshire Water), Duncan Reed (Institute of Hydrology), Gordon Sutcliffe and M. Wilson (Environment Agency) and P. Woodworth (Proudman Oceanographic Laboratory). Any misinterpretations, or other errors are due to me. I am also grateful to Minitab Inc for giving me a copy of Release 11 of Minitab, which I have used extensively. I have appreciated the editorial and production services provided by Peter Waterhouse, Judith Jenkins, Julie Delf, and others at Arnold. Finally I wish to thank Evelyn Kelly for typing the text so quickly and accurately, and for the pleasing layout of the equations.

ANDREW METCALFE
Department of Engineering Mathematics 1997
http://www.ncl.ac.uk/~nengm/

Notation

Introduction

The following is a list of the most commonly occurring symbols, described in a general context. Some occur with different meanings (r in particular), but these should be clear from the application. There is no standard for statistical notation so I have tried to keep to the most common conventions. However, books rarely use identical notation and you do have to read the accompanying definitions.

Generation notation

1. In most places a, b and c are constants.
2. Variables are usually

   $$W, X, Y, Z \text{ or } w, x, y, z \text{ with } T, t \text{ for time}$$

   It is sometimes useful to distinguish explicitly a random variable from the value it takes in a specific instance. Then the upper-case letter is the random variable and the lower-case is the specific value, as in

   $$\Pr(X = x)$$

   Adhering to this rule can become rather awkward, and I have chosen to rely on the context in many places. For example, $\hat{\beta}$ represents both the estimator of a coefficient in a regression model and the estimate for a particular data set.
3. The following are usually integers:

   $$i, j, k, l, m, n$$

   They always are if they appear as subscripts. In time series and spectral analysis r, s, t are usually integers if they are subscripts.
4. e is the base of natural logarithms ($2.71828\ldots$). π is the area of circle with radius one ($3.14159\ldots$). Both notations for the exponential function are used:

   $$e^x, \qquad \exp(x)$$

5. i is the square root of -1.

6. Multiplication is $\times$ or juxtaposition: e.g. 2×5, and ab or $a \times b$. Natural logarithms are used unless base 10 is stated: $\ln(y) = x$ if and only if $y = e^x$. Logarithms base 10 are written $\log(\ )$.

7. Matrices are in bold italic, superscript ^{T} is transpose, det or $|\ |$ stands for the determinant. The latter also denotes absolute value or modulus.

8. $\{x_i\}$ is a sample of size n, and $x_{i:n}$ are the ordered data. That is,
$$x_{1:n} < x_{2:n} < \ldots x_{n:n}$$

9. $[a, b]$ is the interval on the real number line between a and b, i.e. x such that $a \leqslant x \leqslant b$, and (a, b) is the interval such that $a < x < b$

10. A sum of terms is written
$$\sum x_i = x_1 + \ldots + x_n$$
A product of terms is written
$$\prod x_i = x_1 \times \ldots \times x_n$$
If there is any doubt about the range of i it is added to $\sum$ or $\prod$.

11.
MOM	method of moments
MOME	method of moments estimation/estimators/estimates
PWM	probability weighted moments
ML	maximum likelihood
MLE	maximum likelihood estimation/estimators/estimates
LSE	least squares estimation/estimators/estimates
$\mathscr{L}$	likelihood function

12. $\{x_t\}$ represents a time series.

13. ARIMA $\{p, d, q)$ autoregressive integrated moving average model with parameters p, d and q.

14. ∇ is the backward difference operator and B is back shift operator. That is:
$$\nabla x_t = x_t - x_{t-1} \quad \text{and} \quad Bx_t = x_{t-1}$$

15. A dot is used to represent differentiation with respect to time. As in a general linear system
$$\dot{x} = Ax + Bu$$
$$y = Cx$$

Symbols

$[a, b]$	all real numbers x such that $a \leqslant x \leqslant b$
(a, b)	all real numbers x such that $a < x < b$
$\text{Bin}(n, p)$	binomial distribution: n trials, probability of success p
$\text{cov}(X, Y), \widehat{\text{cov}}$	population, sample covariance

E[ ]	expected value
CV, $\widehat{CV}$	population, sample coefficient of variation
E_i, r_i	errors about regression relationship and their estimates (residuals).
f_k	frequency of event referred to by k
$f(y \mid x)$	conditional pdf
$F(x)$, $f(x)(f_X(x))$	cdf (cumultive distribution function), pdf (probability distribution function) (subscript added if necessary)
$F(x, y)$, $f(x, y)$	bivariate cdf, pdf
F_{ν_1, ν_2}	F distribution with ν_1 and ν_2 degrees of freedom for the numerator and denominator respectively.
$h(t)$, $H(\omega)$	impulse response, and transfer function of a linear system (a complex quantity)
$M(\lambda)$	exponential distribution with mean λ^{-1}
N, n	population, sample size
$n!$	n factorial
$_nP_r$, $_nC_r$	permutations (arrangements) and combinations (choices) of r from n
$N(\mu, \sigma^2)$	normal distribution with mean μ and variance σ^2
p, $\hat{p}$	population, sample proportion
ρ, r	population sample correlation
Poisson (λt)	Poisson distribution with rate λ, time interval t
Pr( )	probability of
$\Pr(B \mid A)$	probability of B conditional on A
$P(x)$	probability function (for discrete variable)
$P(x, y)$	bivariate probability function
$P(y \mid x)$	conditional probability function
R^2	'R squared' (proportion of variance accounted for by regression), formally known as the coefficient of multiple determination
R^2_{adj}	adjusted R^2
$R(t)$, $h(t)$	reliability or survivor function, hazard or failure rate function
s	the sample estimate of the standard deviation of the errors in regression, known as the residual mean square
s^2, s	sample estimate of population variance, standard deviation

stdev (x)	standard deviation of x
t_ν	Student's t-distribution with ν degrees of freedom
$U[a, b]$	uniform distribution on the interval $[a, b]$
var(x) or σ_x^2	variance of x
x_ε	upper $\varepsilon \times 100\%$ point of distribution of X
$y(t), Y(\omega)$	time varying signal, Fourier transform of signal
$\bar{y}_{i.}, \bar{y}_{..}$	average values of y_{ij} over j, over both i and j.
$Z \sim N(0, 1)$	Z distributed standard normal
z^*	complete conjugate of z
$z_{a/2}$	percentage points of $N(0, 1)$, for two sided $(1 - \alpha) \times 100\%$ confidence intervals
α_k, a_k	PWM($E[X(1 - F)^k]$) in population, sample estimate
β_k, b_k	PWM$\{E[XF^k]\}$ in population, sample estimate
$\beta_0, \beta_1, \ldots, \beta_k$	population coefficients of multiple regression model
$\hat{\beta}_0, \hat{\beta}_1, \ldots, \hat{\beta}_k$	least squares estimators, and estimates, of $\beta_0, \ldots, \beta_k$
$\gamma, \hat{\gamma}$	population, sample skewness
$\gamma(k), c(k)$	ensemble autocovariance function (acvf), sample acvf calculated from time series
$\gamma_{xy}(\tau), c_{xy}(k)$	cross covariance function (ccvf) for continuous signals, sample estimate from a time series
$\Gamma_{xy}(\omega)$	cross spectrum (a complex quantity)
$\Gamma(\omega), C(\omega), \tilde{C}(\omega)$	spectrum of a random process, sample spectrum, periodogram [double subscript added if necessary, e.g. $\Gamma_{yy}(\omega)$]
$\Gamma(\ )$	gamma function (in particular, $\Gamma(n + 1) = n!$)
$\kappa, \hat{\kappa}$	population, sample kurtosis
$\mu, \bar{x}$	population, sample mean
$\rho(k), r(k)$	ensemble autocorrelation function (acf), sample acf calculated from a time series
σ^2, σ	population variance, standard deviation
σ^2, σ	variance, standard deviation of errors
$\Phi(z), \phi(z)$	cdf, pdf of standard normal distribution
χ_ν^2	chi-square distribution with ν degrees of freedom
$\Upsilon(h)$	variogram
ω	frequency in radians per second
$\sim$	distributed as

1

Statistics in Civil Engineering

1.1 Introduction

Civil engineers have always had to deal with uncertainty, but they are now
expected to do so in more accountable ways. Probability theory provides a
mathematical description of random variation, and enables us to make realistic
risk assessments. Statistics is the analysis of data, and the subsequent fitting of
probability models. Computers continue to push back the limits imposed by the
arithmetic and, to some extent, the algebraic manipulation involved. This
greatly increases the scope of feasible analyses. The range of applications
within civil engineering is vast, and I have aimed to show this through practical
examples. The book is structured around the main application areas, in the
following way.

1.2 Computer simulation

The widespread availability of modern PCs makes computer simulation a
relatively quick means of investigating the performance of dynamic systems.
Simulation is used by transport engineers to investigate the effects of changes
to road networks, the performance of public transport systems, and queueing
situations at quarries. Water resource engineers use simulation to investigate
the efficiency of control rules for reservoir releases. Structural engineers
simulate the responses of structures to extreme events (e.g. Jancauskas *et al*.
1994). Historic records are not long enough to estimate low probabilities. Even
if we have records from hundreds of years, as we do for the River Nile, it is
most unlikely that climatic and other variables have not changed. Simulation
allows us to generate very long records from a random process which has
similar statistical properties to a relatively short historical record.

Simulation is also used as an alternative to theoretical sampling results in
statistical analysis. For example, we can use simulations to construct
confidence intervals for estimates of return periods. This is known as

bootstrapping. The advantages over theoretical results are that simulation can be used in complex situations for which appropriate formulae are not known, and that it does not rely on approximations which are found by letting sample sizes tend to infinity, known as asymptotic results, although these often remain adequate for small samples. The disadvantage is that simulation is far less convenient than an algebraic formula. In Chapter 2 we look at random number generation, which is the basis of any stochastic simulation.

Scale models of large civil engineering structures, such as bridges and offshore platforms, will usually be tested in wind tunnels or wave tanks before construction starts. Random number sequences are often part of the algorithms for driving wavemakers, and fans, so that they provide a realistic environment.

1.3   Quality assurance and reliability

Concrete and steel are the most important materials in civil engineering construction, and it is vital to maintain a high quality. Sampling inspection is no substitute for ensuring that the supplier is capable of producing batches of concrete that are consistently within specification. We cannot separate good from bad on site, and we cannot afford for any part of a structure to be made from defective concrete. However, we must take samples of the concrete supplied for at least two reasons. The first is that an efficient sampling procedure, such as that recommended by Day (1995), will control quality and provide early warning of any undesirable trends, before any structurally defective concrete is produced. The second is that we need to be able to supply impartial evidence that the structure has been built to specification. The CUSUM technique is well suited to this application.

Another important aspect of quality assurance is the separation of variability into its components. This can be used to decide how best to reduce the overall variability of some product, or to highlight some underlying problem. For example, it is not satisfactory if laboratories return substantially different assays of the same material. I describe an inter-laboratory trial of measurements of polished stone values for a roadstone, in which I was involved.

Reliability, as a technical subject, is the study of the failure of systems. Three aspects of reliability are discussed in Chapter 3. The first is lifetime distributions, such as the Weibull, for steel cable and bitumen surfaces on bridge roadways. The second is interference distributions to model the effects of varying cover depth and carbonation in concrete structures. Markov processes, and their application to distributed systems such as water supply, is the third.

1.4   Extreme value distributions

Probability distributions for flood water levels, from the sea or rivers, are needed for the design of defence structures. The relationship between flood severity and return period, up to one thousand years, might be needed for a typical design. This has to be estimated from records that extend over tens, rather than hundreds of years, and inevitably involves extrapolation into the tails of the distribution. In addition to the variety of models for distributions of annual floods, and regionalization methods that aim to make use of data from additional sites, there are peak over threshold methods. It is essential to base estimates on realistic assumptions and to have some measure of the precision of these estimates (e.g. Rooke, 1996). Ang and Tang (1984) discuss the economic value of information which will increase the precision. The possibility of climatic change, on an unprecedented scale (e.g. Whyte, 1995), is an additional challenge for the designer. Statistical methods can be used to make some short term allowance for this, but longer term predictions depend more on scientific arguments.

Another strategy for lessening the effects of extreme river flow is to allow flooding of low lying grassland, often referred to as washlands in the UK. This has advantages over defensive walls, which, if used excessively, tend to channel a river and pass the problem downstream. Also, high walls may look intrusive and spoil the character of the environment near the river. The joint distribution of peak flows and flood volumes is required for the assessment of washland schemes. There are several methods for modelling multivariate extreme value distributions, and it is an active research topic. Gumbel's work in this area, which includes two forms of bivariate extreme value distributions, goes back to the early 1960s.

There are many other applications of extreme value distributions in civil engineering. Examples include allowing for extreme wind speeds, assessing the risk of scouring of pipelines by ice-keels (Wadhams 1983 and Chouinard 1995), and the design of structures to withstand earthquakes (Singhal and Kiremidjian, 1996).

1.5 Relationships between variables

We often wish to find a relationship between variables, so that we can explain, or at least predict, a crucial one (the dependent variable or response) from the others (the predictor variables). The investigation can be: a designed experiment, for example, a project to find how geometry, welding techniques and cathodic protection affect the lifetimes of welded plate joints under test conditions; a survey, for example, how does the number of trips per household depend on car ownership, household size, and social factors, and should improvement of public transport provision be a priority for the local authority; or a monitoring study, for example, can we predict a flood peak flow from rainfall and a measure of catchment wetness, and include this relationship in a flood warning system? The multiple regression model specifies that some known function of the dependent variable is a linear combination of known functions of the predictor variables.

The model is therefore linear in the unknown parameters, and the estimates of the parameters are the solutions of a set of linear equations. Most spreadsheet software will fit the standard multiple regression model, which can be used to analyse all the examples given above. The multiple regression model includes, for example, quadratic surfaces, because squared terms and cross product terms are simply additional predictor variables.

However, there are important applications in which the usual assumptions about the errors in the model are not satisfied, even approximately. One example is when monitoring the effects of road safety measures. Appropriate modifications to the standard analysis should be used. Sometimes we would like to fit a model which is not linear in the parameters, such as that relating subsidence to the dimensions of mining tunnels, and need to solve non-linear equations numerically.

1.6   Bayesian analysis

In a Bayesian approach we interpret probability as our subjective belief about the chances of some given event occurring. However, it is assumed that we will use all available data in a rational, and well defined, manner so it is quite disciplined. In the Bayesian paradigm, all the relevant knowledge we have before some study is summarized by a prior probability distribution. We update this distribution, using data from the study, to obtain the posterior distribution. If all our knowledge, in the context of the study, is incorporated in the prior distribution we should be indifferent to the means of obtaining new information. In principle, we would not insist it is from a random sample. This perspective is realistic when we analyse environmental data. It is not, however, a satisfactory basis for acceptance sampling of a large consignment of pavers. Customer and supplier are unlikely to share the same prior distribution. Randomization should help resolve this conflict. In scientific work we aim to reduce the subjective element as much as possible by, for example, using randomization to assign experimental material to treatments unless there are ethical constraints.

The Bayesian approach seems very suitable for engineering applications, but it does lead to more complicated calculations. Bayesian statistics software is available, sometimes free with books (e.g. Pole *et al.*, 1994). Three specific applications, for which a Bayesian approach has marked advantages, are covered. The first is an overview study of the effects of replacing priority controlled junctions with mini-roundabouts. The second is an application to updating an asset management plan for a water company, and the third is forecasting house sales for a builder when we might wish to include information about special events, such as a change in taxation.

1.7   Time series

There are often competing demands for water resources, which include domestic use, industrial use, agricultural requirements and irrigation schemes. Engineers are expected to supply this water, and maintain flows in rivers for navigation and environmental reasons, despite the random nature of rainfall. This is a challenging task in any climate, and a vitally important one in arid regions. A wide range of methods for modelling stochastic processes have been developed, and these are discussed with an emphasis on hydrological applications, in Chapter 7. The models are used to simulate time series for planning purposes, and for short term predictions in, for example, flood warning schemes. There are many other applications of time series models, which include the modelling of wind, wave and earthquake forces.

The second major topic of Chapter 7 is spatially distributed random processes. Much of the impetus for this work came from the mining industry, but other important applications include the modelling of contaminant flow from accidental land spills or land fill waste disposal schemes.

1.8   Spectral analysis and dynamic systems

The Tacoma Narrows Bridge was by no means the first bridge to collapse because of wind related forces, but it is a particularly well documented case. Fortunately there were no human casualties, but a dog that stayed in a car, despite his owner's efforts to get him out, did perish. The disaster has attracted attention because the wind speed, a mild gale, was not exceptionally high. The torsional oscillations were filmed, and can be seen on the internet (www.wsdot.wa.gov) under Galloping Gertie, the nickname the bridge had already attracted. The collapse is not satisfactorily explained by forced vibration at a resonant frequency (Billah and Scanlan, 1991), but any analysis requires some mathematical description of the wind forces. Wind velocities are not usually constant, and wave heights are not usually realistically described by some periodic pattern. Spectral analysis is a means of describing randomly varying signals in terms of their average frequency composition. This, and the relationship between spectra and dynamic systems, is the main topic of Chapter 8.

Some aspects of signal processing and control theory are also covered. There are many novel civil engineering applications. Satellite global positioning systems are being used to position pile driving rigs, and satellites are also used to obtain accurate wave height data. Active control systems are being used to protect buildings against extreme wind forces and earth tremors. Reinhorn *et al.* (1993) report results from a full-scale test structure, a 22 m tower in Tokyo.

1.9   Design of experiments

Controlled experiments, often carried out in laboratories, can provide information for civil engineering design much quicker than monitoring structures and systems in normal operation. Examples include the specification for welded joints on an offshore structure, choice of flagstones for pavements, and optimum parameter settings for extraction of copper from an ore. Four essential principles are emphasized in Chapter 9. The first is to reduce any extraneous variation as much as is possible, and to measure that which we cannot remove. The second is that it is inefficient, and potentially misleading, to change one variable at a time. Variables often interact, i.e, the effect of one depends on the level of the other. The third is that we should randomize the choice of experimental material, and allocation of that material to treatments, subject to the constraints we impose by the design of the experiment. The fourth is that we should make the experiment as realistic as possible, so that our findings are likely to be relevant to actual operating conditions, but nevertheless monitor field performance closely.

1.10   Computing

Most modern statistical techniques rely on computers for carrying out the calculations. Most people who work in universities will have access to a wide range of mathematical and statistical programs. Consulting engineers may not have access to such a variety of software, and will usually be working with more constraints on their time. This means that learning to use a new package for some specialist application may not be an attractive option. Most of the techniques described in this book can be implemented reasonably quickly in a scientific programming language with good matrix manipulation facilities, and preferably some optimization sub-routines available. Fortran is widely used in the industry and continues to be developed. Matlab, including the toolboxes, is a great asset for work with linear models of systems. I have used Matlab and Fortran, with the NAG subroutine library, for making the calculations for several of the examples in this book.

Algebraic manipulation packages, such as Maple, can be very helpful and I have used Maple for some examples in Chapters 3 and 4. However, most data analysis will be restricted to numerical calculations. I relied on Minitab for routine statistical calculations, and used the more specialist GLIM and MLP in a few places.

1.11   Using the book

This book is an intermediate level text. A summary of the material covered in a typical first statistics course is given in Appendix 1. It has been written for revision, rather than as an introduction. The main chapters are fairly self-contained. Their ordering corresponds, roughly, to distributions of a single variable, relationships between variables which are not time dependent, and variables which change over time. The technical difficulty does not increase as you go through the book, and it is not necessary to have read all the material in preceding chapters before reading the one relevant to your work. In addition, the exercises contain many other useful results.

2

Probability Distributions and Monte Carlo Simulation

2.1 Introduction

Modern instrumentation and data logging lead to increasingly large data sets in a wide range of engineering contexts. Powerful PCs make computationally intensive methods of parameter estimation, and simulation studies of random processes, feasible. There are considerable benefits to be gained from the use of these methods. We start with algorithms for generating random numbers from discrete and continuous probability distributions. Details of the common distributions, which are covered in most introductory statistics courses, are given in Appendix 1, but some other useful distributions are introduced in this chapter together with techniques for estimating their parameters. Appropriate theoretical probability distributions have several advantages over empirical distributions from specific investigations. For example, a simulation study of the response of a structure in high winds would require gust speeds and directions. A theoretical distribution of gust speeds will allow for more extreme events than those that have already been recorded. The parameters of the distribution can be estimated from diverse information, and there is no need for a long record at the site. An algebraic form of the distribution may be necessary for a theoretical analysis of the structure's response.

2.2 Stochastic (Monte Carlo) simulation

Computer simulations of apparently random environments are commonly used to assess the performance of systems. Examples include the modelling of: propagation of cracks in offshore structures (Wu, 1993); queues in construction and mining (Carmichael, 1987); water resource systems (Reitsma *et al.*, 1996); rainfall as an input to hydraulic flow models (Cowpertwait *et al.*, 1996); transportation of radio-nucliides in porous rock (Mackay *et al.*, 1996); and urban traffic networks (Rathi and Santiago, 1990).

Another transport application is bus scheduling. When you have waited a long time for a bus, only to see several turn up together, you may have suspected that the bus company arranges this on purpose. A more prosaic explanation is that, although buses leave the depot at equal time intervals, traffic and variation in the number of passengers boarding soon disrupt this order. If one bus starts catching up with another it will tend to pick up fewer passengers at each stop, which hastens the process. Appleby's (1994) BUSTLE program gives a nice demonstration of this. Senevirante (1990) describes a more detailed computer program for simulating bus services. The purpose of these simulations is to investigate the effects of changes to the system, such as: more buses; route changes; new routes; bus lanes; and employing conductors rather than operating driver only buses.

Simulations are usually set up in terms of discrete time changes. Senevirante, for example, increments the clock each second. At each time step we can simulate a passenger arriving at a bus stop with some given probability. An alternative, which may be more convenient in some cases, is to generate the continuous times between passengers arriving, and round these to an integer number of time steps. In either case we have to start with uniformly distributed random numbers over an interval $[0, 1]$ (see Appendix A1.6). A concise notation for this distribution is $U[0, 1]$.

2.2.1   Generating uniformly distributed pseudo-random numbers

There are electronic devices, such as noise diodes, which use the thermal agitation of electrons (Johnson noise) to approximate white noise. Such devices could be adapted to generate uniform random numbers. However, simulation work usually relies on mathematical algorithms which, although deterministic, produce sequences of numbers that appear to be distributed as independent uniform random variables. These algorithms are properly known as pseudo-random number generators, but this is usually abbreviated by dropping the 'pseudo'. Cooke *et al.* (1990) describe more of the background. The following is a simple example of a *mixed congruential generator*, using *division modulo* 6075

$$I_{j+1} = 106I_j + 1283 \qquad (\text{mod } 6075)$$

where the instruction (mod 6075) means, let I_{j+1} equal the remainder after dividing $(106I_j + 1283)$ by 6075. It will generate a sequence of integers $\{I_j\}$ between 0 and 6074. These integers are divided by 6075 to give a sequence of decimals between 0 and 1

$$r_j = I_j/6075$$

The $\{r_j\}$ appear to be independent uniformly distributed random variables, except for the fact that the sequence must repeat itself after each run of 6075 numbers – if not before. The generator is called 'mixed' because it involves addition as well as multiplication, and 'congruential' because of the modular division which results in numbers differing by multiples of 6075 being equivalent. To see how the calculation works let us start with I_0 equal to 127.

Then

$$I_1 = 106 \times 127 + 1283 = 14745 \ (\text{mod} \ 6075) = 2595$$

$$I_2 = 276353 \ (\text{mod} \ 6075) = 2978$$

$$I_3 = \cdots = 1051$$

$$I_4 = \cdots = 3339 \ \text{etc.}$$

The corresponding $\{r_j\}$ sequence starts: 0.427161, 0.490206, 0.173004, 0.549630, ... A sequence of random digits $\{n_j\}$ can be obtained by multiplying the $\{r_j\}$ by 10 and rounding. The first few are $4, 5, 2, 5, \ldots$.

Fortran and most other languages have a MOD function which will do the division modulo 6075 for you, but you can obtain the same effect by:

$$(i/k - \text{integer part of } i/k) \times k$$

Remember to use integer arithmetic. The uniform distribution of the $\{r_j\}$ can be checked by a histogram, but there are many tests that can be applied to help justify an assumption that the $\{r_j\}$ and $\{n_j\}$ are independent. For example:

(i)   plot r_{j+1} against r_j and check that the points are spread all over the square;
(ii)   calculate the correlations between $\{r_j\}$ and $\{r_{j+k}\}$ for k from 1 upwards, and check that they are within sampling variation of 0;
(iii) check the proportions of two digit combinations in $\{n_j\}$.

In practice you will want a reliable generator with a much longer cycle length, and you will not wish to test it yourself. Reputable mathematical or statistical software should provide good generators, but it is often convenient to write your own within a program. See Appendix 2 for a thoroughly tested algorithm which only needs a few lines of code (Wichmann and Hill, 1982).

In a typical run of BUSTLE, passengers arrive at stops at an average rate of 0.065 per time step. This is realized by recording an arrival if a random number is between 0.000 and 0.065.

2.2.2 Generating random numbers from discrete distributions

The Neyman-Scott rectangular pulses (NSRP) rainfall model (Rodriguez-Iturbe *et al.*, 1987; Cowpertwait, 1995) is built up from five Poisson (Appendix A1.5) processes. In the simplest form of this model, storm origins, which can correspond to the leading edges of cold fronts, occur as a Poisson process with a rate λ per hour. A typical realization is shown on the first line of Figure 2.1. Each storm origin has a random number (C) of rain cells associated with it. The number of rain cells is generated via a Poisson distribution, specifically ($C - 1$) is a Poisson random variable with mean ($v - 1$). The reason for modelling ($C - 1$), rather than C, as a Poisson random variable is to ensure that all storms have rain. The parameter v is the average number of rain cells per storm. The next stage is to position the cell origins. The waiting times from the storm origin to the cell origins are independent exponential random variables with a mean of $1/\beta$ hours. The second line of Figure 2.1 shows the cell origins, together with the original storm origins. The cells are assumed rectangular and their size depends on duration and intensity. The cell durations

Storm origins arrive according to a Poisson process

Each origin generates a random number of rain cells with cell origins at ⋇

The intensity and duration of each rain cell follow exponential distributions – the intensity is constant throughout the duration

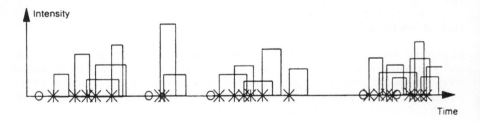

The total intensity at any point in time is the sum of the intensities of all active rain cells at that point

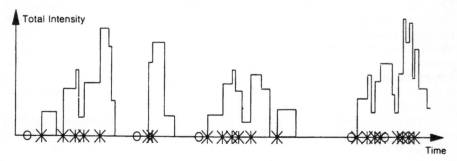

Figure 2.1 A schematic representation of the Neyman–Scott rectangular pulses model (courtesy of P. S. P. Cowpertwait)

are modelled as independent exponential random variables with a mean of $1/\eta$ hours, and the intensities as independent exponential random variables with a mean of $1/\xi$ mm per hour. The cells themselves are included in the third line of Figure 2.1 and summed to give the rainfall intensity at any time, on the bottom line. It follows that the model is built up from five independent Poisson processes and is defined by the five parameters λ, β, v, η and ξ.

The parameters of the model can be estimated from past records, and we return to this in Chapter 7. Simulations of rainfall need random numbers from a Poisson distribution. A Poisson variable X, which is $(C-1)$ in this context, with mean μ has a probability mass function:

$$P(x) = \frac{e^{-\mu}\mu^x}{x!} \qquad \text{for } x = 0, \ldots$$

The cumulative distribution function (cdf) is

$$F(x) = \Pr(X \le x) = \sum_{i=0}^{x} \frac{e^{-\mu}\mu^i}{i!}$$

To generate random numbers from this distribution we first calculate $F(x)$, for $x = 0, \ldots, N$, where N is large enough for $F(N)$ to be almost exactly 1. Having done this, we generate a $U(0,1)$ random number R. The corresponding Poisson variable, X, is such that

$$F(X-1) < R \le F(X)$$

The construction works because the probability that R is between $F(x-1)$ and $F(x)$, which must both be between 0 and 1, is just the difference between $F(x-1)$ and $F(x)$ which equals $P(x)$.

Example 2.1
Generate random numbers from a Poisson distribution with mean 3.5. First calculate:

$$
\begin{aligned}
P(0) &= e^{-3.5} & &= 0.0302 \\
P(1) &= 3.5P(0) & &= 0.1057 \\
P(2) &= 3.5P(1)/2 &&= 0.1850 \\
P(3) &= 3.5P(2)/3 &&= 0.2158 \\
&\;\;\vdots \\
P(13) &= & &= 0.0001
\end{aligned}
$$

Then calculate

$$
\begin{aligned}
F(0) &= 0.0302 \\
F(1) &= 0.1359 \\
F(2) &= 0.3208 \\
F(3) &= 0.5366 \\
F(4) &= 0.7254 \\
&\;\;\vdots \\
F(13) &= 1.0000
\end{aligned}
$$

Now generate uniform random numbers, and locate the corresponding Poisson numbers. For example, the uniform random numbers on the left give the Poisson random numbers on the right:

$$0.19211 \text{ gives } 2$$
$$0.94520 \text{ gives } 7$$
$$0.70986 \text{ gives } 4$$
$$0.65249 \text{ gives } 4$$
$$\text{etc.}$$

2.2.3   Generating random numbers from continuous distributions

Suppose a random variable X has a cdf $F(x)$ and we want to generate random numbers from this distribution. For any a and b

$$\Pr(a < X < b) = F(b) - F(a)$$

Both $F(a)$ and $F(b)$ are probabilities, and must therefore be between 0 and 1. So, if R is distributed $U[0, 1]$ then

$$\Pr(F(a) < R < F(b)) = F(b) - F(a)$$

as well. Now apply the inverse cdf (F^{-1}) to the inequality to obtain the equivalent probability statement

$$\Pr(a < F^{-1}(R) < b) = F(b) - F(a)$$

Since a and b were arbitrary numbers $F^{-1}(R)$ and X have the same distribution with cdf F. All we need do is generate a sequence of uniform random numbers and then apply F^{-1}.

Example 2.2
Port authorities use Monte Carlo simulation methods for planning the numbers of berths, cranes and other facilities. Suppose bulk carriers arrive, according to a Poisson process, at an average rate of 6.4 per twenty-four hours and we wish to generate random times between arrivals.

Let T be the time in hours between arrivals. The distribution of T is exponential, $M(\lambda)$, with a rate parameter (λ) of 0.267 per hour, and the cdf is

$$F(t) = 1 - e^{-0.267t}$$

If

$$F(t) = r$$

then

$$t = -\ln(1 - r)/0.267$$

The sequence of uniform random numbers on the left would correspond to the times on the right:

$$0.73336 \text{ gives } 4.95 \text{ hours}$$
$$0.44451 \text{ gives } 2.20 \text{ hours}$$
$$0.03817 \text{ gives } 0.15 \text{ hours}$$
$$0.79677 \text{ gives } 5.97 \text{ hours}$$
$$\text{etc.}$$

The general result is that if $R \sim U[0, 1]$

$$-\ln(1 - R)/\lambda \sim M(\lambda)$$

The subtraction of R from 1 is not necessary because $(1 - R)$ and R have identical distributions, but it emphasizes the rationale of the method which is shown graphically in Figure 2.2.

Generating random numbers from a standard normal distribution (see Appendix A1.7) is not so easy, because neither Φ, nor its inverse Φ^{-1} can be expressed as a formula involving elementary functions. The options include: use approximations to Φ^{-1}; add independent uniform random numbers; transform uniform random numbers; and acceptance rejection (AR) methods. The first is neither convenient nor efficient, because you would need to use different approximations for different regions in the domain [0, 1]. The second relies on the Central Limit Theorem. The distribution of the sum of n uniform random variables rapidly approaches normality as n tends to infinity. A value of 12 for n will suffice. If

$$R_i \sim U[0, 1]$$

then to a very good approximation

$$\sum_{i=1}^{12} R_i \sim N(6, (1)^2)$$

These normal random numbers can be scaled to any other normal distribution: subtract 6; multiply by the required standard deviation; and add the required mean. The Box–Muller method is an example of the third option. If R_1 and R_2 are independent variables from $U[0, 1]$ then:

$$Z_1 = (-2 \ln R_1)^{1/2} \sin(2\pi R_2)$$
$$Z_2 = (-2 \ln R_1)^{1/2} \cos(2\pi R_2)$$

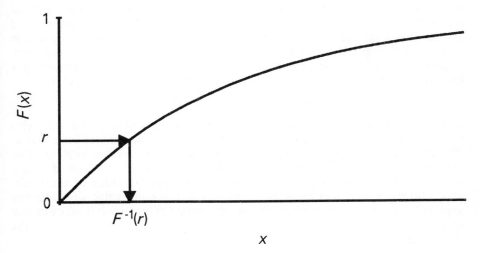

Figure 2.2 Generating exponential random numbers from uniform random numbers

are independent variables from $N(0, 1)$. You will understand the rationale for this construction if you work through Exercise 1. The acceptance–rejection method is described in Appendix 2.

2.3   Applications of Discrete Distributions

2.3.1   The likelihood principle and EM algorithm

Example 2.3
Concrete is used for the foundations of most civil engineering structures, and is frequently the main building material as well. Examples include airport runways, roads, bridges, dams and high rise buildings. Samples will be taken during a project at least twice a day, with the objective of ensuring that all the concrete supplied can safely be incorporated in the structure. There are many variables that can affect the performance of concrete. Day (1995) lists: strength after 28 days; shrinkage; durability; permeability; segregation resistance; setting time; and wear resistance. Some of these can be measured directly on test cubes or cylinders, and others may be related to different variables which can be measured. The specification should include ranges for all the measured variables. Day emphasizes that these ranges must be stringent enough for a clear distinction to be made between contractually defective, in terms of the specification, and structurally defective. In practice, up to 5% of cubes failing one or more of the specified criteria might be acceptable. Contractually defective blocks would be treated as warning signs, and any tendency for their number to increase would lead to a review of the supplier's manufacturing system. The measurements for each variable can be plotted on charts, after each cube is analysed. It might be useful to plot some linear combinations of the variables, known as principal components, as well (see Section 5.3.2 and Chouinard *et al.*, 1995). Here we estimate the proportion (p) of contractually defective cubes in the hypothetical population of all cubes that could be produced under the current system. The answer may seem obvious, but we obtain it in a way that illustrates a powerful general method of estimation. We assume that a sample of n cubes is drawn at random from a large population containing a proportion p of defectives, and that x of the sample are defective because they fail to meet at least one criterion. The probability of obtaining x defectives is then given by the binomial distribution (see Appendix A1.4):

$$P(x \mid p) = {}_nC_x p^x (1 - p)^{n-x}$$

I have emphasized that $P(x)$ depends on p because we will now treat the right-hand side as a function of p, which it is if we have known values for x and n. This is known as the *likelihood function* for p and will be written as $\mathscr{L}(p)$. The *maximum likelihood estimator* (*MLE*) of p is the value at which $\mathscr{L}$ has a maximum. It is often easier to work with the natural logarithm of the likelihood, $\ln(\mathscr{L})$, which will have its maximum at the same value of p. In this application the value of p that minimizes $\mathscr{L}$, written $\hat{p}$ because it is an estimator of p, can be written as an explicit formula in x and n. We start with the likelihood,

$$\mathscr{L}(p) = {}_nC_x p^x (1 - p)^{n-x}$$

then take the log-likelihood,

$$\ln(\mathcal{L}) = \ln(_nC_x) + x \ln p + (n - x)\ln(1 - p)$$

and find the value of the p for which the derivative with respect to p is 0.

$$\frac{d \ln(\mathcal{L})}{dp} = \frac{x}{p} - \frac{(n - x)}{1 - p}$$

Setting

$$\frac{d \ln(\mathcal{L})}{dp} = 0$$

gives

$$\hat{p} = x/n$$

This is the value of p at the unique stationary point of $\ln(\mathcal{L})$, and hence $\mathcal{L}$. It is clear, from the definition of $\mathcal{L}$, that it is a maximum. So, the population proportion is estimated by the sample proportion. The general principle of estimating population parameters by their sample equivalents is known as the method of moments (MOM). In this example the MLE and MOM estimators are the same, but you will see from the next example that this is not always the case.

Example 2.4

In a *compound Poisson process* the occurrences are no longer restricted to being single. One model uses a Poisson process for points at which events occur, and a truncated Poisson distribution for the number of occurrences, X, at those points. This is a relatively straightforward example of a distribution for which the MLE cannot be expressed as an explicit function of sample statistics (Everitt, 1987). The probability mass function for the truncated Poisson distribution consists of Poisson probabilities for $1, 2, \ldots$ scaled so that they add to one.

$$P(x) = \frac{e^{-\theta}\theta^x}{x!(1 - e^{-\theta})} \qquad \text{for } x = 1, 2, \ldots$$

It might be a plausible model for the number of raincells in a storm but, since the numbers of raincells in storms cannot be counted, it is unlikely that rainfall data will show much difference between assuming a truncated Poisson variable and assuming a Poisson variable with one added.

In this example we will fit the truncated Poisson distribution to data from a survey of numbers of people in vehicles, excluding public service vehicles, on business trips (Wood, 1981). The survey site was situated 20 km north of Gretna on the A74 in Scotland. South-bound traffic was surveyed between 6 a.m. and 10 p.m. for one week. Thirty per cent of vehicles were stopped, of which about one third were on business trips. The following data were recorded from 2544 vehicles in the business trip category.

Number of people in car (x_i)	Number of cars
1	1679
2	687
3	127
4	51
5 or more	0

I cannot think of any convincing theoretical reason for assuming the numbers of people to have a truncated Poisson distribution. However, we would have some justification for assuming this distribution, in simulations or cost benefit analyses, if we found that it gave a reasonable empirical fit to data from several surveys.

In general, if there are n cars, and the number of people in the ith car is x_i, the likelihood is:

$$\mathcal{L}(\theta) = \prod_{i=1}^{n} e^{-\theta}\theta^{x_i} \Big/ [x_i!(1 - e^{-\theta})]$$

The log-likelihood is therefore

$$\ln(\mathcal{L}) = -n\theta + \ln\theta \sum_{i=1}^{n} x_i - \sum_{i=1}^{n} \ln(x_i!) - n\ln(1 - e^{-\theta})$$

and its derivative with respect to θ is,

$$\frac{d\ln(\mathcal{L})}{d\theta} = -n + \frac{\sum x_i}{\theta} - \frac{ne^{-\theta}}{1 - e^{-\theta}}$$

So $\hat{\theta}$ is the solution of the equation

$$-n + \frac{\sum x_i}{\hat{\theta}} - \frac{ne^{-\hat{\theta}}}{1 - e^{-\hat{\theta}}} = 0$$

which has no explicit algebraic form. However, it is easy to solve numerically. In this case $n = 2544$, $\sum x_i = 3638$ and a plot of the function on the left against values of $\hat{\theta}$ from 0.5 to 1.5 in steps of 0.1, and then from 0.70 to 0.80, leads to the solution $\hat{\theta} = 0.764$. Everitt (1987) describes more efficient methods for finding the solution than plotting the function. You are asked to compare the MLE of θ with the method of moments estimator (MOME) of θ in Exercise 3, and you will find that they are different. When the MLE and MOME differ it is better to use the MLE because it is, usually, a more precise estimator. You might think, intuitively, that it explicitly uses all the information provided by the data, rather than just summary statistics. However, intuition is not always a good guide and you should read Appendix 3 for more details. Other advantages of MLE are that there is a general method for calculating approximations to their standard deviations; and that the likelihood principle is often the easiest approach to estimation problems, such as allowing for a trend in extreme values of sea levels in Example 4.5.

Now that we have estimated θ, we should see whether there is a reasonable agreement between the observed numbers of cars in each category and their expected values. The expected values are $P(x \mid \theta = 0.764) \times n$ and equal: 1694.7; 647.4; 164.9; 37.0 for 1, 2, 3 and 4 or more people, respectively. The chi-square goodness of fit statistic (Appendix A1.21) is 15.8 and the differences between observed and expected values cannot be convincingly attributed to chance ($\chi^2_{2,0.0005} = 15.20$). However, we should remember that it is a large sample, and the lack of fit may not be practically important.

This example also provides a simple demonstration of the EM algorithm (Dempster *et al.*, 1977) which is one method of maximum likelihood estimation with missing data. I calculated the number of cars with 1, 2, 3 or 4 people in from the table of percentages and the total number of cars given in the report. Up to 12 cars in the 5 or more category might have been rounded to 0%. We will now suppose that '4' represents '4 or more'. The EM algorithm iterates between finding the MLE, the M bit, and replacing missing values by their expected values, the E bit. The $P(x \mid \theta = 0.764)$ for $x = 4$, 5, 6 and 7 are 0.0124; 0.0019; 0.0002 and 0.0000 to four decimal places. If we distribute the 51 cars in the '4 or more' category in proportion to these probabilities we obtain 43.6, 6.7 and 0.7 for 4, 5 and 6 people respectively. If we repeat the MLE procedure we obtain a value of 0.769 for θ. We now recalculate the probabilities for $x = 4$, 5, 6 and 7 and iterate until the algorithm converges, in this case to 0.769. An alternative to the EM algorithm would be to maximize the likelihood:

$$\mathcal{L}(\theta) = [P(1)]^{1679}[P(2)]^{687}[P(3)]^{127}[1 - P(1) - P(2) - P(3)]^{51}$$

2.3.2 Negative binomial distribution

Schemes for reducing traffic accidents are often first implemented in selected areas on a trial basis. Analysts commonly assume that occurrences of accidents can be modelled by a Poisson process. Yearly totals of accidents have a Poisson distribution, even if the underlying rate varies seasonally, if accidents occur randomly and independently. The variance of the Poisson distribution equals the mean. Sometimes the variance of yearly accident totals, after allowing for any systematic changes, is significantly higher than the mean. This is known as *overdispersion*. A plausible explanation is that the underlying rate varies from year to year because of different weather and other factors, which are not included in the analysis.

The number of vehicles involved in accidents each year will not have a Poisson distribution because most accidents involve more than one vehicle. If highway authorities record the number of vehicles involved in accidents they could fit a compound Poisson process (see Example 2.3), but a simpler expedient is to assume a distribution which can accommodate the additional variability. The negative binomial distribution has two parameters which allow the ratio of the variance to the mean to take values greater than one. Miaou (1994) investigates its use for modelling the number of trucks involved in accidents each year.

There are alternative forms of the negative binomial distribution and I use the definitions given by Evans *et al.* (1993). The Pascal variable X is the number of failures before the rth success in a sequence of independent trials in which the probabilities of a success and failure are p and $1 - p$ respectively (*Bernoulli trials*, as in the binomial distribution). The probability mass function of the *Pascal distribution* is therefore

$$P(x) = {}_{r+x-1}C_{r-1}p^r(1 - p)^x \qquad \text{for} \quad x = 0, \ldots$$

This generalizes to the *negative binomial distribution* for non-integer values of r

$$P(x) = \frac{\Gamma(r + x)}{\Gamma(r)x!} p^r(1 - p)^x \qquad \text{for } x = 0, \ldots$$

where $\Gamma(\ )$ is the *gamma function* (see Appendix A1.1 and Table A15.7). The gamma function is a generalization of the factorial function, and $\Gamma(r)$ is $(r - 1)!$. The mean and variance of X are:

$$\mu = r(1 - p)/p$$
$$\sigma^2 = \mu/p$$

Example 2.5
The numbers of cars arriving at a car park during 5 second intervals were recorded. The following observations were made one morning between 8:30 a.m. and 8:55 a.m.

Number of cars arriving in 5 s Interval	Frequency
0	160
1	96
2	28
3	13
4	2
5	1
Total	300

The mean and standard deviation of the number of cars arriving in 5 s intervals are 0.680 and 0.902 respectively. The variance, 0.814, is appreciably greater than the mean and this suggests that the negative binomial distribution may be more realistic than the Poisson distribution. The MOM estimates of r and p are the solutions of the equations

$$\bar{x} = \hat{r}(1 - \hat{p})\hat{p}$$
$$s^2 = \bar{x}/\hat{p}$$

so

$$\hat{p} = 0.680/0.814 = 0.835$$

and

$$\hat{r} = 0.835 \times 0.680/(1 - 0.835) = 3.45$$

Estimated probabilities, and hence expected values, can now be calculated. For example, for $x = 2$

$$P(2) = \frac{\Gamma(5.45)}{\Gamma(3.45)2!} (0.835)^{3.45}(1 - 0.835)^2$$

$$= \frac{4.45 \times 3.45}{2} (0.835)^{3.45}(0.165)^2 = 0.112$$

and the expected count for 2 cars is 300 times $P(2)$ which equals 33.6. If the '4 and 5' car categories are combined into '4 or more', the calculated value of a chi-square goodness of fit statistic is 2.27, which is less than the upper 10% point of a chi-square distribution with $5 - 1 - 2 = 2$ degrees of freedom.

2.4 Applications of Continuous Distributions

2.4.1 Lognormal distribution

Annual maximum floodpeak inflows $\{x_i\}$ to the Hardap Dam in Namibia, for the water years 1962/3 to 1986/7, are given in Table 2.1 (Adamson, 1989). The box plot Figure 2.3 and summary statistics:

$$\bar{x} = 863 \text{ m}^3 \text{ s}^{-1}$$
$$\hat{x}_{0.5} = 412 \text{ m}^3 \text{ s}^{-1}$$
$$s_x = 1310 \text{ m}^3 \text{ s}^{-1}$$
$$\widehat{CV} = 1.52$$
$$\hat{\gamma} = 2.87$$
$$\hat{\kappa} = 11.2$$

emphasize the unpredictable nature of the arid climate in southern Africa. Water is a valuable resource and planners needed to know how frequently they

Table 2.1 *Annual maximum floodpeak inflows to Hardap Dam: catchment area 12 600 km² (courtesy of the Department of Water Affairs, Namibia)*

Year	Inflow ($m^3 s^{-1}$)	Year	Inflow ($m^3 s^{-1}$)
1962–3	1864	1975–6	1506
1963–4	44	1976–7	1508
1964–5	146	1977–8	236
1965–6	364	1978–9	635
1966–7	911	1979–0	230
1967–8	83	1980–1	125
1968–9	477	1981–2	131
1969–0	457	1982–3	30
1970–1	782	1983–4	765
1971–2	6100	1984–5	408
1972–3	197	1985–6	347
1973–4	3259	1986–7	412
1974–5	554		

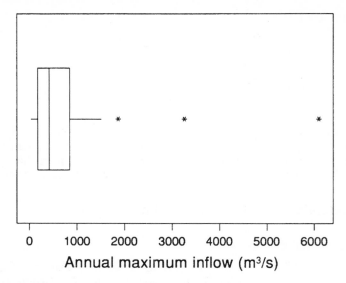

Figure 2.3   Boxplot of annual maximum inflows to Hardap Dam

could expect yearly maximum floodpeaks to exceed $5000 \text{ m}^3\text{s}^{-1}$. If exceedances of this level spill, water is wasted, but retaining the water would require additional construction. The associated cost may not be justified if exceedances are rare. A naive estimate of an exceedance once every 25 years is too unreliable to be useful. The only practical approach with such a small sample is to assume the data are a random sample from a particular form of distribution, estimate its parameters, and hence estimate the probability of exceeding $5000 \text{ m}^3\text{s}^{-1}$. We can either look for a suitable form of distribution, or try transforming the original data so that their transforms can reasonably be assumed to come from a normal distribution. The Box–Cox transform (Box and Cox, 1964)

$$y_i \equiv \begin{cases} [(x_i - L)^\lambda - 1]/\lambda & \text{for } \lambda \neq 0 \\ \ln(x_i - L) & \text{for } \lambda = 0 \end{cases}$$

is often used. The parameter L must be set so that $X - L$ is positive, and zero is often a natural choice. The scaling for non-zero λ is not necessary, but it makes the transformation continuous at 0, with respect to λ, which is useful when deriving mathematical results. The logarithmic transform has some physical justification. The normal distribution is obtained as the sum of a large number of random perturbations (Appendix A1.7), so if a variable arises from multiplicative perturbations its logarithm will be normally distributed. Figure 2.4 is a normal probability plot of the logarithms of the Hardap Dam inflows $\{y_i\}$ and the fit is remarkably good (the Anderson–Darling statistic is discussed in Appendix A1.21). We let X represent annual maximum inflows and assume

$$Y = \ln(X)$$

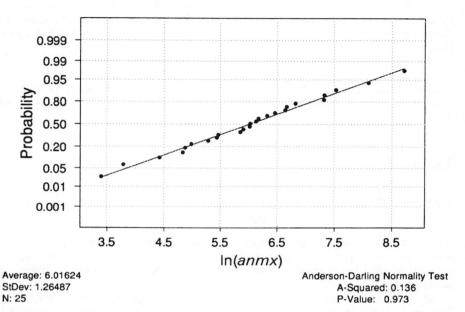

Figure 2.4   Logarithms of annual maximum inflows to Hardap Dam plotted on normal probability paper

has a normal distribution. The mean and standard deviation of the $\{y_i\}$ are

$$\bar{y} = 6.016$$

$$s_y = 1.265$$

and the probability that Y exceeds $\ln(5000)$ which is 8.517, is estimated using Table A15.1 by

$$\Pr\left(Z > \frac{8.517 - 6.016}{1.265}\right) = \Pr(Z > 1.98) = 0.024$$

Now suppose planners ask for an estimate of the inflow with a *return period* of 100 years. That is, the inflow that will be exceeded, on average, once every 100 years – equivalent to the upper 1% quantile of the distribution of yearly inflows. The upper 1% quantile of the normal distribution of Y is

$$\mu_Y + 2.326\sigma_Y$$

where 2.326 is the upper 1% point of the standard normal distribution (final row of Table A15.2). The natural estimate of this quantile is

$$\bar{y} + 2.326 s_y = 8.96$$

but see Appendix A1.16 for a slight refinement. The corresponding 1% point of the distribution of X is $\exp(8.96)$ which equals an inflow of 7773 $\mathrm{m^3\,s^{-1}}$. We can quite easily calculate a, rather approximate, 90% confidence interval for this quantile. The variance of $\bar{Y}$ is σ^2/n, and the variance of S, when sampling from a normal distribution, is very approximately $\sigma^2/(2n)$ (Appendix A1.14). Since $\bar{Y}$ and S are independent

$$\mathrm{Var}(\bar{Y} + 2.326S) \approx \sigma^2/n + (2.326)^2\sigma^2/(2n)$$

The approximate 90% confidence interval is obtained by replacing σ with s_y, and then assuming a near normal distribution, with an upper 5% point of about 1.7, for $\bar{Y} + 2.326S$:

$$\bar{y} + 2.326s_y \pm 1.7\,s_y\sqrt{[1/n + (2.326)^2/(2n)]}$$

This becomes

$$[8.13, 9.79]$$

for Hardap Dam, and returning to $\mathrm{m^3\,s^{-1}}$ by taking the exponential leads to

$$[3396, 17785]$$

The confidence interval for the upper 1% quantile is very wide, but this is an inevitable consequence of the relatively small sample size. An alternative method of calculating confidence intervals, the *bootstrap method*, is described in Appendix 4.

So far we have not required the probability density function (pdf) of the lognormal distribution, but we do need it to find an expression for the mean of X in terms of the mean and standard deviation of Y. This result is of considerable practical importance if we take logarithms before fitting a regression model. The pdf of the lognormal distribution can be found from that of the normal distribution by the following argument, which will be used in other contexts.

Let Y be a random variable with cdf $F_Y(y)$; and $X = g(Y)$ be an increasing function of Y with an inverse function g^{-1}, so that $g^{-1}(X) = Y$. Then, by definition

$$\mathrm{Pr}(Y < y) = F_Y(y)$$

and since $g(Y)$ is increasing

$$\mathrm{Pr}(Y < y) = \mathrm{Pr}(g(Y) < g(y)) = F_Y(y)$$

So we can now write

$$F_X(x) = \mathrm{Pr}(X < x) = F_Y(g^{-1}(x))$$

$$f_X(x) = \frac{\mathrm{d}}{\mathrm{d}x}F_X(x) = \frac{\mathrm{d}}{\mathrm{d}y}F_Y(g^{-1}(x))\frac{\mathrm{d}y}{\mathrm{d}x}$$

$$= f_Y(g^{-1}(x))\bigg/\frac{\mathrm{d}x}{\mathrm{d}y}$$

In the case of the lognormal distribution

$$Y = \ln X \quad \text{and} \quad X = \exp(Y)$$

$$y = g^{-1}(x) = \ln x$$

and Y is normally distributed.

$$Y \sim N(\mu_Y, \sigma_Y^2)$$

$$f_Y(y) = \frac{1}{\sigma\sqrt{2\pi}} \exp\{-[(y-\mu_Y)/\sigma_Y]^2\}$$

$$\frac{dy}{dx} = \frac{1}{x} \quad \text{and} \quad \frac{dx}{dy} = x$$

and, application of the general result gives the pdf of X:

$$f_X(x) = \frac{1}{x\sigma\sqrt{2\pi}} \exp\{-[(\ln x - \mu_Y)/\sigma_Y]^2\} \qquad \text{for } 0 \leqslant x$$

The mean, median, standard deviation and skewness of X are:

$$\mu_X = \exp(\mu_Y + \sigma_Y^2/2)$$
$$\sigma_X^2 = \mu_X^2[\exp(\sigma_Y^2) - 1]$$
$$x_{0.5} = \exp(\mu_Y)$$
$$\gamma = [\exp(\sigma_Y^2) - 1]^3 + 3[\exp(\sigma_Y^2) - 1]$$

Notice that the mean of X is greater than the exponential of the mean of Y. The exponential of the mean of Y is the median value of X. The distribution is positively skewed so the mean is to the right of the median (Figure 2.5).

2.4.2 Rayleigh distribution

Liu (1991) suggests that the Rayleigh distribution gives an adequate approximation to ordinary wind speeds for calculating heat losses through buildings and energy available to wind turbines. However, he emphasizes that it is not adequate for predicting extreme values. The distribution is named after Lord Rayleigh (J.W. Strutt) who derived it in 1919 as part of the solution to an acoustics problem. If a random variable X has a Rayleigh distribution, its probability density function is,

$$f(x) = \frac{x}{b^2} e^{-x^2/(2b^2)} \qquad \text{for } 0 \leqslant x$$

and the cumulative distribution function is

$$F(x) = 1 - \exp[-x^2/(2b^2)] \qquad \text{for} \quad 0 \leqslant x$$

The mean, variance, skewness and kurtosis are:

$$\mu = b(\pi/2)^{1/2}$$
$$\sigma^2 = (2 - \pi/2)b^2$$
$$\gamma = 2(\pi - 3)\pi^{1/2}/(4 - \pi)^{3/2} \approx 0.63$$
$$\kappa = (32 - 3\pi^2)/(4 - \pi)^2 \approx 3.25$$

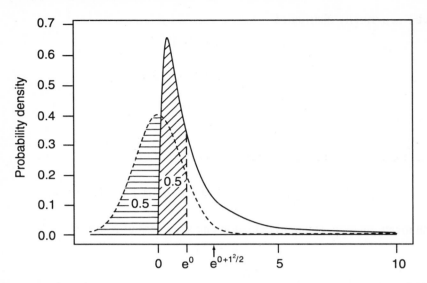

Figure 2.5   Relationship between means and medians of lognormal distribution of X and the normal distribution of $Y = \ln X$ (broken curve)

The method of moments estimate of b follows from the equation for μ, by replacing μ with $\bar{x}$ to obtain

$$\hat{b} = \bar{x}(2/\pi)^{1/2}$$

The maximum likelihood estimate is different, and there are physical reasons for preferring it in many applications. The likelihood function is defined in the same way as for discrete distributions since $f(y_i)\delta y$, for small δy, is the probability of being close to, by which I mean within $\delta y/2$ of, y_i. The exact size of δy does not have any effect on the values of the parameters that minimize the likelihood function, and it is usually omitted from the expression. If there are n data

$$\mathscr{L}(b) = \prod_{i=1}^{n} \frac{x_i}{b^2}\, e^{-x_i^2/(2b^2)} (\delta x)^n$$

The log-likelihood is therefore

$$\ln(\mathscr{L}) = \sum_{i=1}^{n} \{\ln(x_i) - \ln(b^2) - x_i^2/(2b^2)\}$$

where we have dropped the irrelevant $n\ln(\delta x)$. The MLE is the solution of the equation

$$\frac{d\ln(\mathscr{L})}{db} = 0$$

This is

$$\sum \left\{ -\frac{2}{\hat{b}} - \frac{x_i^2}{2}(-2\hat{b}^{-3}) \right\} = 0$$

which is easily rearranged to give

$$\hat{b}^2 = \sum x_i^2 / (2n)$$

We now show that the MLE is equivalent to matching $E[X^2]$ with its sample estimate. To begin with

$$E[X^2] = \mu^2 + \sigma^2$$

for any random variable. For the Rayleigh distribution this

$$= b^2(\pi/2) + (2 - \pi/2)b^2$$
$$= 2b^2$$

The result follows because $\sum x_i^2 / n$ is the sample estimate of $E[X^2]$. In many engineering applications X^2 will be proportional to energy, and it makes more sense to match its expected value with the corresponding sample value than to match means. With only one parameter to be estimated we cannot do both, although it is worth noting that the Rayleigh distribution is a special case of the two parameter Weibull distribution (Chapter 3).

Example 2.6
Staff from the Geography Department of the University of Newcastle upon Tyne, UK, have taken readings from an anemometer on the roof of the Claremont Tower over 19 years from 1969. Sharp (1988) included the following data for winter months (December, January and February) in a detailed report.

Wind speed averaged over 6 hours (km/hr)	Percentage of record (%)
calm	6.2
1–9	17.1
10–19	23.1
20–29	23.8
30–39	15.2
40–49	8.8
50–59	3.4
60 and over	2.4

We can fit a Rayleigh distribution by taking an average calm as 0.5 km/hr, an average wind over 60 as 70 km/hr, and the mid points of the other intervals. Then

$$\sum x_i^2 / n = (0.5^2 \times 6.2 + 5^2 \times 17.1 + \cdots + 54.5^2 \times 3.4 + 70^2 \times 2.4)/100$$
$$= 769$$

The fitted Rayleigh distribution has cdf

$$F(x) = 1 - \exp[-x^2/769]$$

We can calculate expected percentages as: $100 \times F(0.5)$; $100 \times (F(9.5) - F(0.5))$; and so on up to $100 \times (1 - F(60))$. They are 0.03%; 11.1%; 27.9%; 28.7%; 19.1%; 9.0%; 3.1%; and 1.0%. The fit is rather rough[1] (Figure 2.6) but

---

[1] A chi-squared goodness of fit test must not be based on percentages. It would need the number of independent observations which is an unknown, but large, number. Even small discrepancies between observed and expected counts would be statistically significant.

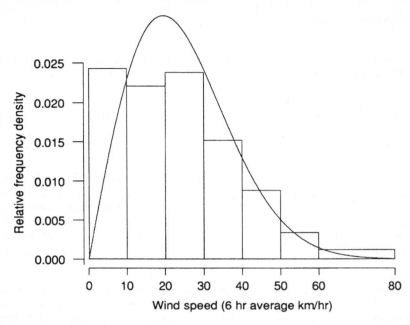

Figure 2.6 Histogram of windspeeds (six hour means) measured on top of Claremont Tower, and pdf of fitted Rayleigh distribution

adequate for calculating heat losses through buildings when we reflect that the effective wind speeds around them will be different from those on the roof of the Claremont Tower.

The Rayleigh distribution is often used as an approximation to the distribution of wave heights in marine applications. The significant wave height is customarily used to describe sea states. It is defined as the mean height of the highest third of waves. Fatigue crack growth can be modelled as being proportional to the product of existing crack length and some power between 2 and 4 of the stress range (Paris and Erdogan, 1963). If the stress range is assumed to be proportional to wave height we have another justification for preferring the MLE estimate of *b*. If we have estimates of *b* from wave buoy records in a sea area, we can interpolate these to find suitable values for other locations where we intend to install offshore structures.

2.4.3 Gamma distribution

There are many other forms of distribution which have a positive skewness. The gamma distribution, also known as the Pearson type III distribution, is one of the more commonly chosen ones, and special cases of it go by yet other names. The Water Resources Council of the USA used to recommend it as a model for the logarithms of annual maximum flows, but nowadays the generalized extreme value distribution is usually preferred (Önöz and Bayazit, 1995; and Chapter 3). The gamma distribution tends towards a normal distribution as the skewness decreases, so the log-Pearson III can approximate a lognormal distribution as a special case. The sum of independent exponential random variables, with the same rate parameter, has a gamma distribution, which is one reason for its use in reliability analysis. Senevirante (1980) uses it to model the times taken for passengers to board a bus, and journey speeds, and refers to other work to justify the choice.

The distribution of the time until the cth occurrence in a Poisson process has a gamma distribution. The pdf (multiplied by δx) can be written down directly as: the product of $c-1$ occurrences in time x with the probability of an occurrence in the interval $(x, x+\delta x)$. Hence the probability density function for a gamma distributed random variable X is

$$f(x) = \frac{\lambda(\lambda x)^{c-1}}{\Gamma(c)}\, e^{-\lambda x} \qquad \text{for } 0 \le x, \text{ with } 0 < \lambda \text{ and } 0 < c$$

The mean, variance, skewness and kurtosis are

$$\mu = c/\lambda$$
$$\sigma^2 = c/\lambda^2$$
$$\gamma = 2/\sqrt{c}$$
$$\kappa = 3 + 6/c$$

The special case of c equal to one is an exponential distribution. The moment generating function (Appendix 1) of the gamma distribution is

$$M(\theta) = (1 - \theta/\lambda)^{-c} \qquad \text{for} \quad \theta < \lambda$$

which is consistent with the gamma variable being the sum of c independent exponential variables. These are also called Erlang distributions, after the Danish mathematician A.K. Erlang (1878–1929) who is renowned for his pioneering work on stochastic processes. The chi-square distribution with ν degrees of freedom is a gamma distribution with $c = \nu/2$ and $\lambda = 1/2$. The cdf has to be evaluated numerically, except for integer values of c which correspond to Erlang distributions and chi-square distributions with even degrees of freedom. Random numbers from Erlang distributions can be generated as the sum of c exponential random numbers. Some typical gamma probability density functions are shown in Figure 2.7.

Method of moments estimates of the parameters are given by:

$$\hat{c} = \bar{x}^2/s^2, \qquad \hat{\lambda} = \bar{x}/s^2$$

Any distribution with a lower bound of 0 can be made into a distribution with a lower bound of L by replacing x with $x - L$, so the pdf becomes:

$$f(x|L, \lambda, c) = \frac{\lambda(\lambda(x-L))^{c-1}}{\Gamma(c)}\, e^{-\lambda(x-L)} \qquad \text{for } L \le x$$

If we wish to fit a gamma distribution to the logarithms of the Hardap Dam inflows $\{y_i\}$ there is no natural value for L; except that it must be less than the logarithm of the smallest inflow which is $\ln(30) = 3.4$. We have to either

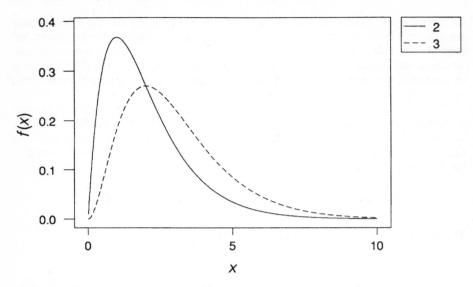

Figure 2.7   Probability density functions of gamma distributions, with $c = 2$ and $c = 3$

choose a reasonable value for L and then use MOM estimates for c and λ, or estimate all three parameters from the data. The latter seems preferable in principle. The likelihood function is defined in the same way as for discrete distributions since $f(y_i)\delta y$, for small δy, is the probability of being close to, within $\delta y/2$ of, y_i. The exact size of δy does not have any effect on the values of the parameters that minimize the likelihood function, and it is usually omitted from the expression. If there are n data

$$\mathcal{L} = \prod_{i=1}^{n} \frac{\lambda(\lambda(y_i - L))^{c-1}}{\Gamma(c)} \, e^{-\lambda(y_i - L)} \, (\delta y)^n$$

The log-likelihood is therefore

$$\ln(\mathcal{L}) = nc \ln(\lambda) - n \ln(\Gamma(c)) + (c - 1) \sum \ln(y_i - L) - \lambda \sum (y_i - L)$$

where we have dropped the irrelevant $n \ln(\delta y)$. The MLE are the solutions of the equations:

$$\frac{\partial \ln(\mathcal{L})}{\partial L} = 0; \qquad \frac{\partial \ln(\mathcal{L})}{\partial \lambda} = 0; \qquad \frac{\partial \ln(\mathcal{L})}{\partial c} = 0$$

which are

$$-(\hat{c} - 1) \sum (y_i - \hat{L})^{-1} + n\hat{\lambda} = 0$$

$$n\hat{c}\hat{\lambda} - \sum (y_i - \hat{L}) = 0$$

$$n \ln(\hat{\lambda}) - n \frac{d}{d\hat{c}} \ln(\Gamma(\hat{c})) + \sum \ln(y_i - \hat{L}) = 0$$

We cannot rearrange these to give explicit solutions for $\hat{L}$, $\hat{\lambda}$ and $\hat{c}$ and have to resort to numerical methods. The third equation includes the di-gamma function $\psi(c)$, which is defined as:

$$\psi(c) = \frac{d}{dc} \ln(\Gamma(c)) = \frac{d}{dc} \Gamma(c)/\Gamma(c)$$

This is available in Abramowitz and Stegun (1965), some computer packages, and as a NAG subroutine, but we can avoid using it if we minimize the log-likelihood directly. A further complication is that there are constraints on the parameters: $L < $ minimum y_i; $\lambda > 0$; and $c > 0$. Everitt (1987) describes various numerical methods for minimizing a function of several variables. He also suggests redefining the parameters as an alternative to using a constrained minimization routine. For example:

$$L = y_{\min} - \exp(u)$$
$$\lambda = \exp(v)$$
$$c = \exp(w)$$

The log-likelihood can then be recast as a function of u, v, and w. The simplex method is a straightforward way of minimizing a function of several variables although it is not particularly efficient. With three variables the simplex is a tetrahedron. You evaluate the function at the four vertices; reflect the vertex with the largest function value in the opposite face; evaluate the function at this mirror image point; continue in this way until the two largest function values are mirror images; then reduce the size of the simplex. The details are in Appendix 5. All numerical minimization routines depend on a sensible choice of starting values, and there is no guarantee they find a global, rather than a local, maximum. A sensible starting point is something slightly less than the minimum y_i value for L and, given this choice, the MOME of λ and c.

Example 2.7
I used the NAG simplex algorithm E04CCF to fit a three-parameter gamma distribution to the natural logarithms of the inflows to the Hardap Dam (Table 2.2). The minimum value, mean and standard deviation of the logarithms are 3.401, 6.016 and 1.265 respectively; and initial values of $L = 3.3$, $c = 3$ and $\lambda = 1$ seemed reasonable. The routine eventually converged to $L = -172.2165$, $c = 20678.97$ and $\lambda = 116.0217$. These values correspond to a gamma distribution which is very close to the normal distribution. Let Y represent the logarithms of inflows. The probability that an inflow exceeds 5000 is identical to

$$\Pr(Y > \ln 5000) = \Pr(Y + 172.217 > \ln 5000 + 172.217)$$

Now $(Y + 172.217)$ has a two parameter gamma distribution ($L = 0$), with $c = 20679.0$ and $\lambda = 116.022$, and the probability

$$\Pr(Y + 172.217 > 180.734) = 0.024$$

This is the same as we calculated for the lognormal distribution in Section 2.4.1.

Table 2.2 Means and standard deviations of the proportions of flow in each month for the River Swat at Kalam, Pakistan, for the years 1961–1993, together with coefficients of the fitted Dirichlet distribution. Months are coded 1 to 12 for January to December. (Summary data courtesy of Sir William Halcrow and Partners)

month i	0	1	2	3	4	5	6	7	8	9	10	11	12
mean		0.0145	0.0118	0.0156	0.0417	0.1201	0.2319	0.2480	0.1677	0.0759	0.0342	0.0214	0.0172
stdev		0.0025	0.0020	0.0028	0.0098	0.0269	0.0246	0.0306	0.0187	0.0136	0.0041	0.0031	0.0028
c_i	1	4	3	4	11	33	63	68	46	21	9	6	5

2.4.4   Vector variables and von Mises distribution

Wind speeds, faults in rock, and many other data needed for engineering work are vectors. Summary statistics of directions need different formulae from the usual definitions. The mean of a set of n angles $\{\theta_i\}$ is defined as the direction of the resultant of n unit vectors in the directions θ_i. This mean $(\bar{\theta})$ can be calculated from the following equations

$$C = \Sigma \cos \theta_i \qquad S = \Sigma \sin \theta_i$$
$$R^2 = C^2 + S^2$$
$$\cos \bar{\theta} = C/R \qquad \sin \bar{\theta} = S/R$$

R is the length of the resultant vector and lies in the range $[0, n]$. The mean resultant length $(\bar{R})$ is defined by

$$\bar{R} = R/n$$

the sample circular variance is defined by

$$V = 1 - \bar{R}$$

and the sample circular standard deviation is defined by

$$v = \sqrt{[-2 \ln(1 - V)]}$$

The von Mises probability distribution is commonly used to model directions in two dimensions if the distribution has a single mode. The form of the generalization to three dimensions, the Fisher distribution, is given in Exercise 2.7 (Mardia, 1972; Fisher *et al.*, 1987). The pdf of the *von Mises distribution* is given below

$$f(x) = \exp[b \cos(x - a)]/[2\pi I_0(b)] \qquad 0 < x \leqslant 2\pi; \text{ with } 0 < a < 2\pi \text{ and } 0 < b$$

The parameter b is called the concentration parameter. The distribution is symmetric about its modal value, $x = a$. This is also the mean direction. The factor $2\pi I_0(b)$ makes the area under the curve equal one. It is called the normalizing factor, and includes the modified Bessel function of the first kind of order 0 $(I_0(b))$. The MLE of the parameters a and b are the solutions of the equations:

$$\tan \hat{a} = S/C$$
$$I_1(\hat{b})/I_0(\hat{b}) = \bar{R}$$

where $I_1(\hat{b})$ is the modified Bessel function of the first kind of order 1. These Bessel functions of order k are defined by,

$$I_k(b) = (b/2)^k \sum_{i=0}^{\infty} (b^2/4)^i \Big/ [i!\Gamma(k + i + 1)]$$

and they are tabled and available in MAPLE, for example. However, it is much easier to use the approximation given in Fisher (1993)

$$\frac{I_0(b)}{I_1(b)} \approx \begin{cases} 2b + b^3 + 5b^5/6 & \text{for } b < 0.53 \\ -0.4 + 1.39b + 0.43/(1 - b) & \text{for } 0.53 \leqslant b < 0.85 \\ 1/(b^3 - 4b^2 + 3b) & \text{for } b \geqslant 0.85 \end{cases}$$

to calculate $\hat{b}$. The normalizing factor can be deduced from the area under $\exp[\hat{b} \cos(x - \hat{a})]$ as x increases from 0 to 2π, if $I_0(\hat{b})$ is not readily available.

Example 2.8
The following table gives directions of waves in Area 11 in the North Sea (British Maritime Technology Ltd, 1986). I have given degrees in the usual convention for bearings, and radians in the mathematical convention of anticlockwise from the positive real axis.

Compass point	Degrees	Radians	waves per thousand
north	0	$\pi/2$	87
north east	45°	$\pi/4$	64
east	90°	0 rad	95
south east	135°	$7\pi/4$	112
south	180°	$3\pi/2$	151
south west	225°	$5\pi/4$	179
west	270°	π	185
north west	315°	$3\pi/4$	127

For convenience in the arithmetic we will take a nominal n of 1000, although the number of waves observed will be much greater than this, and treat the 'per thousand' as frequencies. The formulae for C and S are modified to allow for grouped data in the usual way, i.e. multiply by the corresponding frequencies and sum over the 8 directions. Then:

$$C = -181.924 \qquad S = -134.711$$
$$R^2 = (226.37)^2$$
$$\hat{a} = 3.779 \qquad \hat{b} = 0.8785$$

Then the normalizing factor of 7.55 was found as the area under the curve

$$\exp[\hat{b} \cos(x - \hat{a})]$$

for x between 0 and 2π using Simpson's rule.

2.4.5 Disaggregation and the Dirichlet distribution

We need to simulate rainfall to drive physically based rainfall run-off models of catchments and hydraulic models of urban storm water sewer systems. The flow routing models are quite sensitive to the distribution of total storm rainfall within time intervals as short as five minutes. One of many approaches to modelling rainfall is first to model the duration and total precipitation of storms and then to disaggregate the rainfall within a storm. Woolhiser and Osborn (1985) defined rescaled increments as follows. Divide a storm into m time steps, and define $R(i)$ as the total precipitation at the end of the ith step. Then $R(m) = 1$ and the rescaled increments are defined by

$$x(i) = \frac{R(i) - R(i-1)}{1 - R(i-1)}$$

for i from 1 to m. The rescaled increments must lie in the range $[0, 1]$ and Woolhiser and Osborn (1985) fitted a beta distribution (see Exercise at the end of this chapter) to data from many storms. They allowed the parameters of the distribution for $x(i)$ to depend on the value of $x(i-1)$ by fitting a linear regression of $x(i)$ on $x(i-1)$ (see Chapter 5).

The Dirichlet distribution is a multivariate generalization of the beta distribution. I shall describe an application for disaggregating yearly data to monthly values. The pdf of a standard, or Type I, Dirichlet distribution is

$$f(x_1, \ldots, x_k) = \frac{\Gamma\left(\sum_{i=0}^{k} c_i\right)}{\prod_{i=0}^{k} \Gamma(c_i)} \prod_{i=1}^{k} x_i^{c_i-1} \left(1 - \sum_{i=1}^{k} x_i\right)^{c_0-1}$$

where $(x_1, \ldots, x_l)^{\mathrm{T}}$ is the vector representing the relative monthly amounts. These are converted into proportions by dividing by the sum of x_i, which lies between 0 and 1. The mean and variance of individual elements, as proportions, are given by,

$$\mu_i = c_i/c$$

$$\sigma_i^2 = \frac{c_i(c - c_i)}{[c^2(c + 1)]}$$

where $c = \sum_{i=1}^{k} c_i$.

The covariance between X_i and X_j is given by

$$-c_i c_j / [c^2(c + 1)]$$

The c_i can be estimated by MOM. Random number generation is more convenient if we restrict the c_i to integers. The closest solution to the MOM equations, subject to this constraint is then found. For disaggregation from years to months we take $k = 12$ and set $c_0 = 1$. A program (written by Benoît Parmentier) begins with c equal to 100. The c_i are estimated by MOM using the average monthly proportions. The ratio (η) of the mean of the theoretical variances for each month to the mean of the historical variances is calculated.

$$\eta = \frac{\sum \sigma_i^2/12}{\sum_{\text{months}} \text{variance of proportions for month } i/12}$$

If η is within some chosen range, e.g. $(0.95, 1.05)$, the program stops. If it is not, the program iterates with c set equal to the product of the previous value and $\sqrt{\eta}$. During the calculations the c_i are floating point numbers, but the final values are rounded to the nearest integer.

In order to generate Dirichlet variates, we generate first independent standard Gamma variates. The relationship between variables from these distributions is

$$X_i \sim \frac{\gamma: 1, c_i}{\sum_{j=1}^{k} (\gamma: 1, c_j)} \qquad i = 1, \ldots, k$$

where the c_i of the Dirichlet distribution are also taken as the shape parameters of the Gamma distribution. A standard Gamma variate $\gamma:1,c$, where c is an integer, can be generated using

$$\gamma: 1, c \sim -log\left(\prod_{i=1}^{c} R_i\right) = \sum_{i=1}^{c} - log\, R_i$$

where R_i are independent uniform variates.

Example 2.9
The distribution was fitted to monthly inflows of the Swat River in Pakistan from 33 years. The monthly means and standard deviations, the estimated values of c_i, and a comparison of the monthly means, as proportions of the yearly total, for historic and simulated data are given in Table 2.2 on page 30. We could test the suitability of the disaggregation procedure by comparing historical correlations with those imposed by the distribution.

2.5 Summary

1. Uniform $U[0, 1]$ random numbers $\{r_j\}$ can be generated from congruential generators such as

    $$I_{j+1} = 106I_j + 1283 \qquad (\text{mod } 6075)$$
    $$r_j = I_j/6075$$

    For any serious work you should use reputable software or program the Wichmann and Hill (1982) algorithm, given in Appendix 2.

2. In principle, random numbers from a distribution with cdf F can be calculated from $U[0, 1]$ random numbers (r) using the relationship

    $$F(x) = r \qquad \text{and equivalently} \qquad x = F^{-1}(r)$$

    For discrete distributions this is x such that

    $$F(x-1) < r \le F(x)$$

    An example of a continuous distribution for which F^{-1} is available as a formula is the exponential distribution. Then:

    $$F(x) = 1 - e^{-\lambda x} = r \qquad \text{gives} \qquad x = -[\ln(1-r)]/\lambda$$

3. Suppose we wish to use realizations of a Poisson process, with rate λ, in a simulation study. We can use one of the following procedures, in which r is a random number from $U[0, 1]$.

    (i) The times between occurrences (T) are independent, exponentially distributed random variables. We can generate a sequence of times between occurrences from:

    $$t = -[\ln(1-r)]/\lambda$$

    (ii) Divide time into very small increments, Δt, such that $\lambda \Delta t$ is small (less than 10^{-3} as a guide). An occurrence is recorded if:

$$r < \lambda \Delta t$$

This will give a good approximation to a Poisson process, if $(\lambda \Delta t)^2$ and higher powers are negligible, and is convenient if the rate λ is allowed to vary seasonally throughout the year, for example.

4. F^{-1} is not available as a formula for a normal distribution. A convenient way of generating $N(0, 1)$ random numbers from independent $U[0, 1]$ numbers, paired as (r_1, r_2), is

$$z_1 = \sqrt{(-2 \ln r_1)} \sin(2\pi r_2)$$
$$z_2 = \sqrt{(-2 \ln r_1)} \cos(2\pi r_2)$$

5. Method of moments (MOM) estimates of parameters are obtained by equating sample moments with the equivalent moments of the population, which are expressed in terms of population parameters. Maximum likelihood estimates (MLE) are the values of the parameters that maximize the expression for the probability of the observed data occurring.

6. Discrete distributions. The negative binomial distribution is useful for modelling a count when the variance exceeds the mean. It is often used as an alternative to the Poisson distribution when the assumption of independent occurrences is untenable because of clustering.

$$P(x) = \frac{\Gamma(r+x)}{\Gamma(r)x!} p^r (1 - p)^x \qquad \text{for } x = 0, \ldots$$

where r now represents a parameter of the distribution rather than a uniform random number.

$$\mu = r(1 - p)/p$$
$$\sigma^2 = \mu/p$$

7. Continuous distributions.

    (i) lognormal
    X is lognormal if $Y = \ln X \sim N(\mu_Y, \sigma_Y^2)$

    $$\mu_X = \exp(\mu_Y + \sigma_Y^2/2)$$
    $$x_{0.5} = \exp(\mu_Y)$$
    $$\sigma_X^2 = \mu_X^2[\exp(\sigma^2{}_Y) - 1]$$

    (ii) Rayleigh distribution

    $$f(x) = \frac{x}{b^2} e^{-x^2/(2b^2)} \qquad \text{for } 0 \leqslant x$$

    $$F(x) = 1 - exp[-x^2/(2b^2)] \qquad \text{for } 0 \leqslant x$$

    MLE of b is $\hat{b} = \sum x_i^2/(2n)$.

Random number generation from a $U[0, 1]$ number r is:

$$x = \sqrt{[-2b^2 \ln(1 - r)]}$$

(iii) Gamma distribution

$$f(x) = \frac{\lambda(\lambda x)^{c-1}}{\Gamma(c)} \, e^{-\lambda x} \qquad \text{for } 0 \leqslant x$$

$$\mu = c/\lambda$$
$$\sigma^2 = c/\lambda^2$$
$$\gamma = 2/\sqrt{c}$$

The cdf is not available as a formula. If c is an integer, a random number can be generated from $c \cup [0, 1]$ random number $\{r_i\}$ by calculating

$$\sum_{i=1}^{c} -\ln(r_i)/\lambda$$

This makes use of the fact that a $U[0, 1]$ random number $(1 - r)$ can always be replaced by r, saving computation time. For non-integer c, use the acceptance rejection (AR) method described in Appendix 2.

Exercises

1   Box–Muller method
    Let R_1 and R_2 be independent variables from $U[0, 1]$.

   (i)   Explain why the point with coordinates

$$(\cos(2\pi R_2), \sin(2\pi R_2))$$

   is uniformly distributed on the unit circle.
   (ii) Show that

$$-2 \ln(R_1) \sim \chi_2^2$$

   (iii) If Z_1 and Z_2 are independent $N(0, 1)$, it follows, from the definition of the chi-square distribution, that if

$$W = Z_1^2 + Z_2^2$$

   then $W \sim \chi_2^2$ Deduce that $(X_1/W, X_2/W)$ is uniformly distributed on the unit circle, and so justify the construction:

$$Z_1 = (-2 \ln R_1)^{1/2} \sin(2\pi R_2)$$
$$Z_2 = (-2 \ln R_1)^{1/2} \cos(2\pi R_2)$$

2   The multinomial distribution is a multidimensional generalization of the binomial distribution. Each trial can result in one of k possible outcomes,

outcome i occurring with probability p_i. Let $P(x_1, \ldots, x_k)$ denote the probability of x_i occurrences of outcome i in n trials. Then

$$n = \sum_{i=1}^{k} x_i$$

(i)   Explain why

$$P(x_1, \ldots, x_k) = n! \prod_{i=1}^{k} p_i^{x_i}/x_i!$$

(ii) Explain why the mean and variance of an element X_i are np_i and $np_i(1 - p_i)$.

(iii) The covariance

$$\mathrm{cov}(X_i X_j) = -np_i p_j$$

for $i \neq j$. Calculate the correlation between X_i and X_j. Prove the result for the covariance by using the mgf

$$M(\theta_1, \ldots, \theta_k) = \left[\sum_{i=1}^{k} p_i \exp(\theta_i) \right]^n$$

or otherwise.

3   Show that the expected value of a variable with the truncated Poisson distribution, defined for $x = 1, \ldots$ in Example 2.4, is

$$\theta/(1 - e^{-\theta})$$

Compare the MOME with the MLE found in the example.

4   Significant wave height is defined as the mean wave height of the highest third of waves. Show that if wave heights are assumed to have a Rayleigh distribution, this is close to the product of $\sqrt{2}$ and the root mean square of all wave heights.

5   The *beta distribution* has a pdf

$$f(x) = kx^{\alpha-1}(1 - x)^{\beta-1} \qquad \text{for} \quad 0 \leqslant x \leqslant 1$$

where the normalizing constant, k, is defined in terms of the beta function

$$k = B(\alpha, \beta)$$

(i)   Use the definition of $f(x)$ to write down the definition of $B(\alpha, \beta)$.

(ii) The mean and variance are given by:

$$\mu = \alpha/(\alpha + \beta)$$
$$\sigma^2 = \alpha\beta/[(\alpha + \beta)^2(\alpha + \beta + 1)]$$

Verify this result for the special case of the uniform distribution.

(iii) Plot the pdf for a few choices of α and β.

(iv) Set up an acceptance–rejection algorithm to generate a random sample from a beta distribution.

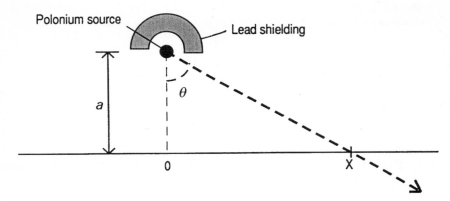

Figure 2.8   The variable *X* has a Cauchy distribution

(v) Show that the MOME of α and β are
$$\hat{\alpha} = \bar{x}\{[\bar{x}(1-\bar{x})/s^2] - 1\}$$
$$\hat{\beta} = (1-\bar{x})\{[\bar{x}(1-\bar{x})/s^2] - 1\}$$

6   Think of some possible graphical representations for circular data. The terms angular histogram, radar plot and rose diagram are used in this context.

7   The pdf of the Fisher distribution is
$$f(\theta, \psi) = C_F\exp[\kappa(\sin\theta\,\sin\alpha\,\cos(\psi-\beta) + \cos\theta\,\cos\alpha)\sin\theta]$$
$$\text{for} \qquad 0 \leqslant \theta \leqslant \pi, \qquad 0 \leqslant \psi < 2\pi$$

There are three parameters, κ, α and β. The location parameters are α and β. The shape parameter κ is called the concentration parameter. Show that the conditional distribution of Ψ, given that Θ equals θ, is a Von Mises distribution.

8   In Figure 2.8 the radioactive source emits α-particles at angles θ which are uniformly distributed over $[-\pi/2, \pi/2]$. A screen is set up at a distance a from the source.
(i) Show that the intercepts on the screen X have a *Cauchy distribution* which has pdf
$$f(x) = \frac{a}{\pi(a^2 + x^2)} \qquad \text{for } -\infty < x < \infty$$

(ii) Show that the Cauchy distribution has an infinite variance if the mean (which is not properly defined) is taken as 0. The central limit theorem does not hold for the Cauchy distribution. The ratio of two standard normal random variables has a Cauchy distribution, and ratios for which the denominator can be very close to zero will behave similarly.

3

Quality Assurance and Reliability

3.1 Components of variance: reducing variability in the strength of concrete

If we are to reduce, or at least control, variation we need to identify its sources. Graham and Martin (1946) describe the methods used to ensure the high quality of concrete paving used for the main runway at Heathrow. Since the compressive strength of test cubes made from the same batch of concrete varies, six cubes were made from each of four different batches of concrete during a working day.

The compressive strengths of two of the cubes in each batch were measured after 7 days, and the strengths of the remaining four were measured after 28 days. Graham and Martin's paper includes results on nearly 3000 cubes, and I have randomly selected two consecutive sets of cubes from every third page. The 28-day compressive strengths are given in Table 3.1, the units have been changed from pounds per square-inch to MPa. We will assume that while consecutive sets of cubes were from two batches made using the same delivery of cement, these pairs of batches were sufficiently separated, in time, for them to have been made from different deliveries of cement. To summarize: from each delivery of cement we have two batches of concrete mix; and from each batch we have the 28-day strengths of four concrete test cubes. This arrangement is shown schematically in Figure 3.1. Denote a typical observation by Y_{ijk},

Cement delivery

Concrete batch

Concrete test cubes

Figure 3.1 Hierarchical structure for concrete test cubes

Table 3.1 *Compressive strengths of 28-day concrete cubes (MPa) (source Graham and Martin, 1946)*

35.6	38.6	30.7	31.7	30.0	27.9	34.3	38.7	33.2	35.8	39.5	38.7
33.6	41.6	30.5	30.0	35.0	27.7	36.4	38.5	35.2	37.1	42.1	36.1
34.1	40.7	27.2	33.8	35.0	29.0	33.4	43.3	37.8	37.1	38.5	35.9
34.5	39.9	26.8	29.6	32.6	32.8	33.4	36.7	35.4	39.5	40.2	42.8

Batch means

34.45	40.20	28.80	31.28	33.15	29.35	34.38	39.30	35.40	37.38	40.08	38.38

Cement delivery means

37.32		30.04		31.25		36.84		36.39		39.23	

Overall mean

35.18

in which the i refers to the delivery of cement, the j refers to the batch, and the k refers to the test. A model for this observation is:

$$Y_{ijk} = \mu + \alpha_i + \beta_{j(i)} + \varepsilon_{k(ij)}$$

$\alpha_i \sim$ mean 0 and variance σ_α^2

$\beta_{j(i)} \sim$ mean 0 and variance σ_β^2

$\varepsilon_{k(ij)} \sim$ mean 0 and variance σ_ε^2

μ is an overall mean.

i runs from 1 to 6 and indicates the cement delivery

j runs from 1 to 2 and is the batch number

k runs from 1 to 4 and is the cube number.

The notation $\beta_{j(i)}$, and $\varepsilon_{k(ij)}$, emphasizes that there is no connection between batch 1 and batch 2 in different cement deliveries, or between the numbering of test cubes in different batches. This is referred to as a *hierarchical design*. If we assume that the three components of variance are independent, then (as shown in Appendix 1)

$$\text{var}(Y_{ijk}) = \sigma_\alpha^2 + \sigma_\beta^2 + \sigma_\varepsilon^2$$

and

$$\text{stdev}(Y_{ijk}) = \sqrt{\sigma_\alpha^2 + \sigma_\beta^2 + \sigma_\varepsilon^2}$$

The result for the standard deviation is an application of Pythagoras' Theorem in three dimensions (Figure 3.2). Our objective is to estimate the three components of variance. Define:

$\overline{Y}_{ij}$. as the mean of the four strength measurements within a batch, the batch mean,

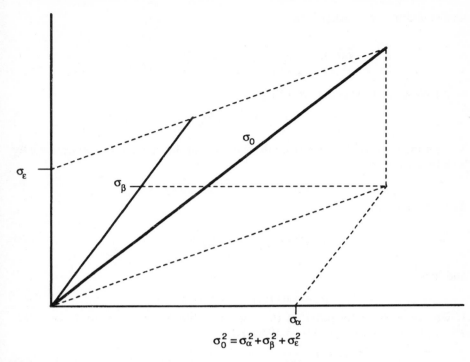

$$\sigma_0^2 = \sigma_\alpha^2 + \sigma_\beta^2 + \sigma_\varepsilon^2$$

Figure 3.2   Independent components of variance are additive

$\overline{Y}_{i..}$ as the mean of the two batch means within a cement delivery, the delivery mean,

$\overline{Y}_{...}$ as the mean of the six delivery means, the overall mean.

In order to generalize the formulae we now assume n cubes within a batch, b batches of concrete made from a delivery of cement and a deliveries of cement:

$$i = 1, \ldots, a; \quad j = 1, \ldots, b \quad \text{and} \quad k = 1, \ldots, n$$

The estimator, $\hat{\sigma}_\varepsilon^2$, of σ_ε^2 is the average of the within batch estimators of the test variance

$$\hat{\sigma}_\varepsilon^2 = \sum_{i=1}^{a} \sum_{j=1}^{b} \left\{ \sum_{k=1}^{n} (Y_{ijk} - \overline{Y}_{ij.})^2/(n-1) \right\} \Big/ (ab)$$

The next step is to estimate σ_β^2 from the standard deviation of the batch means, but we have to take into account the variation of the estimators of the within batch means. That is, for a fixed value of i

$$\operatorname*{var}_{j|i}(\overline{Y}_{ij.}) = \sigma_\beta^2 + \sigma_\varepsilon^2/n$$

where the subtext $j|i$ is to emphasize that it is the variance over batches (j) given a fixed delivery of cement (i). The estimator of the variance of batch means within cement deliveries is

$$\widehat{\text{var}}_{j} (\overline{Y}_{ij.}) = \sum_{j} \left\{ \sum_{i=1}^{a} \left[\sum_{j=1}^{b} (Y_{ij.} - Y_{i..})^2 / (b-1) \right] \right\} \bigg/ a$$

so it follows that the estimator of σ_β^2 is

$$\hat{\sigma}_\beta^2 = \widehat{\text{var}}_{j}(\overline{Y}_{ij.}) - \hat{\sigma}_\varepsilon^2 / n$$

The estimator of σ_α^2 follows from the variance of the cement delivery means by a similar argument

$$\text{var}(\overline{Y}_{i..}) = \sigma_\alpha^2 + \sigma_\beta^2 / b + \sigma_\varepsilon^2 / (nb)$$

The estimator of $(\overline{Y}_{i..})$ is

$$\widehat{\text{var}}(\overline{Y}_{i..}) = \sum_{i=1}^{a} (\overline{Y}_{i..} - \overline{Y}_{...})^2 / (a-1)$$

and hence

$$\hat{\sigma}_\alpha^2 = \widehat{\text{var}}(\overline{Y}_{i..}) - \hat{\sigma}_\beta^2 / b - \hat{\sigma}_\varepsilon^2 / (nb)$$

If we apply these formulae to the data in Table 3.1 we obtain: $\hat{\sigma}_\varepsilon^2 = 4.1797$, $\widehat{\text{var}}_{j}(\overline{y}_{ij.}) = 7.0562$, $\widehat{\text{var}}(\overline{y}_{i..}) = 13.3956$. Hence

$$\hat{\sigma}_\varepsilon^2 = (2.044)^2$$
$$\hat{\sigma}_\beta^2 = 7.0562 - (2.044)^2 / 4 = (2.452)^2$$
$$\hat{\sigma}_\alpha^2 = 13.3956 - (2.452)^2 / 2 - (2.044)^2 / 8 = (3.141)^2$$

The best strategy for reducing overall variability is to concentrate on the factors that make the biggest contributions to the sum of variances. The most important factor appears to be variation in cement between deliveries, but the variation between batches is also quite substantial. Also, our estimators of the variances are not very precise: $\hat{\sigma}_\varepsilon^2$ is based on $(4-1) \times 12 = 36$ degrees of freedom; but σ_β^2 is based only on $(2-1) \times 6 = 6$ degrees of freedom; and σ_α^2 is based only on $6 - 1 = 5$ degrees of freedom. A 90% confidence interval for the ratio of σ_β^2 to σ_α^2 is given by

$$\left[\frac{\hat{\sigma}_\beta^2}{\hat{\sigma}_\alpha^2} F_{5,6,0.95}, \quad \frac{\hat{\sigma}_\beta^2}{\hat{\sigma}_\alpha^2} F_{5,6,0.05} \right]$$

which is

$$[0.123, \quad 2.67]$$

Details of this construction are given in Appendix A1.18. The corresponding 90% confidence interval for the ratio of σ_β to σ_α is [0.35, 1.63], and since this includes 1 we cannot be confident that the between batch variability is less important. It would be prudent to try and reduce both the variation between deliveries of cement, and the variation between batches made from the same delivery.

3.2 Case study: polished stone value inter-laboratory trial

The analysis of an inter-laboratory trial depends on the same statistical principles that were used for identifying components of variance in Section 3.1.

3.2.1 Background to the trial

The friction between a vehicle's tyres and a tarmacadam road is due to the aggregate that is bound with the tar. A good roadstone will maintain frictional forces despite the polishing action of tyres. British Standard BS 812: Part 114: 1989 is a method for measuring the friction between rubber and polished-stone:

> The principle is that the polished-stone value PSV gives a measure of the resistance of roadstone to the polishing action of vehicle tyres under conditions similar to those occurring on the surface of a road. Where the surface of a road consists largely of roadstone the state of polish of the sample will be one of the main factors affecting the resistance of the surface to skidding. The actual relationship between PSV and skidding resistance will vary with the traffic conditions, type of surfacing and other factors.

The test is based on a comparison between the stone to be tested and a control stone, which has been stockpiled by the recognized supplier. However, tests on supposedly identical materials will rarely give identical results. Many factors contribute to this variability, including differences in: test specimens; equipment; operators; calibration; and environmental conditions. *Precision* is a measure of how close together are results of tests on the same material. It is defined as the reciprocal of the variance of test results. The term *accuracy* is reserved for a measure of how close average test results are to the true value. The true value is either established by, or derived from, international conventions and accuracy depends on careful calibration against standards. In the UK, the National Physical Laboratory offers a formal measurement accreditation service to calibration and testing laboratories (NAMAS). In 1994 there were about 30 NAMAS accredited laboratories which offered the PSV test facility. Sixteen agreed to take part in an inter-laboratory trial, which was run according to the British Standard on the precision of test methods (BS 5497: Part 1: 1987).

The objective of the trial was to review the *repeatability* and *reproducibility* of the test method. Repeatability quantifies the precision of tests, on specimens from the same material, under conditions that are as constant as possible: performed by the same operator; using the same equipment; during a short interval of time. Reproducibility refers to tests on the same material in different laboratories with different apparatus and operators. Since reproducibility is a comparison of different laboratories it will take account of inaccurate calibrations: 'Thus repeatability and reproducibility are two extremes, the first measuring the minimum and the second the maximum variability in results' (BS 5497: Part 1: 1987). The 16 laboratories were sent samples of two roadstones referred to as *X* and *Y*. The method for sampling aggregates described in another British standard, BS 812: Part 102: 1989, was used to take bulk samples of X and Y. Each bulk sample was divided into 16 small samples which were randomly assigned to the 16 laboratories. So a sample of X and a sample of Y were dispatched to each laboratory with the following instructions.

3.2.2   Instructions to laboratories

PSV INTER-LABORATORY COMPARISON – INSTRUCTIONS TO LABORATORIES

1. Each of the two samples (X and Y) should be divided into four specimens.
2. For the first run of the test, polishing and friction testing according to BS 812: Part 114: 1989, use two specimens of the control stone C, two specimens of X and two specimens of Y selected as follows.
   (i)   Label the four specimens of X from 1 to 4.
   (ii) Toss two coins, of different values A and B, and assign a number to the outcome according to the rule:

Coin A	Coin B	
Head	Head	1
Head	Tail	2
Tail	Head	3
Tail	Tail	4

   (iii) Repeat, as necessary, to obtain a second, different, number.
   (iv) Use the two corresponding specimens of X.
   (v)   Follow the same procedure to select two specimens of Y.
3. Repeat the whole procedure, including the polishing, for the second run. [Please do not polish all the specimens at the same time!]

IMPORTANT NOTES

(a)   The test (BS 812: Part 114: 1989) consists of two independent runs and includes a check of repeatability (paragraph 11.2). If we wish to deduce the repeatability of the entire test procedure the polishing must be repeated. It would not do to polish all four specimens at the same time.

(b)   The two runs should be carried out by the same operator on the same equipment under similar environmental conditions. The equipment should not be recalibrated, except for the adjustments specified in BS 812: Part 114 such as the height adjustment specified in paragraph 10.5. It is most important that the whole test is performed as usual.

3.2.3   Data from laboratories

The following data were taken from the laboratory reports (indexed by j which runs from 1 to 16) in order to calculate repeatability (r) and reproducibility (R) defined in BS 5497: Part 1: 1987.

(i)   The average of the two friction test readings (FTR) values for the two specimens of X in run 1, denoted by

$$x_{1j}$$

(ii)   The average of the two FTR values for the two specimens of Y in run 1, denoted by

$$y_{1j}$$

(iii)   The average of the two FTR values for the two control specimens in run 1, denoted by

$$z_{1j}$$

(iv)   The average of the two FTR values for the two specimens of X in run 2 (x_{2j}).

(v)   The average of the two FTR values for the two specimens of Y in run 2 (y_{2j}).

(vi)   The average of the two FTR values for the two control specimens in run 2 (z_{2j}).

The original data and these averages are given in Table 3.2.

3.2.4   Analysis of within and between laboratory variation

Roadstone X   For each laboratory, a polished stone value (PSV, v_{ij}) is calculated, for each run, from the formulae (BS 812):

$$v_{1j} = x_{1j} + 52.5 - z_{1j}$$
$$v_{2j} = x_{2j} + 52.5 - z_{2j}$$

The 52.5 is the standard PSV value of the control stone. The estimate of within laboratory variance, s_w^2, is calculated from the formula[1]:

$$s_w^2 = \left[\sum_{j=1}^{16} (v_{1j} - v_{2j})^2 / 2 \right] \Big/ 16$$

Next, the averages of the two PSV measurements from the 16 laboratories are calculated,

$$\bar{v}_j = (v_{1j} + v_{2j})/2$$

followed by the overall average PSV for stone X

$$\bar{v} = \sum \bar{v}_j / 16$$

and then the variance of the 16 averages

$$s_o^2 = \sum (\bar{v}_j - \bar{v})^2 / (16 - 1)$$

The estimate of the between laboratories variance, s_b^2, is calculated from the relation

$$s_o^2 = s_b^2 + s_w^2 / 2$$

---

[1] $[(v_{1j} - \bar{v}_j)^2 + (v_{2j} - \bar{v}_j)^2]/(2 - 1) = (v_{1j} - v_{2j})^2 / 2$

Table 3.2   *(a) Reported friction test readings (FTR) of roadstone from 16 laboratories; (b) averages of two FTR values from the same run*

a (FTR) of roadstone from 16 laboratories

Laboratory	Run 1						Run 2					
	X run 1		Y run 1		Control $r1$		X run 2		Y run 2		Control $r2$	
A	62.3	62.0	55.0	56.0	55.7	55.0	60.7	59.3	54.7	54.7	56.5	54.3
B	52.3	54.7	50.7	50.0	52.0	49.3	53.7	54.7	49.3	48.0	50.3	52.0
C	55.0	55.0	52.0	52.3	53.0	51.3	55.3	55.0	52.3	52.0	53.0	51.0
D	62.0	61.0	54.0	53.0	54.0	54.0	61.3	61.7	52.7	53.7	54.0	54.0
E	62.3	62.3	58.7	59.3	54.7	54.3	62.7	63.0	57.7	58.7	54.3	53.7
F	55.0	58.0	54.0	54.7	52.0	51.7	54.3	55.0	52.0	52.7	51.0	52.0
G	58.0	60.0	55.7	55.3	53.0	52.7	59.3	55.3	52.3	50.0	55.0	54.0
H	56.0	56.0	49.0	50.0	49.0	49.0	54.0	56.0	49.0	57.0	50.0	50.0
I	57.0	59.0	52.0	52.0	53.0	53.0	59.0	58.0	52.0	52.0	52.3	52.0
J	55.3	53.0	50.0	52.0	51.0	51.0	56.0	56.0	52.7	52.3	50.7	50.3
K	55.0	54.3	49.0	49.0	50.3	49.0	54.7	55.3	50.0	48.3	50.3	49.7
L	54.3	55.3	48.0	48.0	50.0	49.0	55.0	52.0	50.0	48.3	46.3	53.0
M	64.0	64.3	54.3	54.0	55.0	55.0	62.0	60.3	55.0	55.0	55.0	55.0
N	59.3	48.7	53.0	53.0	52.0	52.7	60.7	60.7	54.3	53.3	53.7	53.3
O	64.3	64.0	56.7	56.0	53.0	54.3	60.0	62.0	52.3	55.3	50.3	49.7
P	60.0	58.3	49.0	49.3	51.0	52.0	62.0	61.0	51.0	51.3	53.0	52.7

b Averages of two FTR values from the same run

Laboratory	Run 1			Run 2		
	x_1	y_1	z_1	x_2	y_2	z_2
A	62.15	55.50	55.35	60.00	54.70	55.40
B	53.50	50.35	50.65	54.20	48.65	51.15
C	55.00	52.15	52.15	55.15	52.15	52.00
D	61.50	53.50	54.00	61.50	53.20	54.00
E	62.30	59.00	54.50	62.85	58.20	54.00
F	56.50	54.35	51.85	54.65	52.35	51.50
G	59.00	55.50	52.85	57.30	51.15	54.50
H	56.00	49.50	49.00	55.00	53.00	50.00
I	58.00	52.00	53.00	58.50	52.00	52.15
J	54.15	51.00	51.00	56.00	52.50	50.50
K	54.65	49.00	49.65	55.00	49.15	50.00
L	54.80	48.00	49.50	53.50	49.15	49.65
M	64.15	54.15	55.00	61.15	55.00	55.00
N	54.00	53.00	52.35	60.70	53.80	53.50
O	64.15	56.35	53.65	61.00	53.80	50.00
P	59.15	49.15	51.50	61.50	51.15	52.85

The numerical results are:

$$s_o^2 = 4.687 \qquad s_o = 2.16$$
$$s_b^2 = 3.531 \qquad s_b = 1.88$$
$$s_w^2 = 2.314 \qquad s_w = 1.52.$$

A 95% confidence interval for the PSV of stone X can be calculated from

$$\bar{v} \pm t_{15,.025} \quad s_0/\sqrt{16}$$

which is

$$58.3 \pm 1.2$$

Roadstone Y In the same notation, where it is assumed that y_{ij} have replaced the x_{ij}, , the results are:

$$s_o^2 = 2.913 \qquad s_o = 1.71$$
$$s_b^2 = 1.992 \qquad s_b = 1.41$$
$$s_w^2 = 1.843 \qquad s_w = 1.36$$

A 95% confidence interval for the PSV of stone Y is

$$52.8 \pm 0.9$$

3.2.5 *Calculation of repeatability and reproducibility*

Roadstone X In practice, the laboratories would report the average of the two runs to clients (BS 812: Part 114: 1989). Therefore, the within laboratory variance would be decreased by a factor of 2. Call this the adjusted within laboratory variance, estimated by

$$s_{wadj}^2 = s_w^2/2 = 1.16$$

The rationale for defining r as $2.8 s_{wadj}$ is that a 95% confidence interval for the difference between two test results under repeatability conditions is

$$\pm 1.96\sqrt{(\sigma_{wadj}^2 + \sigma_{wadj}^2)}$$

which is estimated by

$$\pm 2.8 s_{wadj}$$

In this case

$$r = 2.8 s_{wadj} = 3.0$$

The between laboratories variance is estimated by s_b^2. The overall variance of a PSV is the sum of the between laboratories variances and the adjusted within laboratory variance. The definition of R includes a factor of 2.8 for the same reason that it appears in the definition or r. So

$$R = 2.8\sqrt{(s_b^2 + s_{wadj}^2)} = 6.1$$

If two laboratories report PSVs for stone from the same source which differ by more than R, some investigation might be warranted.

Roadstone Y In the same notation

$$s_{wadj}^2 = 1.843/2 = 0.9215$$

and it follows that

$$r = 2.7$$
$$R = 4.8$$

3.2.6 Discussion

1. A 95% confidence interval for σ_{wadj}, and hence r, can be constructed if it is assumed that: the within laboratory standard deviation is the same for all laboratories; and test results are near normally distributed. The estimator s_{wadj} has 16 degrees of freedom, because it is the average of 16 independent estimators which all have one degree of freedom. So

$$16s^2_{wadj}/\sigma^2_{wadj} \sim \chi^2_{16}$$

and the 95% confidence interval for σ_{wadj} based on roadstone X is:

$$[0.80, 1.64]$$

The corresponding confidence interval for the population equivalent of r follows from multiplication by 2.8:

$$[2.2, 4.6]$$

The interval based on roadstone Y is:

$$[2.0, 4.1]$$

The confidence intervals for R should be slightly wider, in proportional terms. It follows that both estimates of r and R are well within sampling error of the values given in BS 812: Part 114: 1989, which are 3 and 5 respectively.

2. The standard, BS 5497, assumes that within laboratory variances can reasonably be considered equal, and recommends testing this assumption with Cochran's test. The test statistic (C) is based on a set of p sample standard deviations, all computed from the same number of test results.

$$C = s^2_{max} \bigg/ \sum_{i=1}^{p} s^2_i$$

Critical values for C at the 5% and 1% levels are given for $p = 2$ to 40 and $n = 2$ to 6 in the standard. In our case $p = 16$ and $n = 2$, and several comparisons of within laboratory variances can be made: variability within runs and between runs for roadstone X, roadstone Y and the control stone. The calculated values of C are given in the following table.

	Within 1 run	Within run 2	Between runs
Roadstone X	0.769	0.381	0.513
Roadstone Y	0.432	0.714	0.345
Control stone	0.432	0.746	0.607

The 5% and 1% critical values (Annex A of BS 5497) are 0.452 and 0.553 respectively. The high within run values (X run 1, Y run 2, control run 2) are due to three surprisingly large within run differences which occurred at three different laboratories. There is some evidence that variability between runs differs amongst laboratories (from the control stone results and, to a lesser extent, from roadstone X). However, the largest differences for the control stone and roadstone X are again at different laboratories. In contrast to this, laboratories 3 and 4 returned very consistent FTR for both runs of all three roadstones. If these results are excluded, the estimated within laboratory standard deviation increases by 7%. In my opinion, any differences in within laboratory standard deviations do not invalidate the analysis.

3. Stone Y was from the same source as the control stone, so its PSV should be 52.5. The laboratories did not know the origin of Y at the time of the test. The mean PSV over all laboratories for Y (52.8) is within sampling error (its standard error is $1.71/\sqrt{16}$) of 52.5.

4. It follows from their definitions that any difference between r and R is due to a between laboratory variance. The usual explanation for a between laboratory variance is that there will be small systematic differences between equipment and operators' techniques. But the PSV relies on a comparison with a control stone, and such differences should therefore cancel out for roadstone Y, which is from the same source as the control. The same argument does not apply to roadstone X because different laboratories might deviate in their assessments of actual differences, and this could account for the higher R value for roadstone X. The most likely explanation for a between laboratories variance, when measuring the PSV of Y, is differences in the control stone held by laboratories.

It is rather unlikely that the apparent between laboratories variance, when measuring the PSV of Y, is due to sampling variability. If there is no between laboratories variance, σ_b^2 equals 0, then the expected value of s_0^2 would equal the expected value of $s_w^2/2$, and it would follow that:

$$\frac{2s_0^2}{s_w^2} \sim F_{15,15}$$

The calculated value of this statistic is $2 \times (2.913)/1.843$ which equals 3.16. The probability of a value at least this high, compared with the 'expected' 1, is about 0.02.[2] Formally, the P-value equals 0.02. Furthermore, the largest PSV for roadstone Y is 56.85 from Laboratory 5. This is not identified as an outlier by Dixon's test (BS 5497). The test statistic is:

$$\max\left\{\frac{51.500 - 50.475}{54.250 - 50.475}, \frac{56.850 - 54.250}{56.850 - 51.500}\right\} = 0.486$$

This is less than the 5% critical value, which is 0.546 (Annex B of BS 5497). However, the difference between 52.5 and 56.85 is more than four

---

[2] This statistical test depends on an assumption that the within laboratory variances are equal, but does not seem very sensitive to the assumption. For example, a simulation with laboratory variances ranging from 1^2 up to 16^2 gave the following upper percentage points: 10% 2.4, 5% 3.0, 2% 3.7 and 1% 4.7. Even in this rather extreme scenario, the P-value would be less than 0.05.

within laboratory standard deviations, and such a large discrepancy is most unlikely if there is no between laboratories variance.

3.2.7   Summary

1. The values of r and R calculated from the experiment are consistent with those given in BS 812: Part 114: 1989.

2. The experiment with 16 laboratories is precise enough to have reasonable empirical evidence that R is greater than r, even when comparing a stone (Y) from the same source as the control stones.

3. The natural explanation for this difference between R and r, in a comparative experiment, is that stocks of control stone differ slightly. The experiment is not sufficiently precise to give strong evidence in favour of, or against, this hypothesis.

4. If the difference between r and R is of practical importance, a follow up experiment involving the exchange of control stones could be designed.

5. The estimated value of R for stone X was higher than that for Y. It is quite likely that R does increase with the magnitude of the difference in PSV between roadstone and the control. This would account for the between laboratory variance when comparing stones X and Y directly, without reference to the control stone.

3.3   Control and CUSUM charts

Day (1995) defines *structurally defective* concrete as that which is unable to serve its purpose. We must not allow any such concrete to be produced because we cannot rely on any practical monitoring system to guarantee detection, and even if we could the delays involved would be unacceptable. Day defines *contractually defective* concrete as that which does not quite meet the specification. A small proportion of contractually defective concrete in a structure is tolerable. There is usually a substantial difference between contractually and structurally defective concrete. So a system that detects, and penalizes, production of the former should prevent any of the latter being made.

Even when a process is running well, i.e. *in control*, there will be some inherent variability, which is often called *common cause variation*. A concrete supplier needs to ensure that the common cause variation is low enough to meet the specification. Although it would usually be possible to remove some of this inherent variability by buying more expensive plant and materials, it is unlikely to be economic to do so. We will assume that the standard deviation of samples taken from batches when the process is in control can be estimated from records. The purpose of a control chart is to indicate when a change has taken place, in time for corrective action to be taken. This presents an immediate challenge, because one of the most important measures of the soundness of concrete is its 28-day strength. Day (1995) recommends charting several variables that can be measured more quickly, such as slump and 24-hour density. Slump may not be a reliable absolute measure, but sudden changes in slump measurements made on batches of concrete from the same supplier, on the same contract, should be investigated. It may also be possible to make reasonably precise estimates of 28-day strength from slump, and other variables measured in the first 24-hours, using multiple regression (see Chapter 5).

The most common charts are Shewhart charts for sample means, with a corresponding chart for the sample ranges, and cumulative sum (CUSUM) charts. There are many variations and modifications. A useful recent reference is Caulcutt (1995), and the relevant British Standard is BS5700 15703: 1984.

3.3.1 Shewhart mean and range charts

Suppose we make five slump tests a day on concrete delivered during the construction of a dam. The tests should be spread out throughout the day, but the timing of them should be randomized subject to this restriction. Assume we have considerable experience of slump tests, and records of this supplier's performance on similar contracts. On the basis of this information, we are prepared to assume a standard deviation (σ) of slump tests of 8 mm. The specified slump (τ) is 75 mm (drop of apex of a fresh concrete cone). Although individual slump measurements are not necessarily normally distributed, it is reasonable to expect means of 5 to be well approximated by a normal distribution. The principle, for samples of size n, is that

$$\overline{X} \sim N(\mu, (\sigma/\sqrt{n})^2)$$

If the mean is on target, that is $\mu = \tau$, the probability that $\overline{X}$ exceeds

$$\tau + 3.09\sigma/\sqrt{n}$$

is 0.001, and the probability that $\overline{X}$ is less than

$$\tau - 3.09\sigma/\sqrt{n}$$

is the same.

The procedure for the mean chart is to plot sample means against time and take action if a point falls outside action limits drawn at

$$\tau \pm 3.09\sigma/\sqrt{n}$$

The idea dates back to the work of Shewhart, and others, in the 1920s and the mean chart is also commonly known as a *Shewhart chart*. The rationale behind it is that action will be taken unnecessarily, when μ is on target, at an average rate of 1 in 500 samples but that this is offset by the advantages of detecting changes in μ and taking suitable action.

An increase in process variability is as serious as a change in the mean. It is slightly more convenient to monitor ranges instead of standard deviations for small samples of the same size. Factors of σ from which to calculate upper (1 in 1000) and lower (1 in 1000) action lines for the ranges of samples of size n from a normal distribution are given in Table A15.6. These factors are sensitive to the assumption that the sampling is from a normal distribution. If the assumption that the variable is from a normal distribution is not realistic, the action lines should be treated as a rough guide only. More accurate factors could be obtained from simulation experiments. The data in Table 3.3 are summaries of slump measurements, from samples of 5, over 12 days, and

Table 3.3 *Means and ranges of slump values from samples of size 5*

Sample number	Mean	Range
1	77	24
2	81	39
3	80	11
4	78	17
5	72	43
6	70	16
7	78	12
8	83	27
9	76	35
10	77	9
11	84	25
12	89	15

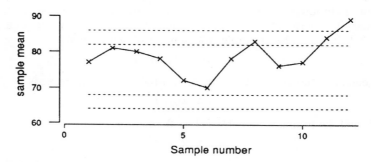

(a)

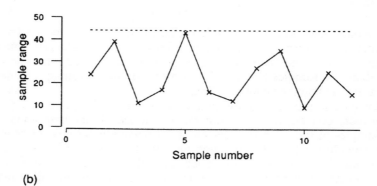

(b)

Figure 3.3 Mean (a) and range (b) charts for concrete slump measurements

Figures 3.3 (a) and (b) are the mean and range charts. The action limits for the mean chart are set at:

$$75 \pm 3.09 \times 8/\sqrt{5}$$

Hence the lower and upper action lines are set at 64 and 86 respectively. The upper action limit for the range is set at:

$$5.48 \times 8 = 44$$

The lower 0.1% limit is

$$0.37 \times 8 = 3$$

There is evidence of an increase in slump on day 12.

3.3.2   CUSUM charts

In Section 3.3.1 we implicitly assumed that variation between days is unacceptable. If we are willing to tolerate some between day variation the mean chart action limits could be increased to $\pm 3.09(\sigma_b^2 + \sigma^2/n)^{1/2}$, where σ_b is the standard deviation between days. If σ_b is of a similar magnitude to σ, it

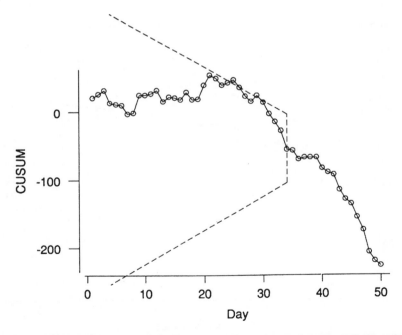

Figure 3.4   CUSUM chart for daily slump measurements. In practice the mask is centred on the current observation. The mask is shown as it would be at day 34 when action is first indicated, because the trace crosses the lines. The width of the front of the mask is 10 standard deviations of the variable plotted (10 in this example). The gradient of the lines is 5 standard deviations per ten sampling units (e.g, width of mask ten days back is 20 standard deviations). With this construction, the false alarm rate is about 1 in 440

Table 3.4 *Daily slump measurements, deviations from 150 mm*

Day	Dev	CUSUM	Day	Dev	CUSUM	Day	Dev	CUSUM
1	21	21	18	−11	18	35	−2	−57
2	5	26	19	1	19	36	−13	−70
3	6	32	20	20	39	37	3	−67
4	−19	13	21	15	54	38	0	−67
5	−2	11	22	−5	49	39	0	−67
6	−1	10	23	−10	39	40	−16	−83
7	−13	−3	24	4	43	41	−6	−89
8	1	−2	25	4	47	42	−3	−92
9	27	25	26	−11	36	43	−22	−114
10	0	25	27	−13	23	44	−14	−128
11	2	27	28	−8	15	45	−7	−135
12	5	32	29	9	24	46	−20	−155
13	−17	15	30	−10	14	47	−19	−174
14	7	22	31	−17	−3	48	−33	−207
15	−1	21	32	−12	−15	49	−13	−220
16	−3	18	33	−13	−28	50	−7	−227
17	11	29	34	−27	−55			

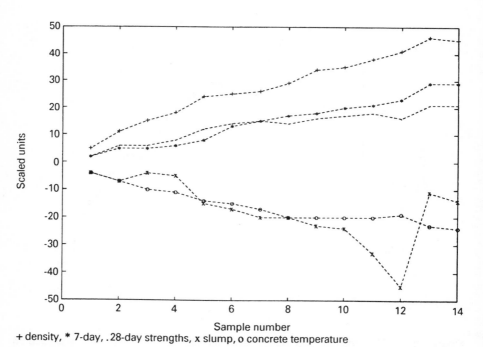

+ density, * 7-day, . 28-day strengths, x slump, o concrete temperature

Figure 3.5 Multivariate CUSUM from CONAD system

may suffice to take one or two samples a day. The cumulative sum chart is a plot of

$$S_t = \sum_{i=1}^{t} x_t - \tau$$

against t, where x_t is the daily slump measurement: or mean of slump measurements if more than one is taken each day. Steep slopes indicate a change in the process mean. A V-mask can be used to decide whether action is justified. The method is demonstrated in Figure 3.4 using the data from Table 3.4 with an overall standard deviation of x_t of 10 mm: corresponding to $\sigma = 8$, $n = 2$, and $\sigma_b = 8.25$. The V-mask can be replaced by a computer algorithm based on some coordinate geometry. The Conad quality control system (Day, 1995) implements this and can plot several CUSUMs on the same graph. A multivariate CUSUM from the Conad system is shown in Figure 3.5. A detailed discussion of CUSUM charts is given in BS5703: Part 1. Their advantages, over Shewhart charts, are that they give a more dramatic indication of a step change in the mean and, on average, lead to quicker control action with the same chance of false alarms. The theory behind the V-mask is given in Appendix 6, and it assumes that the x_t are from a near normal distribution. Day (1995) has found that most important changes in concrete supply are step changes, although he warns that this is not always so.

3.4 Reliability

3.4.1 System reliability in terms of component reliabilities

A complex system can be broken down into components, and its overall reliability expressed in terms of component reliabilities which are usually easier to specify. We define the reliability of the office sprinkler system (Figure 3.6) as the probability it operates if there is a fire. The component reliabilities are

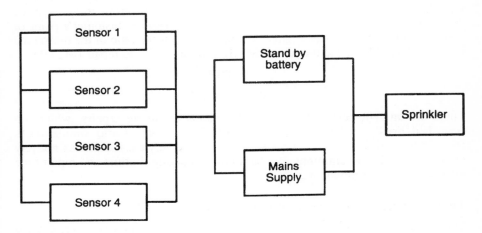

Figure 3.6 Office sprinkler system

the probabilities they function correctly in a fire. The four wireless smoke detectors contain small internal batteries and have reliabilities R_d. The mains supply, back-up power battery, and sprinkler mechanism have reliabilities R_m, R_b and R_k respectively.

The system will operate if: at least one smoke detector works; and the battery or mains power supply works; and the sprinkler mechanism works. The overall reliability (R) follows from the elementary rules of probability (Appendix A1.1). If we assume failures are independent:

$$R = [1 - (1 - R_d)^4] \times [1 - (1 - R_b)(1 - R_m)] \times R_k$$

Suppose R_d, R_b, R_m and R_k have values of 0.95, 0.99, 0.50 and 0.98 respectively. Then R equals 0.975, and the best way to increase the reliability is to improve the sprinkler mechanism. This conclusion would be very different if failures of smoke detectors are not independent. A common cause of failure might be not replacing the small internal batteries in the smoke detectors. One way of allowing for this is to include a common cause element in series with the parallel block of smoke detectors. The probability that a smoke detector fails is assumed to be 0.05. Suppose this can be split up into a common cause failure with probability 0.02 and an independent failure with probability 0.03. The slight implied increase in the probability that a smoke detector fails, 0.02×0.03, is negligible. Then R_d becomes 0.97, but there is an additional factor (R_{cc}) of 0.98 in the expression for R. The overall reliability is now reduced to 0.956. Many systems cannot be reduced to a sequence of series and parallel connections (see Exercises 3 for an example), and there are several techniques for dealing with these including a complete enumeration (e.g. Andrews and Moss, 1993).

A distributed system, such as an electricity or a water supply system (Figure 3.7 after Peel, 1991), is more difficult to analyse because components will be repaired. Both trunk mains may burst at some places during a year, but the supply to the distribution zone will only be cut off if they burst at similar times. We could simplify the system, by assuming all repairs take the rest of a day, and then analyse it in a similar way to the sprinkler system to obtain a reliability for one day, and hence any number of days. However, even if we are prepared to work with the assumption about repairs, this would not give us information about the duration of loss of supplies. The relevant measure from a customer's point of view is availability. In a steady state, availability (A) is the ratio of the mean time before failure (MTBF) to the sum of MTBF and mean time to repair (MTTR)

$$A = \frac{\text{MTBF}}{\text{MTBF} + \text{MTTR}}$$

There are analytic results for idealized systems involving repairs, which are described briefly in Section 3.4.4. Nevertheless, Monte Carlo simulation is frequently used for complex systems, as it is relatively easy and does not rely on any restrictive assumptions. We begin by estimating lifetime and repair time distributions for components.

3.4.2 General results for life distributions

Let the time to failure of a particular component be T with a cdf $F(t)$ defined for positive t. The results are also applicable when time is replaced by

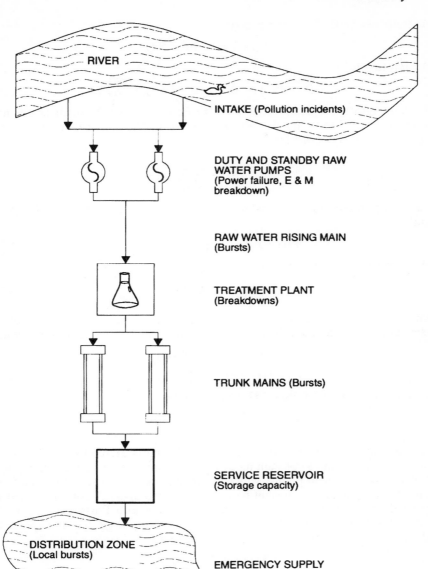

Figure 3.7   Water supply system (after Peel, 1991)

operational time, or cycles to failure, or any non-negative variable such as breaking strain. The *reliability* or *survivor function*, $R(t)$, is defined as the probability that the component does not fail by a time t

$$R(t) = \Pr(T > t) = 1 - \Pr(T < t) = 1 - F(t) \tag{3.1}$$

The probability that a component fails in time $(t, t + \delta t)$ conditional on the fact that it has not failed by time t is given by

$$\Pr(t < T < t + \delta t \,|\, T > t)$$

which formally

$$= \Pr(t < T < t + \delta t \quad \text{and} \quad T > t)/\Pr(T > t)$$

The numerator is logically equivalent to $\Pr(t < T < t + \delta t)$ which is given by $f(t)\delta t$ for small δt. The denominator is $R(t)$. The *hazard* or *failure rate function*, $h(t)$, is defined by

$$h(t) = \frac{f(t)}{R(t)} \tag{3.2}$$

and is proportional to the probability of imminent failure at time t. That is, for small δt

$$\Pr(t < T < t + \delta t \,|\, T > t) = h(t)\delta t \tag{3.3}$$

The *cumulative hazard function $H(t)$* is

$$H(t) = \int_0^t h(u) \, du \tag{3.4}$$

Now f, F, R, h and H are equivalent mathematical descriptions of the distribution of T and the following relationships follow from their definitions.

$$R(t) = \exp[-H(t)] \tag{3.5}$$

$$f(t) = h(t)\exp[-H(t)] \tag{3.6}$$

The mean time before failure MTBF is given by

$$\text{MTBF} = \int_0^\infty t f(t) \, dt = \int_0^\infty R(t) \, dt$$

The last equality follows from integration by parts (Exercise 4). It may be reasonable to assume a constant hazard, λ, for components such as long lengths of main, and treatment plants, which are maintained by localized rehabilitation rather than replaced. A constant hazard would also apply if pollution incidents are random and independent, i.e. occur according to a Poisson process. You may already have guessed that a constant hazard implies T has an exponential distribution. Using equations (3.4) and (3.5) gives

$$R(t) = e^{-\lambda t} \qquad 0 < t$$

However, for many smaller components such as pumps it may be more realistic to assume the hazard rate increases over time. This we do in the next section.

3.4.3 Weibull distribution

The hazard function of the Weibull distribution (Weibull, 1939) is proportional to some power of t, which can be zero giving the exponential distribution as a special case, or even negative. A negative power could correspond to improvements in the operation of the treatment plant or stricter environmental controls over pollution. It is convenient to parameterize the hazard function as

$$h(t) = (\alpha/\beta)(t/\beta)^{\alpha-1}$$

so that the cdf is

$$F(t) = 1 - e^{-(t/\beta)^\alpha} \qquad \text{for } 0 \leqslant t$$

and the pdf is

$$f(t) = \alpha\beta^{-\alpha}t^{\alpha-1}e^{-(t/\beta)^\alpha} \qquad \text{for } 0 \leqslant t$$

The Rayleigh distribution is the special case when α equals 2. The mean and variance of the Weibull distribution are rather awkward functions of the parameters:

$$\mu = \beta\Gamma(1 + 1/\alpha)$$
$$\sigma^2 = \beta^2[\Gamma(1 + 2/\alpha) - \Gamma^2(1 + 1/\alpha)]$$
$$\gamma = \frac{\Gamma(1 + 3/\alpha) - 3\Gamma(1 + 2/\alpha)\Gamma(1 + 1/\alpha) + 2\Gamma^3(1 + 1/\alpha)}{[\Gamma(1 + 2/\alpha) - \Gamma^2(1 + 1/\alpha)]^{3/2}}$$

and

$$E[X'] = \beta'\Gamma(1 + r/\alpha)$$

If we have a sample of n lifetimes $\{t_i\}$, the MLE are the solutions of the simultaneous equations:

$$\hat{\beta} = [\Sigma\, t_i^{\hat{\alpha}}/n]^{1/\hat{\alpha}}$$
$$\hat{\alpha} = n/[\hat{\beta}^{-\hat{\alpha}} \Sigma\, t_i^{\hat{\alpha}} \ln t_i - \Sigma \ln t_i]$$

These can be solved by iteration: choose a plausible starting value for $\hat{\alpha}$; use the first equation to calculate a $\hat{\beta}$; then use the second equation to obtain an improved estimate $\hat{\alpha}$; and so on. A simpler, but less precise, way of fitting the distribution is to use a probability plot.

Suppose we take a random sample of size n and put the variables into ascending order. Denote the ith largest out of n as

$$X_{i:n}$$

For any distribution the proportion of data less than the expected value of the ith largest of n (i.e. average value of the ith largest in imagined repeated samples of n) is very roughly i/n. But this is obviously unrealistic when i equals n, and a better approximation is $(i - 0.4)/(n + 0.2)$. So, in general

$$F(E[X_{i:n}]) = (i - 0.4)/(n + 0.2)$$

and for the Weibull distribution in particular

$$1 - \exp\{-(E[X_{i:n}]/\beta)^\alpha\} = (i - 0.4)/(n + 0.2)$$

This can be rearranged to give

$$\ln(E[X_{i:n}]) = \ln \beta + (1/\alpha)\ln(-\ln(1 - (i - 0.4)/(n + 0.2)))$$

A *Weibull plot* is a plot of the logarithms of the ordered sample values, $\ln(x_{i:n})$, against $\ln(-\ln(1 - (i - 0.4)/(n + 0.2)))$. If the data can reasonably be thought of as coming from a Weibull distribution, they will be scattered about a notional straight line. The intercept and gradient of the line are estimates of ln β and $(1/\alpha)$ respectively. If there is obvious curvature in the plot, a Weibull distribution will not give a realistic fit. The Minitab *Weibull probability plot* labels the scales in original units, so the scales are non-linear, and plots probability vertically. The line is based on the maximum likelihood estimates. The MLE are more precise than graphical estimators.

Example 3.1
The following data are lifetimes (weeks) of 23 submersible pumps, put into ascending order for convenience.

$$
\begin{array}{cccccccc}
18 & 29 & 33 & 41 & 42 & 46 & 48 & 52 \\
52 & 54 & 56 & 68 & 69 & 70 & 72 & 84 \\
93 & 99 & 105 & 106 & 128 & 129 & 173 &
\end{array}
$$

A Weibull plot is shown in Figure 3.8 with the Minitab Weibull probability plot underneath. The ML estimates of α and β are 2.109 and 82.16.

Now suppose that the test had ended after 100 weeks. Our data would be the 18 lifetimes which are less than 100, and five lifetimes which are known only to exceed 100. This is an example of *censored data*. All we need do to estimate the parameters graphically is plot the first 18 against $(i - 0.4)/23.2$ for i from 1 to 18 and draw a plausible line through the points. The ML estimates would be the values of α and β which maximize

$$
\mathcal{L} = \left[\prod_{i=1}^{18} f(x_i) \right] [1 - F(100)]^5
$$

Estimates of α and β would now be 2.94 and 63.97 respectively.

The Weibull distribution can be made into a three-parameter distribution by assuming all lifetimes exceed some lower limit L.

Example 3.2
The following data are cycles to failure (unit of 1000 cycles) for 25 test specimens of mastic asphalt (12% binder) subject to axial load vibration with an initial strain of 200 με (based on a graph in Gurney, 1992). They have been sorted into ascending order.

$$
\begin{array}{cccccccccc}
114 & 143 & 144 & 167 & 170 & 200 & 219 & 246 & 251 & 251 \\
260 & 264 & 269 & 271 & 274 & 312 & 321 & 352 & 362 & 383 \\
389 & 416 & 516 & 624 & 786 & & & & &
\end{array}
$$

Although the Weibull probability plot is not very convincing, the value of the Kolmogorov–Smirnov statistic (Appendix A1.22) with α and β estimated as 2.167 and 349.3 is only 0.154, which is marginally statistically significant ($P \approx 0.10$). If 100 is subtracted from all the data the fit is very good, but a log-normal distribution is an equally impressive fit to the original data. The probability plots are shown in Figure 3.9. It may be worth presenting alternative analyses if several distributions give adequate fits and there is no substantial physical argument for preferring one of them.

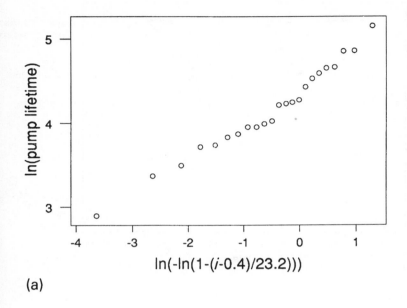

(a)

Weibull Probability Plot for pumplife

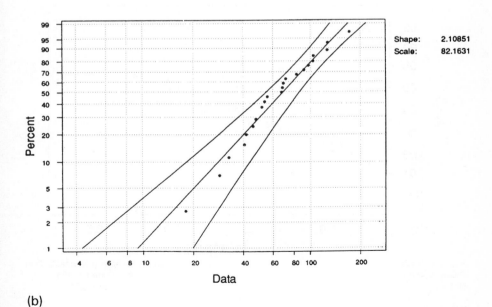

Shape: 2.10851
Scale: 82.1631

(b)

Figure 3.8 (a) Weibull plot, (b) Minitab Weibull probability plot, for lifetimes of pumps

Weibull Probability Plot for asphalt

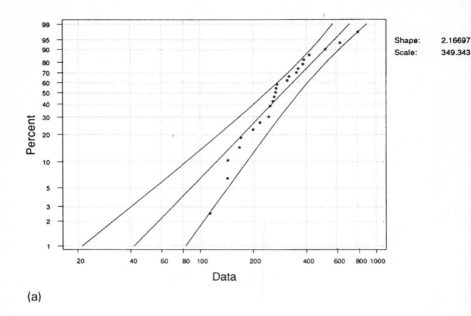

Shape: 2.16697
Scale: 349.343

(a)

Normal Probability Plot

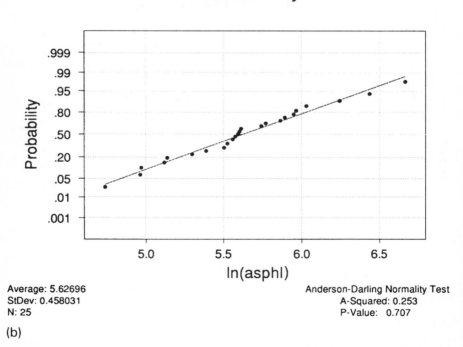

Average: 5.62696
StDev: 0.458031
N: 25

Anderson-Darling Normality Test
A-Squared: 0.253
P-Value: 0.707

(b)

Figure 3.9 (a) Weibull probability plot for cycles (thousands) to failure of asphalt, and (b) normal probability plot for their logarithms

Waloddi Weibull was a Swedish physicist who used the distribution which is named after him to model the strengths of materials. In this context, time (t) is replaced by breaking strength (x). He assumed a *weakest link hypothesis*, which is the notion that a long piece of material consists of independent links, whose weakest member determines its failure. The Weibull distribution is obtained by assuming the survivor function for a single individual link is

$$R_i(x) = e^{-(x/\theta)^\alpha}$$

and applying the weakest link hypothesis to N links

$$R(x) = [R_i(x)]^N = e^{-N(x/\theta)^\alpha} = e^{-(x/\beta)^\alpha}$$

where $\beta = \theta N^{-1/\alpha}$. Crowder *et al.* (1991) give a detailed account of models for strengths of materials.

In the next example we will fit both Weibull and normal distributions to samples of corroded cable and apparently good cable from the George Washington Bridge (Stahl and Gagnon, 1995), and compare the results. One of the principles on which cables and ropes are based is that individual wires will tend to support neighbouring wires, so that the weakest link of one long wire does not undermine the strength of the cable. However, it might still be reasonable to imagine a length of cable to be made up of short independent links of wire rope.

Example 3.3
The following data are breaking loads (kN) of 23 pieces of the new cable tested in 1933.

```
3271  3328  3340  3432  3455  3460  3466  3478  3485  3489
3494  3506  3512  3532  3538  3544  3547  3549  3550  3551
3552  3558  3564
```

The breaking loads (kN) of 18 pieces of corroded cable tested in 1962 were:

```
2406  2651  2661  2691  2750  2799  2808  2828  2858  2897
2966  2995  3054  3162  3172  3201  3221  3260
```

A normal distribution with a mean and standard deviation equal to the sample values of 2910 and 237 appears to be a reasonable fit to the breaking loads of the corroded cable. The ML estimates of a Weibull distribution are 14.56 and 3015 for α and β. The Weibull distribution becomes closest to a normal distribution when α equals 3.6 and its skewness is close to 0. The skewness is -0.78 when α equals 14.56. The lower 1% quantile of the estimated Weibull distribution is 2198 and that of the estimated normal distribution is 2359. The upper 1% points are 3348 and 3461 for the Weibull and normal distributions respectively. Although both distributions seem plausible, I would prefer to rely on the rather more pessimistic Weibull estimate.

Neither the normal nor the Weibull distribution is a convincing fit to the 1933 data. A dot-plot indicates that we need either a distribution with high negative skewness, for which there is no obvious physical justification, or a mixture of two distributions. A mixture might be justified if the cable was from two different manufacturers.

3.4.4   Normal mixture

A *mixture* of two normal distributions,

$$f(x) = pf_1(x) + (1-p)f_2(x)$$

has five parameters to be estimated: two means, two standard deviations and the proportion p. The ML estimates, from a sample of n observations x_i, are the solutions of:

$$\hat{p} = \sum \hat{P}(c_1 \mid x_i)/n \qquad (3.7a)$$

$$\hat{\mu}_j = \sum \hat{P}(c_j \mid x_i)x_i/(n\hat{p}_i) \quad \text{for} \quad j = 1, 2 \qquad (3.7b)$$

$$\hat{\sigma}_j^2 = \sum \hat{P}(c_j \mid x_i)(x_i - \hat{\mu}_j)^2/(n\hat{p}_i) \quad \text{for} \quad j = 1, 2 \qquad (3.7c)$$

where

$$\hat{p}_1 = \hat{p} \quad \text{and} \quad \hat{p}_2 = (1 - \hat{p})$$

$$\hat{P}(c_1 \mid x_i) = \hat{p}f_1(x_i)/f(x_i)$$

and

$$\hat{P}(c_2 \mid x_i) = 1 - P(c_1 \mid x_i)$$

This form of the equations has a neat interpretation. The estimate of p is the mean of the estimated probabilities that each datum came from population 1. These same probabilities are then used to calculate a weighted mean and variance for population 1, and similarly for population 2.

To use these equations: take plausible starting values for all five parameters and work out $\hat{P}(c_1 \mid x_i)$ and $\hat{P}(c_2 \mid x_i)$, which are estimates of the probabilities that an observation x_i came from component distribution j; then re-estimate the five parameters; and repeat until convergence (Everitt, 1987). I started the algorithm by guessing: 0.2 for $\hat{p}$; 3320 and 50 for $\hat{\mu}_1$ and $\hat{\sigma}_1$; and 3520 and 50 for $\hat{\mu}_2$ and $\hat{\sigma}_2$. The algorithm converged after four iterations to give the following estimates: 0.13 for the proportion of distribution 1, which has mean 3313.02 and standard deviation 30.15, and 3513.10 and 39.57 for the mean and standard deviation of distribution 2. Although the algorithm usually converges there is a potential problem because each sample point generates a *singularity* in the likelihood function. For example, if $\hat{\mu}_1$ is set equal to any sample value the likelihood becomes arbitrarily large as σ_1 tends to 0. This feature of the likelihood function is not restricted to mixtures of normal distributions, and I have found it a practical problem if I have attempted to minimize the defining expression for a mixture log-likelihood by the simplex algorithm. One way of avoiding the problem is to use maximum likelihood for grouped data. If there are m classes with n_j data in each class the log-likelihood becomes

$$\ln(\mathcal{L}) = \sum n_j \ln P_j$$

where

$$P_j = \int_{x_{j-1}}^{x_j} f(x) \, \mathrm{d}x = F(x_j) - F(x_j - 1)$$

and x_{j-1} and x_j are the lower and upper limits of the jth class interval. This is straightforward to apply if the cdf can be expressed as a formula (e.g. the

Weibull distribution) and the log-likelihood can be maximized by the simplex algorithm. The formulation is less convenient for the normal distribution, but the fitting algorithms in the Maximum Likelihood Program (MLP), which is supported by NAG Ltd, implement it. The program also calculates approximate standard deviations of the estimators. I compared MLP results with those I obtained from the ML algorithm given by equations (3.7). I specified the right-hand end points of class intervals as: 3300; 3330; ...; 3570. The results are summarized below.

Parameter	Estimate	Standard deviation of estimate
p	0.13	0.07
μ_1	3313.3	16.63
σ_1	25.93	15.2
μ_2	3514.5	8.82
σ_2	38.46	6.43

The only noticeable difference is between the estimates of σ_1 $(25.93 - 30.15 = -4.22)$ and this is small when compared with the standard deviation of the estimate (15.2). Despite the small sample, a mixture of normal distributions is quite a convincing model. A Weibull distribution also happens to be a reasonable fit to $\{y_i\}$, where $y_i = 3566 - x_i$, but I cannot think of any plausible physical justification for the lower limit.

3.4.5 Markov analysis

A water main can be defined as a *process* for delivering water, such that the process is in one of two *states*: working (W) or failed (F). The *Markov property* is that the probability the process changes state in time $(t, t + \delta t)$ depends only on its state at time t. The process is *stationary* if these *transition probabilities* do not vary with time. The process is said to be *memoryless* because the future behaviour depends only on its present state and not on its history. A consequence of the Markov property is that the times between failures, and the times of repairs, are independent random variables from exponential distributions.

We can represent the process by:

$$X(t) = \begin{cases} 0 & \text{if failed} \\ 1 & \text{if working} \end{cases}$$

define the probability of failure in a small time period $(t, t + \delta t)$ if it is working at time t as $\lambda \, \delta t$; and define the probability of repair in time $(t, t + \delta t)$ if it is failed at time t, and therefore under repair, as $\nu \, \delta t$ (see equation (3.3)). The parameters λ and ν are referred to as the failure and repair rates respectively, and if t is in days they will have units of failures, or repairs, per day. Times between failures and times of repairs have exponential distributions with means $(1/\lambda)$ and $(1/\nu)$ respectively (see Section 3.4.2 and Appendix 1). Define $P_0(t)$ as the probability that $X(t)$ equals 0 at time t and $P_1(t)$ similarly. We can obtain differential equations which describe the evolution of these probabilities by considering the relationship between $P_i(t)$ and $P_i(t + \delta t)$, where in this case i is 0 or 1, for a small length of time δt. For example, if the system is failed (state 0) at time $t + \delta t$, it was either under repair at time t and the repair has not been completed, or it was working at time t and there has been a failure since then. We can ignore the possibility that the main has been repaired and failed again within the time increment δt, because the corresponding probability would be of order $(\delta t)^2$. So, if we ignore terms in $(\delta t)^2$ and higher powers

$$P_0(t + \delta t) = P_0(t)(1 - \nu\, \delta t) + P_1(t)\lambda\, \delta t$$
$$P_1(t + \delta t) = P_0(t)\nu\, \delta t + P_1(t)(1 - \lambda\, \delta t)$$

These can be rearranged to give

$$\frac{P_0(t + \delta t) - P_0(t)}{\delta t} = -\nu P_0(t) + \lambda P_1(t)$$

$$\frac{P_1(t + \delta t) - P_1(t)}{\delta t} = \nu P_0(t) - \lambda P_1(t)$$

and if we now let δt tend to zero we obtain two differential equations which can be expressed in matrix notation as:

$$[\dot{P}_0(t) \quad \dot{P}_1(t)] = [P_0(t) \quad P_1(t)] \begin{bmatrix} -\nu & \nu \\ \lambda & -\lambda \end{bmatrix}$$

Terms in $(\delta t)^2$ and higher powers, which correspond to two or more changes of state in time δt, tend to zero in the limit, so the differential equations are exact. The fact that we can ignore the possibility of more than one change of state in a small length of time δt is the key to writing down the differential equations. The square matrix of rates is called the *rate matrix*. The two equations are linearly dependent, but as the system must be in exactly one state any time:

$$P_0(t) + P_1(t) = 1$$

If we substitute $(1 - P_1(t))$ for $P_0(t)$, and assume the main starts in a working condition:

$$\dot{P}_1(t) = \nu(1 - P_1(t)) - \lambda P_1(t)$$
$$P_1(0) = 1$$

which has the solution

$$P_1(t) = \frac{\nu}{\lambda + \nu} + \frac{\lambda \exp[-(\lambda + \nu)t]}{\lambda + \nu}$$

In practice we are usually interested in the value of this probability as t becomes large, because it can also be interpreted as the proportion of time that the system is in the working state. The process is said to be *ergodic* because this limiting probability

$$\frac{\nu}{\lambda + \nu}$$

does not depend on the state the process started from. If you divide by the product $\lambda \nu$ you will see that the expression is equivalent to the availability (A) defined in Section 3.4.1.

Repairable processes which are modelled by exponential failure and repair times are generally ergodic and as the $P_i(t)$ tend to constants the rate of change of these probabilities, $\dot{P}_i(t)$, will tend to zero. The matrix differential equations could be replaced by the algebraic equation:

$$\mathbf{u}\Lambda = 0$$

where Λ is the rate matrix and the elements of $\mathbf{u}$ are

$$u_i = \lim_{t \to \infty} P_i(t)$$

together with the constraint that

$$\sum u_i = 1$$

One disadvantage of this rather neat theory is the restrictive assumptions of exponential distributions for failure and repair times. Bunday (1996) provides a more detailed account which includes an explanation of how a general distribution can be used instead of one of the assumed exponential distributions. A more serious drawback is that we may need as many as 2^n states to describe a system which consists of n components, each of which can work or fail, and simulation is often a more practical option. In the next example we set up the rate matrix for part of the water supply system (Figure 3.7).

Example 3.4
A subsystem in Figure 3.7 is the treatment plant, which can break down, and the two trunk mains, which can burst. We can define six states as follows.

Treatment plant	Mains	State
working	both working	1
working	one failed other working	2
working	both failed	3
failed	both working	4
failed	one failed other working	5
failed	both failed	6

Note that states must be defined so that: the process is always in exactly one state; and a single failure or repair event will result in a change of state. The resulting *state space* is not necessarily unique; here we could have distinguished between east and west mains being the one working and used eight states. A typical column of the rate matrix, the second, is obtained by the following argument. Denote the failure rate of trunk mains by λ_1, and the failure rate of the treatment plant by λ_2. We will assume that repairs are started immediately, using the water company's personnel or contractors, and that the repair rates are ν_1 and ν_2 for mains and treatment plants respectively. If the process is in state 2 at time $t + \delta t$, there are four possibilities for events since time t: it was in state 1 and one of the two mains has failed (remember we can ignore the possibility of both failing in a short length of time δt); it was in state

2 and no events have occurred since; it was in state 3 and one of the two repairs has been completed; it was in state 5 and the treatment plant has since been repaired. It cannot move between state 4 or state 6 and state 2 in time δt because either move would require two events to occur

$$P_2(t + \delta t) = P_1(t)2\lambda_1\delta t + P_2(t)(1 - \lambda_1\delta t - \lambda_2\delta t - \nu_1\delta t)$$
$$+ P_3(t)2\nu_1\delta t + P_5(t)\nu_2\delta t$$

Notice that, for example

Pr(east or west main fails in time δt)
$$= \text{Pr(east main fails in } \delta t) + \text{Pr(west main fails in } \delta t) - \text{Pr(both fail in } \delta t)$$
$$= \lambda_1\delta t + \lambda_1\delta t - (\lambda_1\delta t)^2 \approx 2\lambda_1\delta t$$

Now subtract $P_2(t)$ from both sides, divide by δt, and let δt tend to zero, to obtain:

$$\dot{P}_2(t) = 2\lambda_1 P_1(t) + (-\lambda_1 - \lambda_2 - \nu_1)P_2(t) + 2\nu_1 P_3(t) + \nu_2 P_5(t)$$

The full rate matrix is:

	1	2	3	4	5	6
1	$-2\lambda_1 - \lambda_2$	$2\lambda_1$	0	λ_2	0	0
2	ν_1	$-\lambda_1 - \lambda_2 - \nu_1$	λ_1	0	λ_2	0
3	0	$2\nu_1$	$-\lambda_2 - 2\nu_1$	0	0	λ_2
4	ν_2	0	0	$-2\lambda_1 - \nu_2$	$2\lambda_1$	0
5	0	ν_2	0	ν_1	$-\lambda_1 - \nu_1 - \nu_2$	λ_1
6	0	0	ν_2	0	$2\nu_1$	$-2\nu_1 - \nu_2$

I used MAPLE to solve the system of linear equations:

$$\left. \begin{array}{c} \mathbf{u}\Lambda = 0 \\ \sum u_i = 1 \end{array} \right\} \tag{3.7}$$

The solution is:

$$u_1 = \nu_1^2\nu_2/den; \quad u_2 = 2\lambda_1\nu_1\nu_2/den; \quad u_3 = \lambda_1^2\nu_2/den$$
$$u_4 = \lambda_2\nu_1^2/den; \quad u_5 = 2\lambda_1\lambda_2\nu_1/den; \quad u_6 = \lambda_1^2\lambda_2/den$$

where

$$den = \nu_1^2\nu_2 + 2\lambda_1\nu_1\nu_2 + \lambda_1^2\nu_2 + \lambda_2\nu_1^2 + 2\lambda_1\lambda_2\nu_1 + \lambda_1^2\lambda_2$$

The subsystem functions for a proportion of time given by the sum, $u_1 + u_2$. If λ_1, λ_2, and ν_1 and ν_2 equal 0.02, 0.01, 1 and 0.5 per day respectively, this would be 98% of the time. A useful check of the rate matrix is that rows must add to zero, and you are asked to explain this in Exercise 5. The same principles can be used for analysing queueing systems, three examples are given in exercises 6–8. Carmichael (1987) gives a detailed account of applications in the mining and construction industries.

3.4.5 Interference

If both load X and strength Y are assumed to be independent and to have probability distributions with pdf f_X and f_Y, reliability is defined as

$$R = \Pr(Y > X)$$

$$= \int_{x=0}^{\infty} \int_{y=x}^{\infty} f_Y(y) \, \mathrm{d}y \, f_X(x) \, \mathrm{d}x = \int_{0}^{\infty} (1 - F_Y(x)) f_X(x) \, \mathrm{d}x \qquad (3.8)$$

or

$$= \int_{0}^{\infty} f_Y(y) F_X(y) \, \mathrm{d}y \qquad (3.9)$$

This is known as load–strength interference. In general such integrals may have to be evaluated numerically, but the special case when both X and Y have normal distributions is easy to deal with:

$$R = \Pr(Y > X) = \Pr(Y - X > 0)$$

and

$$Y - X \sim N(\mu_Y - \mu_X, \sigma_x^2 + \sigma_Y^2)$$

This could also be used when X and Y represent logarithms of load and strength.

Example 3.5
A design of bulldozer has buckets with maximum load capacities which are normally distributed with a mean of $6000 \, N$ and a standard deviation of $400 \, N$. Typical full loads are normally distributed with a mean of $4000 \, N$ and a standard deviation of $600 \, N$. If the maximum bucket load is exceeded a safety valve opens, but resetting the valve in working conditions is troublesome and reliability is defined as a successful lift. What is the reliability for one lift by a randomly selected bulldozer?

Let X and Y represent load and strength

$$X \sim N(4000, (600)^2)$$

$$Y \sim N(6000, (400)^2)$$

$$Y - X \sim N(2000, (721)^2)$$

$$\Pr(Y - X > 0) = \Pr(Z > -2.77) = 0.997$$

Crowder (1991) uses a variation of interference estimation to estimate the reliability of steel reinforced concrete. The following example is based on his article which arose from a consultancy.

Example 3.6
When concrete is exposed to the atmosphere a progressive corrosion process known as carbonation can occur. Carbon dioxide is absorbed by the material which, initially alkaline, becomes more acidic. The affected layer diffuses in from the exposed surface, reaches the steel reinforcement, and then corrodes the steel. The depth of carbonation can be measured by drilling cores and using a chemical pH indicator. Both the cover depth, i.e. the distance from the

concrete surface to the steel reinforcement, and the carbonation depth varied throughout the structure. Crowder gives 100 measurements of cover depth, and 20 measurements of carbonation depth after 10 years. These were taken at points which were distributed about the structure in a fairly haphazard manner, and it seemed reasonable to treat them as random samples from their respective populations. The objective of the consultancy was to estimate the proportion of steel reinforcement affected, as a function of time. There was evidence, from other sources, that the average depth of carbonation could reasonably be assumed proportional to the square root of time.

Let Y be the cover depth at a random point and X_t be the carbonation depth at time t. In one analysis he assumes log-normal distributions for X_t and Y.

$$\ln X_t \sim N(\mu_0 + \tfrac{1}{2}\ln(t/t_0), \sigma^2)$$

where μ_0 is the mean at a reference time t_0, and

$$\ln Y \sim N(\nu, \theta^2)$$

The required probability is

$$\Pr(\ln X_t - \ln Y > 0)$$

and this is found from the result that:

$$\ln X_t - \ln Y \sim N(\mu_0 + \tfrac{1}{2}\ln(t/t_0) - \nu, \sigma^2 + \theta^2)$$

The estimates of the parameters are given as:

$$\hat{\mu}_0 = 1.18 \quad \text{when} \quad t_0 = 10, \quad \hat{\sigma} = 0.412$$
$$\hat{\nu} = 2.53, \quad \hat{\theta} = 0.151$$

The estimates of the proportions of affected steel after 20, 50 and 100 years are 0.011, 0.11 and 0.32 respectively. A 95% confidence interval for the 100 year proportion, which could have been constructed using the bootstrap method, is [0.19, 0.49].

Crowder presents an alternative analysis based on the empirical distributions, rather than assumed log-normal distributions. The principle is straightforward. The cdf and pdf of Y and of X_t are estimated directly from the sample, and then equation (3.9) is evaluated numerically. An estimate of a cdf which is made directly from the sample without any assumption about its form is known as an *empirical distribution function (edf)*. The usual definition for calculating the edf from a sample of size n is

$$\hat{F}(x) = (\text{number of data} \leqslant x)/n$$

In this application the measurements of depth were all between 0 mm and 19 mm and recorded to the nearest mm. The data were therefore grouped into intervals of width 1 mm and this naturally led to a slightly smoothed estimator. If f_i is the frequency of depths i in the sample, and r is an integer

$$\bar{F}(r + \tfrac{1}{2}) = (f_0 + \ldots + f_r)/n$$

and $\bar{F}(x)$ is defined by straight line interpolation. The edf of X_0 and Y will be denoted by $\bar{F}_{X_0}(x)$ and $\bar{F}_Y(y)$. The pdf of Y was estimated by the histogram with class width 1. The proportion of concrete affected at the time of the investigation, 10 years after the structure was built, is estimated by

$$1 - \sum_i \frac{f_{i,Y}}{n_Y} \bar{F}_{X_0}(i) \qquad (3.10)$$

where $\bar{F}_{X_0}(i) = (\bar{F}_{X_0}(i + \frac{1}{2}) + \bar{F}_{X_0}(i - \frac{1}{2}))/2$. The summation approximates the integral (3.9). The next step is to scale the cdf of X_0 to that of X_t. The basic assumption is that $X_t/\sqrt{t}$ and $X_0/\sqrt{t_0}$ have the same distribution, that is, X_t increases in proportion to $\sqrt{t}$ but with random variation. Hence

$$\bar{F}_{X_t}(x) = \bar{F}_{X_0}(x(t_0/t)^{1/2})$$

Then $\bar{F}_{X_0}(i)$ in equation (3.10) is replaced by $\bar{F}_{X_t}(i)$, but the expression for this is more complex because of the scaling

$$\bar{F}_{X_t}(i) = (t/t_0)^{1/2} \int_{r_1}^{r_2} \bar{F}_{X_t}(x) \, dx$$

where $r_1 = (r - \frac{1}{2})(t_0/t)^{1/2}$ and $r_2 = (r + \frac{1}{2})(t_0/t)^{1/2}$. Since $\bar{F}_{X_0}$ is defined in terms of straight line segments, this integral is given by the following algorithm where

$$r_3 = [\text{integer value of } (r_2 + \frac{1}{2})] - \frac{1}{2}$$

$$\bar{F}_{X_t}(i) = \begin{cases} H_{12} & \text{if } r_3 \leqslant 1 \\ (t/t_0)^{1/2}[(r_3 - r_1)H_{13} + (r_2 - r_3)H_{23}] & \text{if } r_1 < r_3 \end{cases}$$

and

$$H_{jk} = (\bar{F}_{X_0}(r_j) + \bar{F}_{X_0}(r_k)) \quad \text{for } j, k = 1, 2, 3$$

The results from both analyses are summarized below. Confidence intervals for the empirical distribution estimates were computed by the bootstrap method.

t years	Proportion affected and 95% confidence interval log-normal	Empirical
20	0.011 (0.001, 0.062)	0.0006 (0, 0.002)
50	0.11 (0.04, 0.24)	0.04 (0.002, 0.07)
100	0.32 (0.19, 0.49)	0.26 (0.12, 0.39)

The predictions based on the log-normal distribution are more pessimistic. Both sets of predictions assume that carbonation is proportional to the square root of time.

3.4.6   Kernel density estimation

In the last example the empirical estimate of the pdf of cover depth was just the histogram, but a smooth curve is often needed. The kernel density estimate, $\hat{f}(x)$, is constructed by placing bumps at each of the data points and then adding them. The *kernel*, $K( )$, is itself a probability density function which determines the shapes of the bumps, and the normal pdf is often chosen. If we have a set of n data $\{x_i\}$:

$$\hat{f}(x_i) = \sum_{i=1}^{n} K((x - x_i)/h)/(nh)$$

The *smoothing parameter* h is also called a *window width* or *bandwidth*. It follows from the definition that $\hat{f}(x_i)$ is a pdf, i.e. it is non-negative and has a total area of one, and it has the same continuity and differentiability properties as the kernel. There is a drawback to the use of a constant smoothing parameter. If the data are from a long-tailed distribution, noise can appear in the tails, and increasing h to remove the noise smooths out the main peak too much. A remedy is to use a variable width kernel. Then

$$\hat{f}(x) = \Sigma \ K((x - x_i)/(hd_{ik}))/(nhd_{ik})$$

where d_{ik} is the distance from point x_i to the kth nearest point. The choice of k determines the sensitivity to local detail. An application to the data from the new cable (Example 3.3) is shown in Figure 3.10. Silverman (1986) gives a detailed account of empirical density estimation, and his Applied Statistics algorithm AS176 (Silverman, 1982) is freely available.

3.4.7   Proportional hazards modelling

A company tested n welded joints of tubular steel in a cyclical load testing machine for up to 100 hours. The width of weld (x_i) was measured for each joint before testing, and if it failed the time of failure (t_i) was recorded. The product limit (*PL*) estimate of the survivor function is

$$\hat{R}(t) = \prod_{t_j < t} (1 - 1/n_j)$$

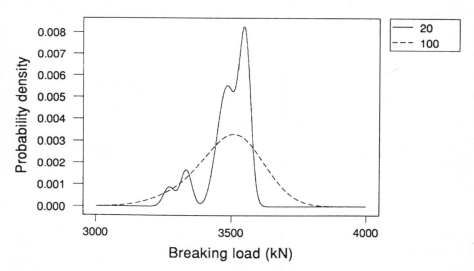

Figure 3.10 Kernel density estimates of strength distribution of new Washington Bridge cable. The smoothing parameter h is 20 (full curve) and 100 (broken curve)

where t_j are distinct failure times, and n_j is the number of joints at risk just before time t_j. Cox (1972) proposed a hazard function which took into account the concomitant variable x

$$h(t) = h_0(t)e^{\beta x}$$

where β is an unknown parameter and h_0 is a baseline hazard function. The likelihood function for β is obtained by noticing that the probability that joint i fails at time t_i given the set of joints at risk is

$$e^{\beta x_i} \Big/ \sum_{\substack{\text{joints} \\ \text{at risk}}} e^{\beta x_k}$$

The joints at risk include joint i. The likelihood function for β is

$$\mathcal{L}(\beta) = \prod_{i=1}^{m} \left[e^{\beta x_i} \Big/ \sum_{\substack{\text{joints} \\ \text{at risk}}} e^{\beta x_k} \right]$$

where m is the number of joints which fail. The remaining $n - m$ joints survive at least 100 hours. The *PL* estimate for the baseline survivor function is (Crowder, 1991)

$$\hat{R}_0(t) = \prod_{t_j < t} \hat{\alpha}_j$$

where

$$\hat{\alpha}_j = 1 - \left\{ e^{\beta x_j} \Big/ \sum_{\substack{\text{joints} \\ \text{at risk}}} e^{\beta x_k} \right\}$$

and $\hat{\beta}$ is the MLE of β. The baseline cumulative hazard function can be estimated from $\hat{H}_0(t) = -\ln \hat{R}_0(t)$. The method can be modified to allow for ties, more concomitant variables can be included, and these can change over time. In the latter case βx is replaced by $x^T \beta$. Kalbfleisch and Prentice (1980) give a detailed account.

Smith and Carr (1986) proposed a variation of the model for short term flood risk prediction, in which $h(t)$ is the rate of occurrence of flows exceeding some threshold, $h_0(t)$ is the seasonally varying baseline rate of exceedances and x^T, which are time dependent, are concomitant variables. The main difference between this application and modelling lifetimes of joints is that a year does not leave the risk set if an exceedance occurs. Futter *et al.* (1991) applied the method with baseflow (x), a measure of catchment wetness, as a concomitant variable. Baseflow is defined as the flow in a river that is due to infiltration. It can be estimated from river flow records using one of a variety of sets of assumptions. Futter *et al.* (1991) approximated baseflow by the minimum flow on the day before an exceedance of the threshold. Smith and Carr (1986) gave the log-likelihood function for β as

$$\ln \mathcal{L}(\beta) = \sum_{i=1}^{n} \sum_{j=1}^{N_i} \beta x_i(t_j) - \sum_{k=1}^{N} \ln \left[\sum_{i=1}^{n} \exp(\beta_{x_i}(t_k)) \right]$$

where t_j are the days on which exceedances occurred in year i, t_k are the days on which exceedances occurred in any of the n years, N_i is the number of exceedances in year i, and N is the total number of exceedances. The estimator used for the baseline rate, over a typical time interval $[t_2 - t_1]$ during which it is assumed constant, was

$$\hat{h}(t) = \left\{ \sum_l \left[\sum_{i=1}^{n} \exp(\hat{\beta}_{x_i}(t_l)) \right]^{-1} \right\} \bigg/ (t_2 - t_1)$$

where the outer summation is over l such that $t_1 < t_l < t_2$. Futter *et al.* (1991) modified the equations by removing a year from the risk set for two days after an exceedance, to allow the recession curve to fall to a baseflow level (Futter *et al.*, 1990), and followed Peto's recommendations for dealing with ties (discussion of Cox, 1972, see exercise 10). Data were available from Burn Hall gauging station on the River Browney in County Durham, UK. The catchment is an area of 180 km^2. It varies in height between 100 m and 300 m above sea level and has a geology dominated by carboniferous coal measures. Daily maximum and minimum flows from 18 years (1968–1985) were used to fit the model. Estimation of β relies on a comparison of baseflow before an exceedance with baseflows on the same day in other years. The estimates for different thresholds are given in Table 3.5. The standard deviations were estimated by bootstrapping on years.

The probability of an exceedance in a short time interval Δ is approximately

$$h(t)\Delta$$

The application described by Smith and Carr (1986) was to reservoir operating policies implemented over one week intervals, during which their concomitant variables did not change much. If the model is to be used for flood risk estimation over longer periods, baseflows must be forecast. Risks are then calculated for each day and combined under an assumption of independence once baseflow has been allowed for. That is

Pr(exceedance of the critical level within next 30 days) =

$$1 - \prod_{k=1}^{30} (1 - h(t_k)\Delta)$$

where t_k is the day k days ahead, $h(t_k)$ is the estimated risk per day on that day,

Table 3.5 *Estimates of the base flow covariates for the Smith and Karr model applied to the River Browney with various critical flows*

Critical flow, m^3/s	Number of events	Regression coefficient (b)	Standard deviation
7	75	0.577	0.112
10	51	0.704	0.131
16	32	0.753	0.161
20	25	0.434	0.150

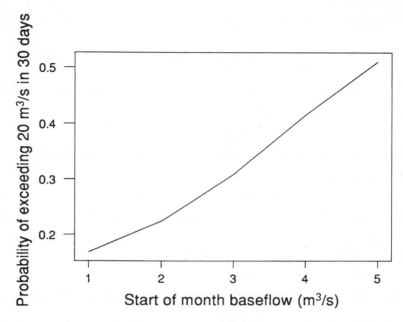

Figure 3.11 Risk of exceeding a flow of 20 m³/s within 30 days, plotted against start of period baseflow, for summer months at Burn Hall gauging station on the River Browney

and Δ is the time increment of 1 day. The inter-event baseflow series was modelled by a constant change scheme. A fixed amount was removed from, or added to, the mean corrected baseflow on day t to bring the value for day $t + 1$ closer to the mean. In symbols, with w representing baseflow

$$(w_{t+1} - \bar{w}) = (w_t - \bar{w}) + \delta w_t$$

until it equals $\bar{w}$ when it remains at $\bar{w}$. Some typical results are shown in Figure 3.11.

3.5 Summary

1 (a) Several batches of concrete are made from one delivery of cement. Several tests are made on each batch. Let Y_{ijk} be the result of test k on batch j of delivery i

$$\mathrm{var}(Y_{ijk}) = \sigma^2_{\text{delivery}} + \sigma^2_{\text{batch | delivery}} + \sigma^2_{\text{test | batch and delivery}}$$

The most efficient way of reducing the overall variance is to reduce the largest component.

(b) In an inter-laboratory trial, the reproducibility standard deviation, σ_R, in the root of the sum of the between laboratory variance, σ^2_L, and the repeatability variance, σ^2_r. That is

$$\sigma_R = \sqrt{\sigma^2_L + \sigma^2_r}$$

2.   A CUSUM is a plot of

$$S_t = \sum_{i=1}^{t} (x_i - \text{target})$$

for some variable x_i. Changes from the target will show up as steep slopes.

3.   The hazard function is defined by

$$h(t) = f(t)/R(t)$$

The Weibull distribution has a hazard which is proportional to t^{a-1}.

4.   The Weibull distribution has a cdf

$$F(t) = 1 - e^{-(t/\beta)^a} \quad \text{for} \quad 0 \le t$$

It can also be derived from the weakest link hypothesis.

5.   In a Markov process $P(t)$ is a column of probabilities that the process is each of a finite number of states at time t

$$P(t + \delta t) = P(t)(I + \Lambda \delta t)$$

where $(I + \Lambda \delta t)$ is known as the transition matrix and Λ is the rate matrix

$$\frac{P(t + \delta t) - P(t)}{\delta t} = P(t)\Lambda$$

and letting δt tend to zero gives

$$\dot{P}(t) = P(t)\Lambda$$

the long term proportions of time spent in each state

$$u_i = \lim_{t \to \infty} P_i(t)$$

are given by the solution of

$$u\Lambda = 0$$

$$\sum u_i = 1$$

6.   Interference.
     If load X and strength Y are independent, the reliability

$$R = \Pr(Y > X) = \int_0^\infty f_y(y)F_X(y)\, dy$$

Exercises

1   In order to investigate the accuracy and precision of a colourmetric method for estimating lead content of water, 15 litres of a carefully prepared 12 ppm solution were made up. The solution was thoroughly shaken and poured into 15 one litre jars. Sets of three jars were chosen at random and sent to five randomly selected laboratories. The results are given below.

Lab A	Lab B	Lab C	Lab D	Lab E
9	12	13	9	13
11	14	13	12	11
10	10	16	12	12

Estimate the within laboratory and between laboratories variances, and show that the estimated between laboratories standard deviation is 1.18 (to 2 decimal places).

2 The amount of impurity (X) in a diesel fuel oil varies from batch to batch. For each batch the amount of impurity (M) is measured. The measurement technique is subject to random error (E) which has a zero mean and is independent of the actual amount of impurity present. If the standard deviations of M and E are 25 ppm and 15 ppm respectively what is the standard deviation of X? Show that the correlation between M and E is 0.6.

3 Reliability for more complex configurations.
The following configuration is not made up of elements in series or parallel.

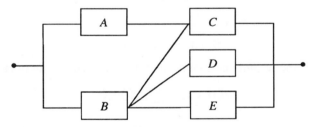

Possible methods of analysis include a complete enumeration or an argument using conditional probability. Show that the overall reliability (R_0) is given by:

$$R_0 = R_B R_C R_D R_E - R_A R_B R_C - R_B R_C R_D$$
$$- R_B R_C R_E - R_B R_D R_E + R_A R_C + R_B R_C$$
$$+ R_B R_D + R_B R_E$$

4 The mean time before failure is

$$E[T] = \int_0^\infty t f(t) \, dt$$

Use integration by parts to show that this is equivalent to

$$\int_0^\infty R(t) \, dt$$

5 The rates (λ_{ij}) of a Markov process can be defined by

$$\lambda_{ij} = \left[\frac{d}{dt} p_{ij}(t) \right]_{\text{evaluated at } t = 0}$$

where $p_{ij}(t)$ is the probability of going from state i to state j in time t. The rate matrix Λ is defined by

$$\Lambda = (\lambda_{ij})$$

(i)   Explain why

$$\sum_j p_{ij}(t) = 1,$$

and hence show that the sum of the elements in any row of the rate matrix is 0.

(ii)   Write the derivative in the definition explicitly as a limit

$$\lambda_{ij} = \lim_{\delta t \to 0} \frac{p_{ij}(\delta t) - p_{ij}(0)}{\delta t}$$

and use this to show that for small δt

$$p_{ij}(\delta t) = \delta t \lambda_{ij}$$
$$p_{ii}(\delta t) = 1 + \delta t \lambda_{ii}$$

(iii)   Give two reasons why

$$\lambda_{ij} \geq 0 \quad \text{if} \quad i \neq j \quad \text{and} \quad \lambda_{ii} \leq 0$$

(iv)   Consider the continuous time form of the Chapman–Kolmogorov equations

$$p_{ij}(t + \tau) = \sum_k p_{ik}(t)p_{kj}(\tau)$$

Differentiate both sides partially with respect to τ, and evaluate these derivatives at $\tau = 0$ to obtain

$$p'_{ij}(t) = \sum p_{ik}(t)\lambda_{kj}$$

with $p_{ij}(0) = 0$ if $i \neq j$ and $p_{ii}(0) = 1$.

(v)   Verify that the matrix form of these equations is

$$\mathbf{P}'(t) = \mathbf{P}(t)\Lambda$$
$$\mathbf{P}(0) = \mathbf{I}$$

6    $M/M/1$ queue.

Lorries arrive at a quarry, randomly and independently, according to a Poisson process with a rate λ per hour. The mean time between arrivals is therefore $1/\lambda$ hours. The service times are very variable and we will approximate their distribution by an exponential distribution with mean $1/\nu$ hours. There is only one server. The notation $M/M/1$ refers to arrival times having the Markov property, i.e. they have an exponential distribution, exponential service times, and 1 server. Let $X(t)$ be the number of lorries in the system at time t.

(i)   Explain why the rate matrix is

$$\Lambda = \begin{vmatrix} -\lambda & \lambda & 0 & 0 & 0 & \cdots \\ \nu & -(\nu+\lambda) & \lambda & 0 & 0 & \cdots \\ 0 & \nu & -(\nu+\lambda) & \lambda & 0 & \cdots \\ 0 & 0 & \nu & -(\nu+\lambda) & \lambda & \cdots \\ \cdot & \cdot & \cdot & \cdot & \cdot \end{vmatrix}$$

(ii)   Hence obtain the normal equations

$$-\lambda u_0 + \nu u_1 = 0$$

$$\lambda u_0 - (\lambda + u)w_1 + \nu u_2 = 0$$

etc

and show that

$$u_1 = (\lambda/\nu)u_0$$
$$u_2 = (\lambda/\nu)^2 u_0$$
$$\vdots$$
$$u_i = (\lambda/\nu)^i u_0$$

(iii) The sum of a geometric progression is given by

$$a + ar + ar^2 + \ldots = a/(1-r)$$

for $|r| < 1$ (see Chapter 7, footnote 3). Deduce that if $\lambda < \nu$ then

$$u_0 = 1 - \lambda/\nu$$

(iv)   Show that the queue increases without limit if $\lambda \geqslant \nu$.
(v)   Show that the average number of lorries in the system, $\sum i\mu_i$, equals $\lambda/(\mu - \lambda)$.
(vi)   Deduce that the average waiting time is $\lambda/(\mu(\mu - \lambda))$, and that the average time spent in the system is $l/(\mu - \lambda)$.

7 *M/M/*1 queue with bounded storage.
   A small ship repair yard has two dry docks for two ships including the one being worked on. Local ships need repair at a rate of λ per day. There is one repair team, and the repair time is exponentially distributed with mean $1/\nu$ days. If a ship needs repair and the yard has a spare dry dock the ship will be sent to the yard, but if there is no free dock it will be sent elsewhere. Show that the proportion of time that both docks are in use is,

$$(\lambda/\nu)^2/[1 + (\lambda/\nu) + (\lambda/\nu)^2]$$

   and deduce the average number of lost repair jobs per day.

8 *M/M/r* queue with bounded storage.
   The ship repair yard, described in Exercise 7, has expanded. There are now

two repair teams and four dry docks. Explain why the rate matrix is:

$$
\begin{array}{c@{\quad}ccccc}
 & 0 & 1 & 2 & 3 & 4 \\
0 & -\lambda & \lambda & 0 & 0 & 0 \\
1 & \nu & -(\nu + \lambda) & \lambda & 0 & 0 \\
2 & 0 & 2\nu & -(2\nu + \lambda) & \lambda & 0 \\
3 & 0 & 0 & 2\nu & (-2\nu + \lambda) & \lambda \\
4 & 0 & 0 & 0 & 2\nu & -2\nu
\end{array}
$$

9 Fourteen pieces of electric cable were subjected to an accelerated life test for 100 days. Three survived 100 days of testing, but the rest failed after

$$21 \quad 33 \quad 39 \quad 46 \quad 57 \quad 58 \quad 67 \quad 71 \quad 78 \quad 89 \quad 93$$

days respectively. Use graph paper to estimate the parameters of a Weibull distribution fitted to these data. Assuming the Weibull model, how many days would you expect the lifetimes of 99% of such test pieces to exceed?

Compare your estimates with the MLE obtained by minimizing the likelihood function:

$$\prod_{i=1}^{11} f(x_i)[1 - F(100)]^3$$

10 Explain why, for example,

Pr(exceedances on day t_j in years 1 and 6 | exactly
two exceedances on day t_j in n years of record) =

$$e^{\beta x_1(t_j)} e^{\beta x_6(t_j)} \left[{}_nC_2 \left(\sum_{i=1}^{n} e^{\beta x_i(t_j)} \middle/ n \right)^2 \right]^{-1}$$

4

Extreme Value and Related Distributions

4.1 Introduction

A flood defence project, which might be a washlands system for river flooding or a barrier for sea flooding, has to be justified by a cost–benefit analysis. There are many factors to be estimated, but one of the most crucial is the distribution of floods. If their severity and frequency is underestimated the project may be curtailed. Further investment would have been justified by the potential benefits. If the flood risk is overstated, the project will be on too large a scale and public funds will have been diverted from more useful schemes.

River floods are generally characterized by peak flows, and floods from the sea are recorded as heights above some ordnance datum. The durations and volumes above specific thresholds are also important, but we begin by looking at univariate distributions for extreme values. Other civil engineering applications include: earthquakes (Nordquist, 1945 and Burton and Makropoulos, 1985); windspeed (Liu, 1991 and ANSI ASCE7, 1988); and depths of ice keels with reference to burial depths of pipelines (Wadhams, 1983 and Chouinard, 1995).

4.2 Extreme value type 1 distribution (Gumbel distribution)

Imagine that we take repeated samples of size N from some parent distribution with an unbounded upper tail, which decays at least as fast as an exponential distribution. The distribution of the greatest values from these samples tends towards an *extreme value type 1* (EV1) distribution as N tends to infinity. The cumulative distribution function (cdf) of the EV1 distribution is,

$$F(x) = e^{-e^{-(x-\xi)/\theta}} \qquad \text{for} \qquad -\infty < x < \infty$$

and the probability density function (pdf), follows from differentiation of the cdf

$$f(x) = \frac{1}{\theta} \exp\{-(x-\xi)/\theta - \exp[-(x-\xi)/\theta]\}$$

I give an outline of the derivation in Appendix 7. The parameter ξ is the mode of the distribution, and θ is a scale factor which is proportional to the standard deviation. Some other features of the distribution, in terms of these parameters, are:

$$\text{median} = \xi + [-\ln(\ln 2)]\theta \approx \xi + 0.36651\theta$$
$$\text{mean } (\mu) = \xi + \in{}^1\theta \qquad \approx \xi + 0.57722\theta$$

$$\text{standard deviation } (\sigma) = \pi\theta/\sqrt{6} \approx 1.28255\theta$$
$$\text{skewness } (\gamma) \approx 1.14$$
$$\text{kurtosis } (\kappa) \approx 5.4$$

Whilst the theoretical conditions for the distribution can never be satisfied exactly, it often fits data sets rather well and these instances provide empirical justification for its use. An example is the set of annual maximum discharges $(m^3 s^{-1})$ of the River Thames at Kingston for 113 years (water years between 1882/83 and 1994/95) in Table A16.1. The data have been naturalized i.e. allowances have been made for water abstraction. Each annual maximum discharge is the greatest of 365 daily mean values, but daily discharges are not stochastically independent. The distributional result can be generalized for the greatest values from correlated sequences of size N, although the convergence to the EV1 distribution is slower. Three hundred and sixty five is not necessarily large enough for the EV1 to be a good model. The assumption that daily discharges come from a distribution with an unbounded upper tail which decays exponentially seems a reasonable approximation, although the unboundedness might be questioned. The easiest way to estimate the parameters of the EV1 distribution is to equate the sample mean and standard deviation with the theoretical population values. The statistics, $\bar{x}$ and s, are 324.5 $m^3 s^{-1}$ and 118 $m^3 s^{-1}$ respectively. The method of moments (MOM) parameter estimates, $\hat{\xi}$ and $\hat{\theta}$, are found by replacing μ and σ in the preceding equations with $\bar{x}$ and s,

$$324.5 = \hat{\xi} + 0.57722\hat{\theta}$$

$$118.8 = 1.28255\hat{\theta}$$

The solutions are 271.0 for $\hat{\xi}$ and 92.6 for $\hat{\theta}$. The EV1 pdf, with these parameter values, is superimposed on the histogram in Figure 4.1, and there appears to be a good correspondence. The sample skewness of the Thames' maxima is 1.04, which is close to the theoretical value of 1.14 for the EV1 distribution and provides further evidence of a good fit. We can make more precise estimates of the parameters of the EV1 distribution if we use maximum likelihood (ML) estimators. The ML estimates are given as the solutions of the

---

[1] In this context $\in$ is Euler's constant, which is defined as the limit as m tends to infinity of

$$1 + \frac{1}{2} + \ldots + \frac{1}{m} - \ln(m) = 0.577215\ldots$$

Lower case Greek gamma is commonly used in the literature for Euler's constant, as well as for population skewness. The distinction is usually clear from the context.

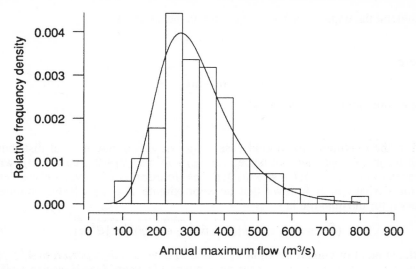

Figure 4.1   EV1 pdf and histogram of Thames maxima

following equations (see Exercise 6).

$$\left.\begin{aligned}\hat{\theta} &= x - \sum x_i\, exp(-x_i/\hat{\theta})/\sum exp(-x_i/\hat{\theta}) \\ \hat{\xi} &= -\hat{\theta} \ln \left\{\sum exp(-x_i/\hat{\theta})/n\right\}\end{aligned}\right\} \tag{4.1}$$

The first formula can be solved iteratively for $\hat{\theta}$ using the MOM estimate as a starting value. When $\hat{\theta}$ has converged to a limit, the second equation gives $\hat{\xi}$ directly.

Example 4.1
We now find the MLE of an EV1 distribution fitted to the Thames' maxima. The MLE of θ converged to 95.5 from a starting value of 92.6 within 20 iterations. The corresponding MLE estimate of ξ is 270.7.

The asymptotic distribution of the ML estimators is bivariate-normal with mean and covariance matrix:

$$\begin{pmatrix} \xi \\ \theta \end{pmatrix}, \quad \left(\frac{\theta^2}{n\pi^2}\right)\begin{pmatrix} \pi^2 + 6(1-\epsilon)^2 & 6(1-\epsilon) \\ 6(1-\epsilon) & 6 \end{pmatrix}$$

Our usual reason for fitting an EV1 distribution will be to estimate some upper quantile of the distribution. For example, the upper 1% quantile, $x_{0.01}$, is the solution of the equation

$$F(x_{0.01}) = 0.99$$

which is

$$x_{0.01} = \xi + [-\ln(-\ln(0.99))]\theta$$

In general the upper $(\varepsilon \times 100)\%$ quantile is the solution of

$$x_\varepsilon = \xi + \psi\theta$$

where

$$\psi = -\ln(-\ln(1 - \varepsilon))$$

The estimator of this quantile is

$$\hat{\xi} + \psi\hat{\theta}$$

and in the maximum likelihood case it has an asymptotic normal distribution with mean $(\xi + \psi\theta)$ and variance, var $(\hat{\xi}) + \psi^2$ var$(\hat{\theta}) + 2\psi$ cov$(\hat{\xi}, \hat{\theta})$, which reduces to $[1 + (6/\pi^2)(1 - \in + \psi)^2](\theta^2/n)$. Johnson *et al.* (1995) gives details of the derivation. It follows that an approximate $(1 - \alpha) \times 100\%$ confidence interval for the quantile is

$$(\hat{\xi} + \psi\hat{\theta}) \pm z_{\alpha/2}\sqrt{\{[1 + 6(1 - \in + \psi)^2/\pi^2]\theta^2/n\}} \qquad (4.2)$$

In the context of annual maxima x_ε is said to have a *return period* of $1/\varepsilon$ years because, on average, it will occur once every $1/\varepsilon$ years (see Exercise 1) if the physical conditions stay the same.

Example 4.2
The estimated upper 1% point of the EV1 distribution fitted to the Thames' maxima is

$$270.7 + (-\ln(-\ln(0.99))) \times 95.5 = 710.0$$

with an estimated standard deviation of 36.3. An approximate 90% confidence interval for the 100 year flood is therefore

$$710 \pm 1.645 \times 36.3$$

$$[650, 770]$$

An approximate 90% confidence interval for the 1000 year flood is 930 ± 86. The confidence intervals only allow for uncertainty in the parameter estimates and it is implicitly assumed that the underlying distribution of annual maximum flows is unchanging. We will analyse the time series of monthly average flows in a later chapter. There is no clear evidence of a trend, but this is no guarantee of what may happen in the long term.

Despite its apt genesis, the EV1 is just one of several distributions which can be used to model extreme values. It has only two parameters and hence fixed values for skewness and kurtosis. The generalized extreme value distribution has an additional parameter which affects the skewness and kurtosis. In practice we often have to select a suitable distribution on the basis of data sets which are considerably shorter than the Thames' record. Design decisions will be made by extrapolating into the tails of the theoretical distributions, so it is important to make an appropriate choice and then to estimate the parameters in an efficient way. Probability plots, which depend on the type of distribution, are a useful means for assessing how good a fit is. A natural way of estimating the parameters of a three-parameter distribution would be to use the MOM and equate the sample mean, standard deviation and skewness to the theoretical population values. However, the sample skewness has undesirable properties. It is highly variable and may be seriously biased. The bias depends on the population distribution as well as the sample size. These results are not surprising, because sample skewness is based on cubed deviations from the mean which may allow outlying values to have disproportionate effects. The sample kurtosis is even less reliable. L-statistics, which are linear functions of order statistics, can be used instead, and their sampling distributions are preferable.

4.3 Order statistics and their uses

4.3.1 Definition and distribution of order statistics

Suppose that we intend to take a random sample of size n from some continuous probability distribution with cdf $F(x)$ and pdf $f(x)$. This can be denoted by

$$\{X_i\} \qquad \text{for } i = 1, \ldots, n$$

where the X_i are independent random variables from the distribution represented by F. Because the X_i are independent their position in the sequence tells us nothing about their relative sizes. However, they can be sorted into ascending order:

$$X_{1:n} < X_{2:n} < \cdots < X_{n:n}$$

where $X_{i:n}$ is now the ith largest in the random sample of n. This is known as the *ith order statistic*. The sample median is an example of an order statistic when n is odd, and an average of two order statistics when n is even.

$$\text{sample median} = \begin{cases} X_{(n+1)/2:n} & \text{when } n \text{ is odd} \\ \dfrac{1}{2} \, (X_{n/2:n} + X_{n/2+1:n}) & \text{when } n \text{ is even} \end{cases}$$

The probability distributions of order statistics are needed to obtain expressions for their expected values (the expected value of the ith order statistic is the average value of the ith largest from a hypothetical infinitely large number of random samples of size n) and other theoretical results. Let $f_r(x)$ be the pdf of the rth order statistic. Then for small δx

$$f_r(x)\delta x = \Pr(x < X_{r:n} < x + \delta x)$$

The event, $x < X_{r:n} < x + \delta x$, corresponds to :$r - 1$ of the X_i being less than x; one X_i being between x and $x + \delta x$; and $n - r$ of the X_i being above $x + \delta x$. The number of ways we can choose one X_i for the middle category is n. There are $_{n-1}C_{n-r}$ choices for the highest category and the remaining $r - 1$ of the X_i go in the lowest category. So the probability can be written:

$$n \, \frac{(n-1)!}{(n-r)! \, (n-1-(n-r))!} \, [F(x)]^{r-1} f(x) \delta x [1 - F(x + \delta x)]^{n-r}$$

Now let δx tend to zero to obtain

$$f_r(x) = \frac{n!}{(n-r)!\,(r-1)!}\,[F(x)]^{r-1}\,[1-F(x)]^{n-r}f(x) \tag{4.3}$$

The r was only introduced as the arbitrary index to avoid confusion with X_i in the explanation, so the expected value of the ith order statistic is

$$E[X_{i:n}] = \int xf_i(x)\,\mathrm{d}x$$

Example 4.3
We will find the expected values of the largest value and the median in samples of size 3 from an exponential distribution. The pdf of the distribution of the largest value is given by equation (4.3) with: $n = 3$, $r = 3$; $F(x) = 1 - \exp(-\lambda x)$; and $f(x) = \lambda \exp(-\lambda x)$. Therefore

$$f_3(x) = 3[1 - e^{-\lambda x}]^2 \lambda e^{-\lambda x}$$

and

$$E[X_{3:3}] = \int_0^{\infty} 3x[1 - e^{-\lambda x}]^2 \lambda e^{-\lambda x}\,\mathrm{d}x$$

This is a straightforward integral, but it is laborious to find without the aid of a computer algebra package. I used Maple to obtain $11/(6\lambda)$. To find the pdf of the median we set $r = 2$, and

$$f_2(x) = 6[1 - e^{-\lambda x}][e^{-\lambda x}]\lambda e^{-\lambda x}$$

The expected value of the median, in samples of size 3, is $5/(6\lambda)$. It is therefore a slightly biased estimator of the median of the distribution, which is $\ln 2/\lambda$.

We can now define the *L*-moments.

4.3.2   Definition of *L*-moments

The *L-moments* exist if X has a finite mean, and they then characterize the distribution (Hosking, 1990). The rth *L*-moment λ_r is defined by

$$\lambda_r = r^{-1} \sum_{k=0}^{r-1} (-1)^k\,_{r-1}C_k E[X_{r-k:r}] \qquad \text{for } r = 1, \ldots$$

Only the first four are commonly used in applications.

$$\lambda_1 = E[X]$$
$$\lambda_2 = \tfrac{1}{2}\{E[X_{2:2}] - E[X_{1:2}]\}$$
$$\lambda_3 = \tfrac{1}{3}\{E[X_{3:3}] - 2E[X_{2:3}] + E[X_{1:3}]\}$$
$$\lambda_4 = \tfrac{1}{4}\{E[X_{4:4}] - 3E[X_{3:4}] + 3E[X_{2:4}] - E[X_{1:4}]\}$$

The first is the mean of the distribution and the second is a measure of the spread, or scale, of the distribution. The third is a measure of asymmetry, and the fourth a measure of how sharply peaked and heavy tailed a distribution is

(Figure 4.2). Dimensionless analogues of skewness and kurtosis are:

$$L\text{-skewness} = \tau_3 = \lambda_3/\lambda_2$$
$$L\text{-kurtosis} = \tau_4 = \lambda_4/\lambda_2$$

The L-skewness is restricted to the range

$$-1 < \tau_3 < 1$$

and L-kurtosis to the range

$$\tfrac{1}{4}(5\tau_3^2 - 1) \leqslant \tau_4 < 1$$

This makes them easier to interpret than the conventional skewness and kurtosis, which can take arbitrarily large values.

Example 4.4
We will find the first four L-moments of the EV1 distribution, and hence calculate the L-skewness and L-kurtosis. The following expected values of order statistics from an EV1, with $\xi = 0$ and $\theta = 1$, distribution are an excerpt

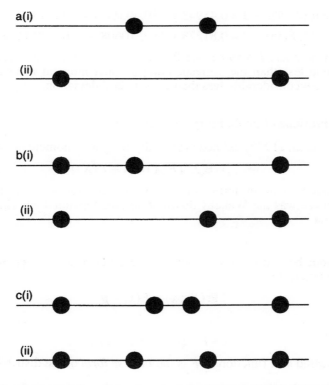

Figure 4.2 Visual interpretation of L-moments (after Hosking, 1990). Expected values of order statisics for (a) $n = 2$ and distribution (ii) is more dispersed than (i); (b) $n = 3$ and distribution (i) is positively skewed whereas (ii) is negatively skewed; (c) $n = 4$ and distribution (i) is more peaked than distribution (ii)

from a table in Johnson *et al.* (1995). They have been calculated from the defining integral.

n	i	Expected value
1	1	0.57722
2	1	-0.11593
2	2	1.27036
3	1	-0.40361
3	2	0.45943
3	3	1.67583
4	1	-0.57351
4	2	0.10608
4	3	0.81278
4	4	1.96351

From the definitions, and a rescaling of the tabled values, we have:

$$\lambda_1 = \xi + 0.57722\theta$$

$$\lambda_2 = (\theta/2)\{1.27036 - (-0.11593)\} = 0.69314\theta$$

$$\lambda_3 = (\theta/3)\{1.67583 - 2 \times 0.45943 + (-0.40361)\} = 0.11779\theta$$

$$\lambda_4 = (\theta/4)\{1.96351 - 3 \times 0.81278 + 3 \times 0.10608 - (-0.57351)\} = 0.10423\theta$$

The *L*-skewness and *L*-kurtosis are 0.17 and 0.15 respectively, nearly an order of magnitude less than the corresponding product moment measures. The *L*-scale λ_2 is also considerably less than the standard deviation.

4.3.3   Definitions of probability weighted moments

Greenwood *et al.* (1979) defined probability weighted moments (PWM) as

$$M_{p,r,s} = E[X^p\{F(X)\}^r\{1 - F(X)\}^s]$$

and expressed them in terms of the parameters for various distributions. Amongst these was the Wakeby distribution (see Section 4.7). This is defined by a formula

$$x = x(F)$$

which cannot be explicitly inverted to give $F(x)$ in an elementary algebraic form. In general, if we write

$$\Pr(X < x) = F(x) = p$$

then

$$x = F^{-1}(F(x)) = F^{-1}(p)$$

and x expressed as a function of F is the *inverse form* of the distribution.

Example 4.5
The exponential distribution has cdf

$$F(x) = 1 - \exp(-\lambda x) = p$$

If we make x the subject of the formula $x = -\ln(1 - p)/\lambda$, and the inverse form is

$$x = -\ln(1 - F)/\lambda$$

The relationship between the PWM and the parameters of a distribution can be used to estimate the parameters in a similar fashion to MOM. The PWM can be expressed as a linear combination of *L*-moments, and vice versa, so PWM are themselves linear combinations of order statistics. We use special cases of the PWM to estimate *L*-moments: $p = 1$, $r = 0$, and r is written in place of s, to give

$$\beta_r = \mathrm{E}[X\{F(X)\}^r]$$

PWM with $p = 1$ and $s = 0$

$$\alpha_r = \mathrm{E}[X\{1 - F(X)\}^r]$$

are used to estimate the parameters of the Wakeby distribution directly.

4.3.4   Estimation of *L*-moments

The natural way to estimate λ_2, from a sample of size n, is to take all possible sub-samples of size 2 from the n and average the $_nC_2$ values of:

$$(x_{2:2} - x_{1:2})/2$$

The first four *sample L-moments* are defined as:

$$l_1 = n^{-1} \sum_i x_i$$

$$l_2 = \frac{1}{2} \, _nC_2^{-1} \sum_{i>j} \sum (x_{i:n} - x_{j:n})$$

$$l_3 = \frac{1}{3} \, _nC_3^{-1} \sum_{i>j>k} \sum \sum (x_{i:n} - 2x_{j:n} + x_{k:n})$$

$$l_4 = \frac{1}{4} \, _nC_4^{-1} \sum_{i>j>k>l} \sum \sum \sum (x_{i:n} - 3x_{j:n} + 3_{k:n} - x_{l:n})$$

These are not in a convenient form for computation. Hosking (1990) gives equivalent forms using sample probability weighted moments (b_r)

$$b_0 = \bar{x}$$

$$b_r = n^{-1} \sum_{i=1}^{n} \frac{(i-1)(i-2)...(i-r)}{(n-1)(n-2)...(n-r)} x_{i:n} \qquad \textit{for } r = 1, ...$$

$$l_{r+1} = \sum_{k=0}^{r} (-1)^{r-k} \, _rC_k \, _{r+k}C_k \, b_k \qquad \textit{for } r = 0, ...$$

The first four are

$$l_1 = b_0$$
$$l_2 = 2b_1 - b_0$$
$$l_3 = 6b_2 - 6b_1 + b_0$$
$$l_4 = 20b_3 - 30b_2 + 12b_1 - b_0$$

The foregoing formulae define unbiased estimators. If you refer back to the definition of β_r you will see that more immediately intuitive estimates are given by the *plotting-position estimates*:

$$b_r \approx \sum x_{1:n} \{F(\mathrm{E}[X_{1:n}])\}^r / n$$

where

$$F(\mathrm{E}[X_{1:n}]) \approx (i - 0.35)/n$$

Plotting positions are discussed in detail in the next section, but $(i - 0.35)/n$ is a plausible enough estimate of the probability that X is less than $\mathrm{E}[X_{i:n}]$. For large n the unbiased and plotting position estimates are almost the same, although the individual factors

$$[(i-1)/(n-1)] \times \cdots \times [(i-r)/(n-r)] \quad \text{and} \quad [(i-0.35)/n]^r$$

differ noticeably when i is large or small in comparison with n, and r is greater than 1. Hosking and Wallis (1995) concluded that for general use the unbiased estimators should be preferred.

Vogel *et al.* (1993) calculated the *L*-skewness and *L*-kurtosis from annual peak flow data at 61 sites in Australia. They plotted the pairs for each site on an *L-moment diagram* (Vogel and Fennessey, 1993) and compared the points with the theoretical relationship between τ_3 and τ_4 for various distributions (reproduced as Figure 4.3). They concluded that the generalized extreme value (GEV) distribution (Section 4.4) and Wakeby (Section 4.7) distributions were more realistic for predicting floods in South West Australia than the log-gamma (log-Pearson 3, LP3). They recommended the use of the Wakeby distribution or the generalized Pareto distribution (usually used for modelling peaks over thresholds as described in Section 4.5) for other regions in Australia.

4.3.5   Plotting against expected values of order statistics

If X has an EV1 distribution with mode ξ and scale factor θ, the *reduced variable Y* defined by

$$Y = \frac{X - \xi}{\theta}, \qquad \text{and equivalently } X = \xi + \theta Y$$

has an EV1 distribution with mode 0 and scale factor 1. The reduced variable has the more convenient cdf,

$$F(y) = e^{-e^{-y}}$$

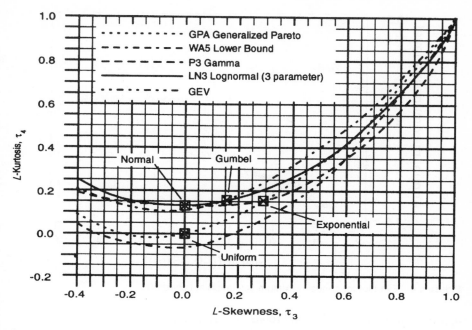

Figure 4.3   L-moment diagram (after Vogel and Fennessey, 1993)

Suppose we have a sample of size n from an EV1 distribution. Since X is just a rescaling of Y,

$$\mathrm{E}[X_{i:n}] = \xi + \theta \mathrm{E}[Y_{i:n}]$$

If we plot the ordered sample $x_{i:n}$ against $\mathrm{E}[Y_{i:n}]$, we would expect to see points scattered about some notional straight line. The slope and intercept of this line are θ and ξ respectively, and these parameters can be estimated by drawing a plausible line through the points. However, even if we use generalized least squares, to allow for different variances and covariances of order statistics (Johnson *et al.*, 1995, give a convenient table of these), such estimators are less precise than those based on L-moments. The value of the plot is that it is a fairly sensitive check whether the EV1 distribution is an appropriate model. As it is a somewhat subjective graphical method, it is usual to use approximate values of $\mathrm{E}[Y_{i:n}]$ rather than the accurate values. The probability of a random variable being less than the average value of the *i*th largest, in a sample of size n, from any distribution can be approximated by an expression of the form:

$$\frac{i+\gamma}{n+\delta} \qquad \text{with} \qquad -1 < \gamma < \delta < 1$$

This can be thought of as i/n, with an adjustment to avoid a probability of 1 when i equals n. A reasonable general approximation for any distribution is $(i-0.4)/(n+0.2)$, but Hosking recommends $(i-0.35)/n$ as the best

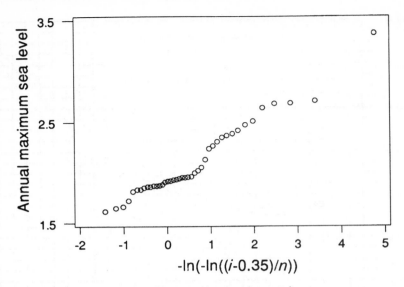

Figure 4.4   EV1 plot for annual maximum sea levels at Lowestoft

approximation for the EV1 and related distribution. So

$$F(E[Y_{i:n}]) \approx (i - 0.35)/n$$

and since $F(y)$ is $\exp(-\exp(-y))$

$$E[Y_{i:n}] \approx -\ln(-\ln((i - 0.35)/n))$$

This procedure, which uses ordinary graph paper, is equivalent to plotting $x_{i:n}$ against $(i - 0.35)/n$ on Gumbel probability paper. A point p on the probability scale is related to the linear scale y by:

$$p = F(y) = e^{-e^{-y}}$$

A value used for $F(E[x_{i:n}])$, such as $(i - 0.35)/n$, is known as a *plotting position*.

Example 4.6
Annual maximum sea levels at Lowestoft between 1953 and 1994 (metres relative to Ordnance Datum Newlyn) are given in Table A16.2. They have been sorted into ascending order and plotted against $-\ln(-\ln((i - 0.35)/42))$ in Figure 4.4. The plot indicates that an EV1 distribution will be a good fit. Data for Sheerness are also given, and you are asked to investigate these in Exercise 3.

4.4   Generalized extreme value distribution

The generalized extreme value (GEV) distribution is a way of writing the Type 2 and Type 3 extreme value distributions (as defined in Exercises 4 and 6), in terms of a parameter k, so that they approach an EV1 distribution as k tends towards zero. The cdf is

$$F(x) = \begin{cases} \exp\{-[1 - k(x - \xi)/\theta]^{1/k}\} & \text{for } x > \xi + \theta/k & \text{when } k < 0 \\ \exp\{-\exp[-(x - \xi)/\theta]\} & \text{for } -\infty < x < \infty & \text{when } k = 0 \\ \exp\{-[1 - k(x - \xi)/\theta]^{1/k}\} & \text{for } x < \xi + \theta/k & \text{when } k > 0 \end{cases}$$

If k is less than zero it is the Type 2 extreme value distribution. This arises in a similar way to the EV1 distribution; the difference being that the tails of the parent distribution do not decay as fast as a negative exponential. The Pareto distributions (with $k < 0$), covered in the next section, have this property. The Type 3 extreme value distribution arises when the parent distribution is bounded above, and this corresponds to k greater than zero. To obtain the EV1 form, use a Taylor expansion for

$$\begin{aligned}[1 - k(x - \xi)/\theta]^{1/k} = {}& 1 + (-k)(1/k)(x - \xi)/\theta \\ & + (-k)^2 (1/k)(1/k - 1)[(x - \xi)/\theta]^2/2! \\ & + \cdots\end{aligned}$$

and then let k tend to zero

$$1 - (x - \xi)/\theta + [(x - \xi)/\theta]^2/2! \ldots = \exp[-(x - \xi)/\theta]$$

The GEV distribution of X can be modified to give the limiting distribution of the sample minimum, W say, by writing down the cdf of $(-X)$ which is then defined as W. The Weibull distribution is an important special case. In Exercise 6, you are asked to show that if a random variable X has an extreme value Type 3 distribution, then $(-X)$ has a Weibull distribution.

Hosking *et al.* (1985) recommend PWM estimators for the GEV distribution as their biases and variances compare favourably with ML estimators in small samples. The PWM estimates are given by the solutions of the equations:

$$\hat{k} = 7.8590c + 2.9554c^2 \quad \text{where} \quad c = (2b_1 - b_0)/(3b_2 - b_0) - \ln 2/\ln 3$$

$$\hat{\theta} = (2b_1 - b_0)\hat{k}/\{(1 - 2^{-\hat{k}})\Gamma(1 + \hat{k})\}$$

$$\hat{\xi} = b_0 + \hat{\Theta}\{\Gamma(1 + \hat{k}) - 1\}/\hat{k}$$

If $\hat{k}$ turns out to be fairly close to 0, and there is no obvious curvature in the Gumbel plot (see Exercise 5), I would prefer to set k equal to zero and revert to the EV1 distribution. The phrase 'fairly close' can be formalized by a statistical hypothesis test. Suppose that we have a sample from an EV1 distribution. This supposition is the hypothesis to be tested. Then it can be shown that, to a reasonable approximation

$$\hat{k} \sim N(0, 0.5633/n)$$

If $|\hat{k}|$ exceeds $z_{a/2}\sqrt{(0.5633/n)}$, we have evidence against our supposition (hypothesis) at the $\alpha \times 100\%$ level. If we do not have evidence against our hypothesis, at a 10% level, we may decide to act as though the hypothesis were true. This should not be misinterpreted as proof that the hypothesis is true; at best it is a reasonable approximation to a complex physical process.

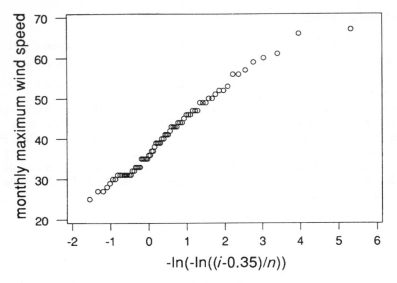

Figure 4.5   EV1 plot for monthly maximum wind speeds at Weston Park

The asymptotic variances of the PWM estimators can be estimated from quite complex expressions in the paper by Hosking *et al.* (1985) (also available in Johnson *et al.*, 1995). The alternative is to use a bootstrap method.

Example 4.7
Table A16.3 gives monthly maximum wind speeds (knots) at Weston Park in Sheffield from January 1990 until December 1995. The EV1 plot (Figure 4.5) is slightly concave down, and we expect a positive estimate of k. The plotting position estimates of the PWM b_0, b_1 and b_2 are: 41.18, 23.4219, and 16.6946 respectively. Hence $\hat{k}$ is 0.0408, and $\hat{\theta}$ and $\hat{\xi}$ are 8.4733 and 36.6187. The standard deviation of $\hat{k}$ is approximately $\sqrt{(0.5633/72)}$ which equals 0.088, so there is no substantial evidence that the GEV is an improvement on an EV1. The estimated upper 0.1% quantile of the GEV distribution is the solution of

$$F(x_{0.001}) = \exp\{-[1 - 0.0408(x_{0.001} - 36.6187)/8.4733]^{1/0.0408}\} = 0.999$$

which is 87.6. For comparison, the MLEs of the parameters θ and ξ for an EV1 distribution fitted to the same data are 7.886 and 36.60 respectively, leading to an estimated 0.1% quantile of 91.1.

4.5   Generalized Pareto distribution and peaks over threshold (POT) analysis

A variant on analysing annual maximum flows is to take all peak flows which exceed some threshold. The increased number of data is a potential advantage, but some criterion for independence needs to be defined.

4.5.1 Definition of generalized Pareto distribution and estimation of parameters

The generalized Pareto distribution includes Pareto, exponential and uniform distributions as particular cases. Its properties, and relationship with the GEV distribution, make it a consistent choice for analysing peak flows above some threshold (L). Another application is in reliability studies, partly because it offers considerable flexibility for modelling the way in which the probability of failure varies with time. The cdf is

$$F(x) = \begin{cases} 1 - (1 - kx/\eta)^{1/k} & \text{for } 0 \leqslant x < \infty & \text{when } k < 0 \\ 1 - \exp(-x/\eta) & \text{for } 0 \leqslant x < \infty & \text{when } k = 0 \\ 1 - (1 - kx/\eta)^{1/k} & \text{for } 0 \leqslant x \leqslant \eta/k & \text{when } k > 0 \end{cases}$$

The pdf, for $k \neq 0$, is given by

$$f(x) = \eta^{-1}(1 - kx/\eta)^{1/k-1} \qquad 0 \leqslant x \leqslant \eta/k \text{ or } \infty$$

If $k < -\frac{1}{2}$ the distribution has an infinite variance. The special case with $k = 0$ is the exponential distribution with pdf

$$f(x) = \eta^{-1}\exp(-x/\eta) \qquad 0 \leqslant x < \infty$$

If k exceeds $\frac{1}{2}$ the distribution has a finite end point, and the special case with $k = 1$ is a uniform distribution on the interval $[0, \eta]$. For most practical applications, we can assume that $-\frac{1}{2} < k < \frac{1}{2}$.

The generalized Pareto distribution has two properties which make it a reasonable choice for peak over threshold analyses. The first is that if a random variable X has a generalized Pareto distribution, then the conditional distribution of $X - L$, given that L exceeds t, also has a generalized Pareto distribution with the same value of k. The second is that the distribution of the maximum of N generalized Pareto variables has a GEV distribution with the same value of k, if N is a random number drawn from a Poisson distribution. An annual flood is the maximum of a number of floods arising from storms during the year. If occurrences of floods are modelled by a Poisson process, and their magnitudes are assumed independent, a choice of a generalized Pareto distribution for flood peaks, over some threshold, is consistent with the use of a GEV distribution for the annual maxima. As usual, a plausible theoretical argument needs to be backed up with some empirical evidence. The moments of the distribution can be obtained quite easily if you notice that

$$E[(1 - kX/\eta)^r] = 1/(1 + rk)$$

provided that the denominator is positive. If $k \leqslant -1/r$ the rth moment does not exist. The mean, variance, skewness and kurtosis follow from this expression; with r equal to one, two, three and four respectively

$$\mu = \eta/(1 + k)$$
$$\sigma^2 = \eta^2/[(1 + k^2)(1 + 2k)]$$
$$\gamma = 2(1 - k)(1 + 2k)^{1/2}/(1 + 3k)$$
$$\kappa = 3(1 + 2k)(3 - k + 2k^2)/[(1 + 3k)(1 + 4k)]$$

Hosking and Wallis (1987) recommend the method of moment estimators of η and k if it is reasonable to suppose k is positive. These are:

$$\hat{\eta} = \tfrac{1}{2}\bar{x}(\bar{x}^2/s^2 + 1), \qquad \hat{k} = \tfrac{1}{2}(\bar{x}^2/s^2 - 1)$$

where $\bar{x}$ and s^2 are the sample mean and variance. If $k < -0.2$, the PWM estimators:

$$\hat{\eta} = 2a_0a_1/(a_0 - 2a_1), \qquad \hat{k} = a_0/(a_0 - 2a_1) - 2$$

where

$$a_r = \sum x_{i:n}\{1 - F(\mathrm{E}[X_{i:n}])\}^r/n$$

are far preferable. In practice it is probably best to use PWM estimators if $\hat{k} < 0$.

4.5.2   Estimation of quantiles

It is straightforward to estimate quantiles and exceedance probabilities of the Pareto distribution, but the frequency of occurrence must be allowed for when transforming these to risk estimates. We will often require quantiles and exceedance probabilities for annual maximum floods. Suppose that exceedances of the threshold (L) follow a Poisson process with a rate of λ per year. Let exceedances, X, of the threshold (L) have a Pareto distribution with cdf $F(x)$, and let Y represent annual maximum floods. If c is some critical level, greater than L

$$
\begin{aligned}
\Pr(Y < c) = {}& \Pr(0 \text{ exceedance of } L \text{ in a year})\\
& + \Pr(1 \text{ exceedance of } L \text{ and exceedance} < c)\\
& + \Pr(2 \text{ exceedances of } L \text{ and both exceedances} < c)\\
& + \Pr(3 \text{ exceedances of } L \text{ and all 3 exceedances} < c)\\
& + \cdots\\
= {}& \mathrm{e}^{-\lambda} + (\lambda\mathrm{e}^{-\lambda})F(c-L) + (\lambda^2\mathrm{e}^{-\lambda}/2!)F^2(c-L)\\
& \hphantom{= \mathrm{e}^{-\lambda} + } + (\lambda^3\mathrm{e}^{-\lambda}/3!)F^3(c-L) + \cdots\\
= {}& \mathrm{e}^{-\lambda}(1 + \lambda F(c-L) + \lambda^2 F^2(c-L)/2!\\
& \hphantom{= \mathrm{e}^{-\lambda}(1} + \lambda^3 F^3(c-L)/3! + \cdots)\\
= {}& \mathrm{e}^{-\lambda}\mathrm{e}^{\lambda F(c-L)} = \mathrm{e}^{-\lambda(1-F(c-L))}
\end{aligned}
$$

The probability that c will be exceeded at least once during a year is therefore

$$\Pr(c < Y) = 1 - \mathrm{e}^{-\lambda(1-F(c-L))} \tag{4.4}$$

The return period of an exceedance of c is $[\lambda(1 - F(c-L)]^{-1}$ years. The relationship between the quantiles of the distributions of Y and X can be found by considering the cdf of Y. We have just shown that

$$\Pr(Y < L) = \mathrm{e}^{-\lambda}$$

$$\Pr(Y < y) = \mathrm{e}^{-\lambda(1 - F(y-L))} \qquad \text{for } L < y$$

Now let y_ε be the upper $\varepsilon \times 100\%$ quantile of the distribution of Y, where $\varepsilon < 1 - \mathrm{e}^{-\lambda}$. Then

$$\Pr(Y < y_\varepsilon) = \exp[-\lambda(1 - F(y_\varepsilon - L))] = 1 - \varepsilon$$

and rearrangement leads to

$$F(y_\varepsilon - L) = 1 + \lambda^{-1} \ln(1 - \varepsilon)$$

Therefore

$$y_\varepsilon = L + F^{-1}(1 + \lambda^{-1} \ln(1 - \varepsilon)) \tag{4.5}$$

and the upper $\varepsilon \times 100\%$ quantile of the distribution of Y is the sum of L and the upper $(-\lambda^{-1} \ln(1 - \varepsilon)) \times 100\%$ quantile of the distribution of X. The upper $\delta \times 100\%$ quantile of the distribution of X is given by

$$F(x_\delta) = \begin{cases} \alpha\{1 - \delta^k\}/k & k \neq 0 \\ -\alpha \ln(\delta) & k = 0 \end{cases}$$

The parameter λ is estimated by the average number of independent exceedances of L per year.

Hosking and Wallis (1987) give the following expression for the asymptotic variances and covariance of $\hat{\alpha}$ and $\hat{k}$

$$\text{var} \begin{bmatrix} \hat{\alpha} \\ \hat{k} \end{bmatrix} \simeq \frac{1}{n(1 + 2k)(1 + 3k)}$$

$$\begin{bmatrix} \alpha^2(7 + 18k + 11k^2 + 2k^3) & \alpha(2 + k)(2 + 6k + 7k^2 + 2k^3) \\ \alpha(2 + k)(2 + 6k + 7k^2 + 2k^3 & (1 + k)(2 + k)^2(1 + k + 2k^2) \end{bmatrix}$$

Approximate confidence intervals for x_ε can be constructed from these results (Exercise 8). Confidence intervals for y_ε also use the variance of $\hat{\lambda}^{-1}$ which follows from the assumption that exceedances of L have a Poisson distribution (Exercise 9). The alternative approach is to construct a confidence interval using the parametric bootstrap method. This has an advantage of avoiding the assumption that the estimator of the quantile has a normal distribution.

Example 4.8
The Environment Agency provided a 21 year series of average daily flows $(m^3 s^{-1})$ of the River Coquet at Rothbury in Northumberland, UK. The first step is to define independent peaks in this series. Hydrologists in the UK will be guided by the Flood Studies Report (Natural Environment Research Council, 1975) which contains the following recommendation. When two or more peaks occur close together in time, the highest one is considered as independent while the others must satisfy the two following independence criteria if they are to be included in the analysis

1. Independent in discharge: the trough between two peaks must be less than two thirds of the earlier peak value.
2. Independent in time: the two peaks must be separated by at least three times the average time to rise to a peak, the latter defined as the mean of the times to rise from five clean typical flood hydrographs in the record.

We adapted this advice slightly, and used a minimum time of 3 days between peaks. Also, we only took peaks above the series mean, and thereby obtained the 485 peaks given in Table A16.4. The results for different thresholds, and for GEV and EV1 distributions fitted to annual maxima can be seen in Table 4.1. The parameters were estimated by PWM, since the MOM estimates of k were negative. The variation in the estimated quantiles of the distribution of annual maxima is a reminder of the lack of precision of the estimators. The slightly anomalous result with the threshold of 40 appears to be due to a decrease in the coefficient of variation, and it is worthwhile investigating the sensitivity of estimates to the threshold. We should also remember that the standard deviation of the EV1 estimator of the upper 1% quantile, in samples of size 21, is roughly 12%. We have also ignored seasonal variation in this analysis.

The Rothbury data are average daily flows, which are appropriate for flood defence measures such as washlands. However, if we wish to estimate instantaneous peak flows from these data, we must make some allowance for the difference between average daily flows and instantaneous peak flows within days. In general, I have found instantaneous peak flow data to be scarcer than daily average flows. We were given the maximum instantaneous flow for each month in the record, but daily peak values were not available. However, the maximum instantaneous flow appeared to correspond clearly to the maximum daily average flow in the month. The average value of the 252 ratios

$$\frac{\text{maximum instantaneous flow in month}}{\text{maximum daily average flow in month}}$$

is 1.122 and their standard deviation is 0.203. Their distribution is well approximated by an exponential distribution (with 1 as the lower limit). There was no distinctive relationship between the ratio and the maximum daily average flow. The Coquet is a small river and the ratios are variable and large enough to be of practical importance. A formal analysis would involve finding the probability that $(\ln Y + \ln(\text{ratio}))$ exceeds some critical value, or estimation of the quantiles of that variable.

Table 4.1 *Results of fitting the generalised Pareto distribution to peak over threshold data from the River Coquet, and the GEV and EV1 to annual maxima.*

Threshold	Number of exceedances	Estimated parameters			Estimated quantiles of the distribution of annual maxima		
		$\hat{\lambda}$	$\hat{\alpha}$	R	$\hat{y}_{0.1}$	$\hat{y}_{0.01}$	$\hat{y}_{0.001}$
AM-EV1	21			0	143	228	312
AM-GEV	21			−0.10	144	255	393
60	25	1.19	24.53	−0.21	137	261	458
50	37	1.76	25.70	−0.13	137	239	375
40	51	2.43	29.78	−0.02	135	210	288
30	85	4.04	20.29	−0.21	142	277	496
20	167	7.95	14.33	−0.29	146	322	669
15	232	11.04	13.80	−0.27	143	303	598
10	326	15.52	13.29	−0.24	141	286	540

4.6   Seasonality, trends and other explanatory variables

In Section 4.5.2 we modelled the occurrences of floods by a Poisson process with a constant parameter λ. We also assumed that the parameters of the distribution of exceedances of the threshold did not vary throughout the year. In most countries there is marked seasonal variation, and ignoring it may lead to misleading estimates of flood risk from the POT analysis. Another reason for considering seasonal differences is that we may need estimates of seasonal flood risk to assess likely damage to crops.

4.6.1   Seasonality

There are three approaches to seasonal flood frequency analysis. The first is to analyse seasons separately, and if a few distinct seasons can be identified this is the most convenient. The second is to allow the parameters of the model to vary throughout the year. There are many variations on this theme, North (1980) for example. The third is to remove estimates of the seasonal component of variation from the original daily series, before carrying out the extreme value analysis. However, this does assume that seasonal effects identified from the central portion of the data remain valid in the tails, and Davison and Smith (1990) suggest that it is best confined to applications where the physical origin of the seasonal component is well understood. There is also the complication of calculating exceedance probabilities, and percentiles, of the reseasonalized distribution of yearly maxima. An example is daily flow data D_t, modelled as the sum of a harmonic seasonal component with an amplitude A and a deseasonalized series S_t:

$$D_t = S_t + A \cos(2\pi t/365 + \phi)$$

Now let S be a peak which exceeds a threshold L, and let X be the exceedance $(S - L)$. From equation (4.4)

$$\Pr(S < c) = F(c - L)$$

but we want the corresponding probability for the reseasonalized D

$$\begin{aligned}
\Pr(D < c) &= \Pr(S + A \cos(2\pi t/365 + \phi) < c) \\
&= \Pr(S < c - A \cos(2\pi t/365 + \phi)) \\
&= F(c - L - A \cos(2\pi t/365 + \phi))
\end{aligned}$$

If we assume S has a stationary distribution, this time varying probability can be averaged over a year,

$$\overline{F}(c - L) = \frac{1}{365} \int_0^{365} F(c - L - A \cos(2\pi/365 + \phi)) \, \mathrm{d}t$$

and substituted into equation (4.4). The rate λ is estimated by the average number of independent exceedances of L by S_t per year. The less elegant alternative is to rely on simulation.

4.6.2   Two component extreme value distribution, mixtures of extreme value distributions, and zeros

Two component extreme value distribution   Many environmental extremes could be the result of one of two, or more, distinct physical processes, e.g.

storms in Florida and the Gulf Coast can be classified as tropical or non-tropical; extreme rainfall can occur in convective or frontal storms; annual maximum floods can be a result of summer storms or winter rain augmented by snow melt. The *two component extreme value distribution* (TCEV) for annual maxima series (Gumbel, 1958 and Rossi *et al.*, 1984) may then be apt. Let the annual maximum wind speed from tropical storms be X_1, the annual maximum wind speed from non-tropical storms be X_2, and the annual maximum wind speed from either source be X. Then

$$\Pr(X < x) = \Pr(X_1 < x \quad \text{and} \quad X_2 < x)$$

If X_1 and X_2 are assumed to have independent EV1 distributions, the cdf of X is

$$F(x) = \exp[-e^{-(x-\xi_1)/\theta_1}] \exp[-e^{-(x-\xi_1)/\theta_1}]$$
$$= \exp[-\phi_1 e^{-x/\theta_1} - \phi_2 e^{-x/\theta_2}]$$

where

$$\phi_j = e^{\xi_j/\theta_j} \qquad \text{for} \quad j = 1, 2$$

The pdf is obtained by differentiating with respect to x

$$f(x) = F(x)\psi(x)$$

where

$$\psi(x) = (\phi_1/\theta_1)e^{-x/\theta_1} + (\phi_2/\theta_2)e^{-x/\theta_2}$$

The log-likelihood is

$$\ln(\mathcal{L}) = \sum \ln(F(x)) + \sum \ln(\psi(x))$$

Notice that it is not necessary to know from which distribution an annual maximum has arisen. You are asked, in Exercise 11, to suggest a means of modifying the analysis when you do know the origin of the maximum. The data are extreme wind speeds from Jacksonville, Florida, given in Table A16.5 (Kinnison, 1985).

Example 4.9
The data in Table A16.6 are annual maximum flows at Skelton, near York, UK, on the River Ouse. They were provided by the Environment Agency. Snow melt may have made a major contribution to some of the peak flows. I used a simplex algorithm (NAG E04CCF) to maximize the log-likelihood, as a function of ϕ_1, θ_1, ϕ_2 and θ_2, from initial guesses of 70, 70, 30 and 150 respectively. It converged to 88.75, 63.75, 11.25 and 143.75, so the corresponding estimates of the ξ_j and θ_j are: $\xi_1 = 286$, $\theta_1 = 64$, $\xi_2 = 348$ and $\theta_2 = 144$. The TCEV is not a convincing model for these data. The maximized log-likelihood is -172.8 which is less than that for a single EV1 distribution (-163.0). This is feasible because the single EV1 distribution is not a special case of the TCEV. Rossi *et al.* (1984) mention that parameter estimates are imprecise if there are no obvious outliers from an EV1 plot, and there are none in this case.

Mixture of extreme value distributions   A different scenario assumes an alternative source of extreme values, which occurs only rarely, but gives rise to the annual maximum when it does occur. A mixture of two EV1 distributions would then be more appropriate. The pdf is

$$f(x) = pf_1(x) + (1 - p)f_2(x)$$

where f_1 and f_2 are pdf of EV1 distributions with modes ξ_1, ξ_2 and scale factors θ_1 and θ_2 respectively. If we know which of two physical processes produced each yearly maximum we can estimate p directly, but in other cases we may either not know which process was involved or not be able to make any definite classification, e.g. a flood might be mainly due to extreme rain but augmented by a small amount of snow melt. The likelihood function for estimating all five parameters is

$$\mathscr{L}(p, \xi_1, \xi_2, \theta_1, \theta_2) = \Pi f(x_i)$$

Maximizing it raises the same issues that we considered for a mixture of normal distributions (Section 3.4.4). In particular, each sample point generates a singularity in the likelihood function unless we restrict θ_2 to be a specified multiple of θ_1.

Zero flows   In arid climates there are years during which some river beds remain dry. The Department of Water Affairs in Namibia has provided annual flood data from 44 stations, and these are given in Table A16.7. For a single site, we can estimate the probability of a year with no flow with a distribution of annual flood conditional on some flow. This is a mixture of a spike at zero and an extreme value distribution. The probability of zero can be estimated by the proportion of years with no flow. An extreme value distribution can be fitted to the annual floods from the other years in the usual way.

Example 4.10
There are 17 annual floods from the River Khan at Ameib. There was no recorded flow in two of these water years. The fitted pdf is of the form

$$f(x) = \begin{cases} p & \text{if } x = 0 \\ (1 - p)f_1(x) & \text{if } 0 < x \end{cases}$$

For this station $\hat{p} = 2/17$ and a lognormal distribution is plausible for f_1. The mean and standard deviation of the 15 logarithms of the non-zero annual floods are 4.51 and 1.57 respectively.

4.6.3   Time or other covariates

Mean sea levels on the south-east coast of England have increased over the past 20 years, because of a slight downwards tilting of the land mass together with global rises in sea levels. We could allow for this when fitting an extreme value distribution. For example we could fit an EV1 distribution, assuming a linear trend in the location parameter, and a constant ratio of location parameter to scale parameter. That is:

$$\xi = \xi_0 + at$$
$$\theta = b\xi$$

The likelihood function will be a function of the three parameters ξ_0, a and b.

Another example is the annual maximum flows from Skelton (Example 4.9). A time series plot suggests that the maxima are becoming more variable, and possibly more extreme as well. However, there is no agreed physical explanation for this, and it may be reasonable to attribute the apparent trend to chance. We can use the likelihood ratio test (see Appendix 3) to test this. Minus twice the natural logarithm of the ratio of likelihoods has an approximate chi-square distribution with degrees of freedom equal to the difference in the number of parameters.

Example 4.11
I started by fitting the, unmodified, EV1 distribution to the annual maximum flows at Skelton, by minimizing the log-likelihood function with the simplex algorithm (NAG E04CCF). The estimates of ξ and θ were 310.02 and 67.56 respectively, and the log-likelihood was -163.051. I used the estimate of ξ, $a = 0$ and $b = 310.02/67.56$ as starting values for the modified EV1 distribution. The simplex algorithm converged to give: $\xi_0 = 305.04$; $a = 4.6472$, where $t = $ year $- 1980$; and $b = 0.19188$. The log-likelihood was -158.238. The chi-square statistic is

$$-2(-163.051 - (-158.238)) = 9.63$$

This exceeds $\chi^2_{3-2, 0.01}$ (which is 6.635) and we have convincing evidence that the mode and scale factor of the assumed EV1 distribution have increased over this period. I found that the convergence of the algorithm to these estimates was sensitive to the initial values. If the initial values were chosen arbitrarily, the routine converged to give different final values with an associated log-likelihood which was less than that for the two-parameter case. If θ was estimated as a constant, rather than as proportional to the mode, the log-likelihood decreased to -160.260.

4.7 Wakeby distribution

The form of this distribution was proposed by H.A. Thomas in 1976. It is said that he was looking across Wakeby Pond, near Cape Cod, at the time. The versatility of the Wakeby distribution is a consequence of its having five parameters. Let $F(x)$ represent the cdf. Then, although F cannot be written as an explicit function of x, x can be expressed in terms of F and the five parameters. The inverse cdf formula is

$$x(F) = \xi + \theta\beta^{-1}[1 - (1 - F)^{\beta}] - \phi\delta^{-1}[1 - (1 - F)^{\delta}] \qquad \text{for} \quad \xi < x$$

where there are some restrictions on the parameter values for $x(F)$ to be strictly increasing with F. In particular, $x(F)$ is properly defined if θ, $\beta > 0$ and either ϕ, $\delta < 0$, in which case x tends to infinity as F tends to 1, or $\delta > 0$ and $\phi < 0$, in which case it is bounded above by $\xi + \theta/\beta - \phi/\delta$. Landwehr *et al.* (1979) list other valid parameter combinations.

Hydrologists were curious about the Wakeby distribution because it can satisfy the *condition of separation*. That is, the skewness statistics of random samples from the distribution can be as variable as sample skewness calculated from hydrologically similar annual maxima records. However, another explanation for the excess variability in sample skewnesses is that the population skewness is not constant within the hydrologically similar region.

The definition may seem rather odd, but it is a convenient form for calculating quantiles once the parameters have been estimated. Landwehr *et al.* (1979) recommend using the method of probability weighted moments. Recall that the α_r are defined by

$$\alpha_r = E[X\{1 - F(X)\}^r]$$

By the definition of expectation, the α_r are given by the integral

$$\int_{-\infty}^{\infty} x(1 - F(x))^r f(x)\, dx$$

But since

$$f(x) = \frac{dF(x)}{dx}$$

the variable of integration can conveniently be changed to F. So

$$\alpha_r = \int_0^1 x(F)(1 - F)^r\, dF$$

which can be shown to be

$$= (r + 1)^{-1}[\xi + \theta(r + \beta + 1)^{-1} + \phi(r - \delta + 1)^{-1}]$$

These PWM are estimated by

$$\alpha_r = \sum [1 - (i - 0.35)/n]^r x_{i:n}/n$$

The parameters themselves are found by equating the population PWM, α_r, with the sample PWM, a_r, for r from 0 to 4. Expressions for the parameter estimators in terms of a_r are given in Appendix 8, but the estimates will not necessarily be real numbers which lead to a valid $x(F)$. Assuming a lower bound ξ, which may be 0 in many applications, reduces the equations to four, for r from 0 to 3. Examples are given in Appendix 8. I have only come across the Wakeby distribution in hydrological applications.

4.8   Improving the precision of flood frequency analysis

4.8.1   Regional flood frequency analysis

Estimates of floods with high return periods, made from data collected at a single site, lack precision. This can be seen from the widths of the associated confidence intervals. There have been many suggestions for improving matters by incorporating more data. One method is to make comparisons with neighbouring sites, if it is reasonable to define regions within which the statistical behaviour of floods is similar. The basic assumption behind regional flood frequency analysis (RFFA) is that the flood peak variable, scaled by its expected value, has the same distribution at every site in the region. In particular, this implies that the coefficient of variation is constant. Formally, if there are M sites and the annual maximum variable at a site k is X_k then

$$Q = X_k/E[X_k] \qquad \text{for} \quad k = 1, \ldots, M$$

has the same distribution at all sites. Cunnane (1988) gives a good review of the method, which is particularly useful for fitting the GEV or Wakeby distributions which require more parameters than the EV1. It may be easier to define hydrologically similar regions in large continents than in the UK, and Wallis and Wood (1985), for example, recommend the use of RFFA in the USA. The procedure is to fit a distribution to Q by calculating a weighted average of scaled PWM from the different sites. Suppose that there are annual maxima from n_k years at site k, and that the PWM and mean at site k are $a_{r,k}$, $b_{r,k}$ and $\bar{x}_k$. Then the averaged scaled PWM ($\bar{a}_r$ and $\bar{b}_r$) are defined by:

$$\bar{a}_r = \sum (a_{r,k}/\bar{x}_k)(n_k/n)$$
$$\bar{b}_r = \sum (b_{r,k}/\bar{x}_k)(n_k/n)$$

where $n = \sum n_k$, and all summations are over k from 1 to M. The site estimat are obtained from the quantiles of the distribution of Q by multiplying by site mean. The report by Dales and Reed (1989), of a three year investigat into regional flood and storm hazard over reservoired catchments, is a detai study of regional flood risk.

Example 4.12
If years with no flow are ignored, the coefficients of variation of annual p flow at the 44 Namibian sites (Table A16.7) vary from 0.48 to 1.90. average value of the CV is quite high 1.038, as we might expect in an region. The averaged estimates of the scaled PWM, $\bar{b}_1$ and $\bar{b}_2$ are 0.75907 0.627526. Application of the equations given in Section 4.4, with $b_0 = 1$, $\hat{k} = -0.3389$, $\hat{\theta} = 0.4861$ and $\hat{\xi} = 0.4776$ for the GEV distribution of Q conditional on a non-zero flow.

Now suppose we wish to estimate the flow that has a 100 year return period at Karris on the River Packriem. Two of the 11 years were dry, and the mean annual maximum for the nine years with some flow is 187.83. We first need to estimate the upper $0.01/(9/11) = 0.0122$ point of the conditional distribution of Q. This equals 5.4114, and the corresponding flow at Karris is $5.4144 \times 187.83 = 1016$.

4.8.2   The *r* largest events method

The motivation for this method is that the r largest, independent, values within a year should provide more information than the annual maximum. The same principle is behind the peak over threshold procedure, and the two methods are equivalent in many respects (Tawn, 1988). However, the r largest events method has some advantages. Records of the r largest events per year may be available more often than daily records. Some variables such as wind speeds, and instantaneous peak flows, would really need a continuous record to apply a peak over threshold method. The r largest events method also provides a direct estimate of the parameters of the distribution of yearly maxima without the need to assume exceedances follow a Poisson process. Weissman (1978) derived an expression for the joint distribution of the r largest in a random sample. Smith (1986) adapted the result to include a trend and analysed sea level data from Venice. Let ξ and θ be the parameters of the EV1 distribution of yearly maxima. Suppose the r largest values are available for n years. Then the likelihood function to be minimized is

$$\mathcal{L}(\xi, \theta) = \theta^{-nr} \exp\left\{-\sum_{j=1}^{n}\left[\exp(-(x_{r,j} - \xi)/\theta) + \sum_{i=1}^{r}(x_{i,j} - \xi)/\theta\right]\right\}$$

4.8.3 Incorporating the greatest known value

It is quite common to know of a few extraordinary extreme values from historical records without having the intervening data. J.R. Hosking (unpublished note) has given a justification for treating an extreme value that occurred T years before an unbroken record of n years as the maximum of a sample of size $n + T$. However, we should remember that we are assuming the extreme value distribution is stationary over this long period if we incorporate the extra point in our analysis. The point can easily be added to the probability plot but it is less straightforward to include it in the likelihood function. We would have to allow for $T - 1$ censored values, i.e. all we know is that they were less than the maximum of all $n + T$ values.

4.9 Multivariate extreme value distributions

Dangerous conditions often arise from a critical combination of extreme environmental variables. Offshore structures such as oil rigs can overturn because of the combined effects of exceptionally large surface waves and violent winds. The peak water level at a dam wall during a storm depends on rainfall and wind as well as the initial volume of water in the reservoir (Anderson and Nadarajah, 1993). Designers of flood alleviation schemes, such as retention tanks and washlands, need to predict volumes of flood water, as well as peak flows. The environmental variables cannot usually be treated as independent, and a variety of analytic techniques, which allow for dependence, have been proposed.

The simplest method for dealing with several dependent environmental variables is to reduce the problem to a single *structure variable* (a term used by Coles and Tawn, 1994). In the offshore context, Cavanié (1985) suggests defining a structure variable y by a *boundary function* of the form

$$y(x_1, \; x_2) = ax_1 + bx_2^2 - d$$

where x_1 and x_2 represent wave heights and wind speeds; and the a, b and d are design parameters. The structure is assumed to fail if y exceeds some critical value. The structure variable approach is to model extremes of y directly. A disadvantage of this method, compared with modelling the joint distribution of x_1 and x_2, is that the whole analysis must be repeated if we wish to investigate different boundary functions. It may also be hard to find a realistic probability distribution for y. If x_1 and x_2 are well modelled by standard distributions, a non-linear function of these correlated variables will not necessarily have any standard form.

A common device for dealing with multivariate data is to find some transformation that makes it plausible to assume they are a random sample from a multivariate normal distribution. Taking logarithms is a familiar example. However, this is not satisfactory if we need to predict extreme values. Coles and Tawn (1994) point out that extremes from a multivariate normal distribution tend to behave like extremes from independent normal distributions, with consequentially misleading results. We need to use multivariate extreme value distributions. Johnson and Kotz (1972) give details of two bivariate extreme value distributions which they attribute to Gumbel. However, there is some ambiguity over the definition of a multivariate extreme value. The example we will refer to throughout this section is peak discharge and volume at Concordia on the Rio Uruguay. A bivariate annual maximum could be defined in any one of three ways: the annual maximum flow and the associated volume; the annual maximum volume and the associated flow; or the annual maximum flow and the annual maximum volume, although they did not necessarily occur with the same flood event. The only readily available data are annual maximum flows with the associated volumes, but the third definition is often used despite the fact that some of the pairs may not correspond to an observed event (Coles and Tawn, 1991).

4.9.1   Bivariate Gumbel distributions

In the following we assume that the marginal distributions for X and Y are reduced Gumbel distributions. That is, they have the same cdf, defined for X by:

$$F(x) = \exp(-e^{-x}) \qquad \text{for} \quad -\infty < x < \infty$$

Type A bivariate Gumbel distribution   The bivariate cdf is

$$F(x, y) = \exp[-e^{-x} - e^{-y} + \theta(e^x + e^y)^{-1}] \qquad \text{for} \quad -\infty < x, \ y < \infty$$

where θ is a parameter which can vary from 0 to 1 giving an increasing positive correlation between X and Y. The median of the reduced Gumbel distribution, $x_{0.5}$ is $-\ln(-\ln(0.5))$ and substitution into the cdf gives

$$F(x_{0.5}, \ y_{0.5}) = (\tfrac{1}{4})^{1-\theta/4} \tag{4.6}$$

So the proportion of (X, Y) pairs for which both variables are less than their medians can range from 0.250, if they are independent, up to 0.354. If the sample proportion $(\hat{p})$ is in this range we can equate it to the right-hand side of equation (4.6) to estimate the association parameter (θ). This gives

$$\hat{\theta} = 4 + 2 \ln(\hat{p})/\ln(2)$$

and Johnson and Kotz (1972) give the following approximation for its sampling variance.

$$\widehat{\text{var}}(\hat{\theta}) = 8.33(1/(2\hat{p}) - 1)/n$$

The original data pairs can be scaled to reduced values by estimating the parameters of the marginal EV1 distributions using the methods of Section 4.1.

The formula for the bivariate pdf can be found by partial differentiation:

$$f(x, y) = \frac{\partial^2 F}{\partial x \partial y}$$

So, the pdf is:

$$f(x, y) = e^{-(x+y)}[1 - \theta(e^{2x} + e^{2y})(e^x + e^y)^{-2} + 2\theta e^{2(x+y)}(e^x + e^y)^{-3}$$
$$\theta^2 e^{2(x+y)}(e^x + e^y)^{-4}] \exp[-e^{-x} - e^{-y} + \theta(e^x + e^y)^{-1}]$$

Similar results can be found for the other extreme value distributions.

Example 4.13
Table A16.8 contains the maximum yearly peak discharges (u) and associated
flood volumes (v) at Concordia on the Rio Uruguay for 91 years (Adamson,
1994) plotted in Figure 4.6. The median discharge is 17.2 $10^3 m^3 s^{-1}$ and the
median volume is 69 $10^9 m^3$. The proportion of pairs for which both u and v are
below their medians is $35\frac{1}{2}/91$ which equals 0.39. This is out of the range of $\hat{p}$
for the Type A distribution, so we can either set θ to its maximum value of 1
(Figure 4.7(a)) or try the Type B distribution.

Type B bivariate Gumbel distribution   The bivariate cdf is

$$F(x, y) = \exp[-(e^{-mx} + e^{-my})^{1/m}]$$

where the association parameter m is greater than or equal to, in the case of
independence, 1

$$F(x_{0.5}, y_{0.5}) = (\tfrac{1}{2})^{2^{1/m}}$$

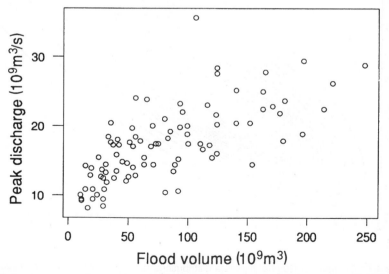

Figure 4.6 Annual maximum peak discharge and associated flood volume at Concordia on
the Rio Uruguay

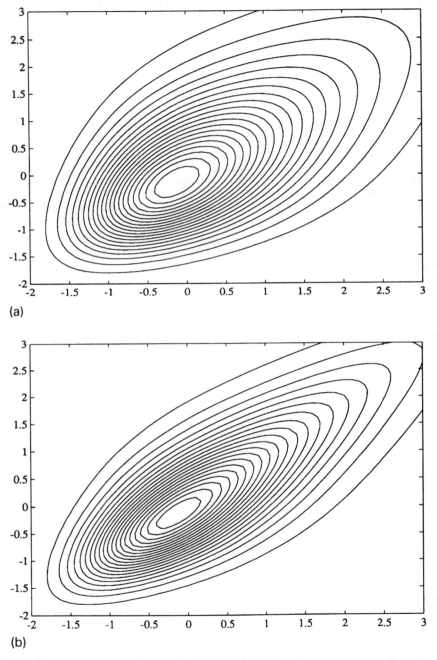

Figure 4.7 Contours of Gumbel bivariate distributions (a) Type A with $\theta = 1$; (b) Type B with $m = 2.2626$; for reduced annual maximum peak and reduced flood volume

and

$$\hat{m} = \ln(2)/[\ln(-\ln(\hat{p})/\ln(2))]$$

$$\widehat{\text{var}}(\hat{m}) \approx 2.08(1/(2\hat{p}) - 1)/[n(-\ln(\hat{p}))^2]$$

The bivariate pdf of the Type B distribution is

$$f(x, y) = e^{-m(x+y)}(e^{-mx} + e^{-my})^{-2+1/m}$$
$$[m - 1 + (e^{-mx} + e^{-my})^{1/m}]\exp[-(e^{-mx} + e^{-my})^{1/m}]$$

Example 4.13 (continued)
If $\hat{p} = 0.39$ them $m = 2.2626$, and contours of the Type B distribution are shown in Figure 4.7(b). The MOME of the mode and scale factor of the marginal distributions of volumes and peaks are: 56.19 and 43.94, and 14.86 and 4.26 respectively. A Gumbel distribution for volume is not a very convincing fit.

4.9.2 Simulation of bivariate distribution from conditional distributions

This method allows us to use detailed models for the conditional distributions, but has the drawback that the end result is an empirical distribution rather than an algebraic function. The principle is that we set up a regression of x and y, and a regression of y on x. (See Section 5.2 for details of regression methods.) We take a starting value of $x(x_0)$ at some arbitrary point within the range of x values. Then, using a random number from the distribution of the errors about the regression line, given x_0, we obtain a randomly generated y_0 to pair with it. Conditional on this y_0, we generate a new x value (x_1) and then given this x_1 we generate a y_1 to pair with it. The iteration can be repeated 1 million times on a modern PC in a few minutes. The bivariate distribution is estimated by the empirical bivariate distribution of 1 million pairs (x_i, y_i). In practice, it is usual to jettison the first few hundred pairs because of the non-random start.

If we assume the errors in the regressions are normally distributed with equal variances we will end up with a very close approximation to a bivariate normal distribution, and the simulation is quite unnecessary. It becomes interesting when we need to use a bivariate distribution which cannot be expressed as a formula. In the next example we assume the residuals have EV1 distributions with means of zero and scale factors that depend on the value of the conditioning variable. A consequence of this assumption is that the marginal distributions will not be exactly EV1. The method is a simple application of the *Gibbs sampler*, which is clearly explained by Casella and George (1992). In general, it does not necessarily converge, but failure to do so should be clear from the results. It must converge if the conditional distributions are bounded.

Example 4.14
Refer to the Concordia data in Table A16.8. The regression of the flood volumes (v) on the yearly peak discharges (u) is

$$v = -43.41 + 7.217u$$

The residuals were divided into four groups according to the associated value of u.

	Group 1	Group 2	Group 3	Group 4
Mid-group value of u	10.75	15.30	18.80	28.00
Standard deviation of residuals	21.02	37.32	43.85	54.00

A regression of standard deviation against mid-group value gives

$$\sigma_{v|u} = 6.26 + 1.80u$$

The regression of peak discharges on volumes is

$$u = 11.790 + 0.06776v$$

and a similar analysis of the residuals leads to

$$\sigma_{u|v} = 2.94 + 0.0094v$$

The simulation used EV1 distributions, with mean 0 and θ equal to $\sigma\sqrt{6}/\pi$, for errors about the regressions. A three dimensional histogram of the simulation is shown in Figure 4.8(a), and a contour plot is shown in Figure 4.8(b). The program (written by Benoit Parmentier) can be adjusted to give the proportion of pairs satisfying any chosen criterion (Adamson *et al.*, 1997).

4.9.3   Point process theory of multivariate extremes

A sketch of the theory for the bivariate case is given here, but it generalizes to multivariate data with any number of components (Coles and Tawn, 1991). Pairs $\{(X_{i1}, X_{i2})$ for $i = 1, \ldots, n\}$ are independent random 'vectors' from a bivariate distribution which has unit Fréchet (EV2) marginal distributions. That is

$$\Pr(X_{ij} < x) = \exp(-1/x) \qquad \text{for} \quad 0 < x$$

This is not a theoretical restriction because other distributions can be transformed to the unit Fréchet. For example, Coles and Tawn (1994) transform data from a generalised Pareto distribution to the unit Fréchet (see Exercise 12). The key result is that the set of n points

$$\{(X_{i1}/n, X_{i2}/n)\}$$

converges to an inhomogeneous Poisson process over the positive quadrant of the plane as n tends to infinity. The intensity function of the Poisson process models the dependence structure between the two components of the random vectors.

The intensity function of the spatial Poisson process can be expressed in a simpler form if we transform to *pseudo-radial* and *pseudo-angular* coordinates. These are defined by:

$$r_i = (X_{i1} + X_{i2})/n$$

and

$$w_i = X_{i1}/(nr_i)$$

respectively, and it follows that $0 \leqslant w_i \leqslant 1$. The intensity then takes on a form

$$\lambda(r, \theta) = \frac{\mathrm{d}r}{r^2}\, h(w)\mathrm{d}w$$

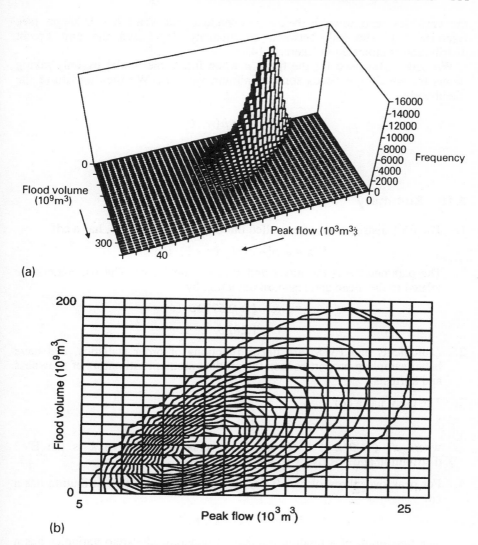

Figure 4.8 Simulated distribution of annual maximum peak and flood volume: (a) histogram, (b) contour plot

(de Haan and Resnick, 1977; Resnick 1987). The $h(w)$ determines the dependence structure. In the case of bivariate data $h(w)$ is itself univariate and the logistic model

$$h(w \mid \theta) = \frac{(\theta - 1)}{[w(1 - w)]^{1 + \theta}} \{w^{-\theta} + (1 - w)^{-\theta}\}^{1/\theta - 2}$$

for $0 \leqslant w \leqslant 1$ and $1 < \theta$ is one, relatively simple, choice (Anderson and Nadarajah, 1993). The larger θ, the greater the degree of dependence. As θ tends to 1 the variables tend towards being independent and $h(w)$ has U-shape (see Exercise 13). The link between an intensity dr/r^2 and the unit Frécht distribution is indicated in Exercise 14.

We make allowance for the finite n when fitting the model by only taking points for which r_i exceeds some minimum value r_0. We then maximize the likelihood

$$L(\theta) = \prod_{\substack{\text{points} \\ \text{beyond} \\ r_0}} h(w_i|\theta)$$

4.10 Summary

1. The EV1 distribution, often called the Gumbel distribution, has a cdf

    $$F(x = exp[-exp[-(x - \xi)/\theta]\}$$

    The parameter ξ is the mode and θ is a scale factor. The parameters are related to the mean and standard deviation by

    $$\mu = \xi + 0.5772\theta$$
    $$\sigma = 1.282\theta$$

2. *L*-moments and PWM are linear functions of order statistics, and have better sampling properties than the product moment estimators of skewness and kurtosis.

3. The GEV has a cdf

    $$F(x) = exp\{-[1 - k(x - \xi/\theta]^{1/k}\}$$

    where $x > \xi + \theta/k$ if $k < 0$ and vice versa. So negative k is the EV2 distribution and positive k is bounded above.

4. For peak over threshold analysis, the generalized Pareto distribution has a cdf

    $$F(x) = 1 - (1 - kx/\eta)^{1/k}$$

    The maximum of a random number of generalized Pareto variables has a GEV distribution with the same k.

5. There are several techniques for modelling multivariate extremes. The bivariate Gumbel distributions are probably the most convenient. Simulation methods are versatile, and relatively simple to implement. The point process characterization is a more sophisticated approach which is already being used in engineering applications.

Exercises

1. Suppose that annual maximum floods (X) have a distribution with cdf F.

The upper 1% point of this distribution is $x_{0.01}$ such that

$$F(x_{0.01}) = 0.99$$

The probability that $x_{0.01}$ is exceeded in any one year is 0.01 and the return period is the reciprocal of this probability, that is 100 years.

(a) Find the probabilities that the '100 year flood' ($x_{0.01}$) is exceeded at least once in:

    (i)   10 years;
    (ii) 50 years
    (iii) 100 years

(b) Find the number of years for which the probability of at least one exceedance of the '100 year flood' is 0.5.

(c) Write down the probability of at least one exceedance in n years of the flood with return period of T years, in terms of n and T.

2. If Y is a reduced Gumbel variable find the pdf of

$$W = -Y$$

by:

    (i)   writing down $\Pr(Y < y)$;
    (ii) and hence finding an expression for

$$\Pr(W > w)$$

    where $W = -Y$ and $w = -y$, in terms of w;
    (iii) and finally differentiating the result.

3. Draw an EV1 plot for the annual maximum sea levels at Sheerness given in Table A16.2.

4. The extreme value type 2 distribution (EV2), also known as the Fréchet-type distribution, has the cdf

$$F(x) = \begin{cases} 0 & x < \lambda \\ \exp[-((x - \lambda)/\beta^{-\alpha}] & \lambda \leqslant x \end{cases}$$

where $\alpha > 0$. Show that this can be rewritten as the GEV with $k < 0$.

5. If Y has a reduced EV1 distribution

$$F(y) = \exp(-\exp(-y))$$

and a plot of y against $(-\ln(-\ln(F)))$ is a straight line.
    (i)   Now suppose that F is

$$\exp(-[1 - kx]^{1/k}) \quad \text{for } 0 < x \text{ and } k < 0$$

    By using a Taylor series as far as the term in x^2 show that a plot of y against $(-\ln(-\ln(F)))$ is concave up.
    (ii) Show that the plot will be concave down if $x < 0$ and $0 < k$.

6. Write down the likelihood function, and the log-likelihood, for estimating the parameters of the EV1 distribution. Hence obtain equations (4.1).

Compare the efficiency of solving these iteratively with maximization of the log-likelihood with the simplex algorithm. What additional information does the latter give you?

7. Show that the GEV distribution with $k > 0$ is equivalent to the extreme value type 3 (EV3) distribution defined by the cdf:

$$F(x) = \exp\{-[(\lambda - x)/\beta]^{\alpha}\} \quad \text{for } x < \lambda, \text{ and } \alpha > 0$$

Now show that if X has an EV3 distribution, and λ is taken as 0, then $-X$ has a Weibull distribution.

8. Show that if X has an EV3 distribution, then $W = -\ln(\lambda - X)$ has an EV1 distribution.

9. (a) A flood peak over threshold discharge (X) is assumed to have an exponential distribution with mean $1/\theta$. The past n peaks over the threshold have been noted, but m of these exceeded the maximum discharge that could be measured (c). So we have $n - m$ known discharges $\{x_i\}$ and m discharges which exceeded c. Show that the maximum likelihood estimator of θ is

$$\hat{\theta} = (n - m)\Big/\left(\sum_{i=1}^{n-m} x_i + mc\right)$$

(b) Use the approximation

$$\text{var}(f(X, Y)) = \left(\frac{\partial f}{\partial x}\right)^2 \text{var}(X) + \left(\frac{\partial f}{\partial y}\right)^2 \text{var}(Y)$$

(see Appendix 1) to obtain an approximate confidence interval for x_δ in a generalized Pareto distribution (refer to Section 4.5.2).

10. Suppose X and Y are Poisson variables with means μ and ν respectively. If μ and ν exceed at least 5 we can approximate the Poisson distributions by normal distributions, so $X \sim N(\mu, \mu)$ and $Y \sim N(\nu, \nu)$. An approximate 95% confidence interval for $\mu - \nu$ is

$$x - y \pm 2\sqrt{(\mu + \nu)}$$

and the usual device of replacing μ and ν by their estimates x and y gives a practical construction for a 95% confidence interval as

$$x - y \pm 2\sqrt{(x + y)}$$

(i) The numbers of electrons emitted from two cathode ray tubes, A and B, over 1 hour under test conditions, are 17 and 32, respectively. Construct a 95% confidence interval for the difference in mean emission rates.

(ii) A mini-roundabout was installed at a crossroads two years ago. In the previous ten years there had been 37 accidents. There have been three accidents since the change. Assume there have been no other significant changes to the junction and construct a 90% confidence interval for the reduction in yearly accident rates.

I suggest you start by letting X represent the number of accidents in the ten-year period before the change and μ_B represent the underlying yearly average. Then, if you assume accidents occur as a Poisson process

$$X \sim N(10\mu_B, \; 10\mu_B)$$

and

$$X/10 \sim N(\mu_B, \; \mu_B/10)$$

[The answer is 2.2 ± 1.8.]

11. Fit the TCEV distribution to the Jacksonville data ignoring the information on storm type. Now analyse the data, making use of the additional information.

12. Distribution of the maximum of N independent GEV variables. If X_i, for $i = 1, \ldots, N$ are independent GEV variables show that $\max(X_1, \ldots, X_n)$ has a GEV distribution with

$$\theta_N = \theta N^{-k}$$
$$\xi_N = \xi + (\theta/k)(1 - N^{-k})$$

Deduce that for the EV1 distribution θ is unchanged and $\xi_N = \xi + \theta \ln N$. [Hint: If $G(x)$ is the cdf of $\max(X_1, \ldots, X_n)$ then $G(x) = [F(x)]^N$. Now take natural logarithms (after Dales and Reed, 1989).]

13. Let u be some threshold value and $Y = X - u$ be a peak over the threshold. Assume Y has the generalized Pareto distribution with cdf

$$F_Y(y) = 1 - (1 - ky/\eta)^{1/k}$$

show that the variable W defined by

$$W = \begin{cases} -(\ln[1 - \{1 - F_X(u)\}\{1 - k(X - u)/\eta\}^{1/k}])^{-1} & u < X \\ -(\ln F_X(X))^{-1} & X < u \end{cases}$$

where F_X, the cdf of X, has a unit Fréchet distribution.

14. Plot $h(w \mid \theta)$ for $\theta = 1.1$ and $\theta = 3$.

15. If a Poisson process has an intensity function $\lambda(r)$ the probability of no occurrence in the time interval $[t_1, \; t_2]$ is,

$$\exp\left[-\int_{t_1}^{t_2} \lambda(r)\,dr\right]$$

Show that if

$$\lambda(r) = 1/r^2$$

then the probability of no occurrence in the interval $[t, \infty)$ is $\exp(-1/t)$.

16. Fit the model of Section 4.9.3 to the Concordia data (Table A16.8).

5

Relationships Between Variables

5.1 Introduction

In this chapter we will review statistical techniques for investigating, and modelling, relationships between variables. Examples abound in civil engineering textbooks and journals, and include: estimating the correspondence between shear resistance of soil and normal stress from results of a shear-box experiment; investigating the relationship between ground movement and excavation depth (Hashash and Whittle, 1996); analysing the effect of increasing beam depth on flexural ductility of seismic resistant steel (Roeder and Foutch, 1996); and producing tables of stopping distance against vehicle speed. The usual objective is to predict values of one variable, the *dependent variable* or *response*, from measurements of other variables which we will refer to as *predictor variables*. Some examples with more than one predictor variable are: forecasting the number of trips per household from car ownership, household size, and socio-economic indicators; predicting 28-day strength of concrete from site measurements of slump, temperature, and mix suitability factor (MSF, see Day, 1995); and predicting peak flow in a river from rainfall and baseflow.

Statistical models are empirical and do not imply that the predictor variables necessarily cause changes in the dependent variable. Causal explanations have to be based on physical reasoning. In most applications there will be some natural explanation for a general relationship between the variables, if not for its precise form, but it need not be causal. For example, there has been a reduction in annual totals of road traffic accidents in the UK over the last 30 years, despite the increase in vehicle registrations. The reduction in accidents over time is not caused by the years passing; it is a consequence of the resources that have been invested in improving road safety over this period. A related issue is that statistical models have been fitted to data, and cannot be relied on for predictions outside the range of values used for the fitting. In some cases extrapolation might lead to disaster. Extrapolation of an assumed linear relationship between extension and load, beyond the elastic limit is a prime example. In other situations we may have to extrapolate, but we must remember that this assumes that the model remains valid. For instance, a mining company may need to estimate the demand for its ore over the next 12 months by assuming present trends continue. Finally, we should remember that models are only approximations to complex physical processes and that there is no true model. A good model will be as simple as possible, while including the essential features for our application. The success of a model is usually judged in terms of the accuracy of predictions made using it, but we also try to capture something, at least, of the way we imagine the world to be.

5.2   The standard multiple regression analysis

5.2.1   The multiple regression model

The term 'regression' is used because *regression towards the mean* is a property of any bivariate, or multivariate, population. The positive association between density, measured at 7 days, and the 28-day strength of concrete cubes provides an illustration. An explanation for the relationship is that cement is the heaviest ingredient in the mix. However, the cement content alone does not determine the compressive strength. The type of sand, type of aggregate, water content of the original mix, and the method of curing also affect the strength. Furthermore, density is not an exact measure of cement content. Now suppose that the density of a particular cube is two standard deviations above the mean density of all cubes. Our best estimate of the strength of the cube is not as high as two standard deviations above the mean strength of all cubes. This is because we would not expect all the other contributory factors to be as far as two standard deviations from their means. In particular, if these other factors are independent of density, we expect them to equal their means. Regression towards the mean is a natural property of a probability distribution; it does not imply that the population will become less diverse over time.

We will start with the theory for predicting a dependent variable (y) from a linear combination of k predictor variables (x_j). If k is equal to one, we have the special case of linear regression on a single variable. In a typical application we will have n data 'vectors' (y_i, x_{1i}, ..., x_{ki}), for i running from 1 to n, from which to estimate the relationship. We can then predict future values of y from measurements of the x_j. The multiple regression model is

$$Y_i = \beta_0 + \beta_1 x_{1i} + \cdots + \beta_k x_{ki} + E_i \qquad for\ i = 1, ..., n \qquad (5.1)$$

The β_j are referred to as *parameters* or *regression coefficients*. The errors, E_i, account for the discrepancies between values of the variable Y_i and the deterministic part of the model. They can include inherent variation in the population and errors in measuring the dependent variable. The assumptions about the errors in the standard form of the model are as follows:

A1   The E_i have an expected value of 0.

A2   The E_i are independent of the values of the x_{ji}.

A3   The E_i are independent of each other.

A4   The E_i have a constant variance σ^2, regardless of the values of the x_{ji}.

A5   The E_i are normally distributed.

The first assumption cannot be checked from the data. Any non-zero expected value, perhaps due to a zero error on the equipment for measuring y, will be indistinguishable from β_0. The second assumption is also fundamental. If, for instance, there is a tendency for errors to increase from negative to positive as x_{ki} increases, the estimate of β_k will be biased upwards. It is not possible to detect this from the data, and measuring equipment must be properly calibrated. Assumption A2 also imposes a requirement that the errors in measuring the x_{ji} should be negligible (see Section 5.8). However, we can use the data to investigate whether the form of the model and the other assumptions are reasonable.

The model is called a *linear model* because it is linear in the unknown parameters $\beta_0, \ldots, \beta_k$. This enables us to give a general matrix formula for their least squares estimates.

5.2.2   Estimating the coefficients

Equation (5.1) can be written in matrix form as

$$
\begin{bmatrix} Y_1 \\ \vdots \\ Y_n \end{bmatrix} = \begin{bmatrix} 1 & x_{11} & \cdots & x_{k1} \\ \vdots & & & \vdots \\ 1 & x_{1n} & \cdots & x_{kn} \end{bmatrix} \begin{bmatrix} \beta_0 \\ \vdots \\ \beta_k \end{bmatrix} + \begin{bmatrix} E_1 \\ \vdots \\ E_n \end{bmatrix}
$$

or, more concisely

$$ Y = XB + E $$

where Y, X, B and E are $n \times 1$, $n \times (k+1)$, $(k+1) \times 1$ and $n \times 1$ matrices respectively. The sum of squared errors is

$$ \psi = (Y - XB)^{\mathrm{T}}(Y - XB) $$

The *principle of least squares* is to estimate the parameters B by the values which minimise ψ (see Exercise 1). The first explicit account of the method was published by A.M. Legendre in 1805 in a paper describing the estimation of comet orbits. Legendre acknowledged Euler's contributions to the method, which had been used by Karl Gauss, working independently, from about 1795. Continuing:

$$
\begin{aligned}
\psi &= (Y - XB)^{\mathrm{T}}(Y - XB) \\
&= Y^{\mathrm{T}}(Y - XB) - B^{\mathrm{T}}X^{\mathrm{T}}(Y - XB) \\
&= Y^{\mathrm{T}}Y - Y^{\mathrm{T}}XB - B^{\mathrm{T}}X^{\mathrm{T}}Y + B^{\mathrm{T}}X^{\mathrm{T}}XB
\end{aligned}
$$

the terms on the right-hand side are 1×1 matrices, that is to say scalars, and therefore equal their transpose. In particular

$$ Y^{\mathrm{T}}XB = (Y^{\mathrm{T}}XB)^{\mathrm{T}} = B^{\mathrm{T}}X^{\mathrm{T}}Y $$

Therefore

$$ \psi = Y^{\mathrm{T}}Y - 2B^{\mathrm{T}}X^{\mathrm{T}}Y + B^{\mathrm{T}}X^{\mathrm{T}}XB $$

We now use standard results from the calculus of several variables (see Exercises 2 and 3), which, noting that $(X^{\mathrm{T}}X)^{\mathrm{T}} = X^{\mathrm{T}}X$, give

$$ \frac{\partial \psi}{\partial B} = -2X^{\mathrm{T}}Y + 2(X^{\mathrm{T}}X)B $$

Provided X^TX has an inverse, the *normal equations* are

$$\hat{B} = (X^TX)^{-1}X^TY \qquad (5.2)$$

5.2.3 Properties of the estimators

The addition of more explanatory variables can only decrease the residual sum of squares. The practical question is whether this decrease is sufficient to give more accurate predictions. The properties of the estimators are needed to answer this. To begin with, the least-squares estimators are unbiased. This can be proved by taking expectations and using the assumption that the E_i are independent of the explanatory variables

$$E[\hat{B}] = (X^TX)^{-1}X^TE[Y] = (X^TX)^{-1}X^TXB = B$$

The *variance–covariance* matrix (often abbreviated to *covariance matrix*) of the estimators is

$$C = \begin{bmatrix} \text{var}(\hat{\beta}_0) & \text{cov}(\hat{\beta}_0, \hat{\beta}_1) & \text{cov}(\hat{\beta}_0, \hat{\beta}_2) & \cdots \\ \text{cov}(\hat{\beta}_0, \hat{\beta}_1) & \text{var}(\hat{\beta}_1) & \text{cov}(\hat{\beta}_1, \hat{\beta}_2) & \\ \vdots & & \text{var}(\hat{\beta}_2) & \\ & & & \ddots \\ & & & & \text{var}(\hat{\beta}_k) \end{bmatrix}$$

Remembering that

$$\text{cov}(\hat{\beta}_i, \hat{\beta}_j) = E[(\hat{\beta}_i - E[\hat{\beta}_i])(\hat{\beta}_j - E[\hat{\beta}_j])]$$

C can be written concisely as

$$C = E[(\hat{B} - B)(\hat{B} - B)^T]$$

since

$$E[\hat{B}] = B$$

We now show that this equals

$$(X^TX)^{-1}\sigma^2$$

The proof begins with:

$$C = E[(\hat{B} - B)(\hat{B} - B)^T]$$
$$= E[(X^TX)^{-1}X^T(Y - E[Y])(Y - E[Y])^TX(X^TX)^{-1}]$$

Now note that $Y - E[Y]$ is a column of E_i. $E[E_i^2]$ is σ^2 and $E[E_iE_j]$ is 0 if i is different from j. So

$$C = (X^TX)^{-1}X^T\sigma^2IX(X^TX)^{-1}$$

As σ^2 is a scalar it can take any position in the matrix product. Hence

$$C = (X^TX)^{-1}\sigma^2$$

In most applications σ^2 will not be known, but it can be estimated from the data in the following way.

5.2.4   Estimating the variance of the errors

The residuals (r_i) are estimates of the values taken by the error variables, and are defined as the differences between the observed responses (y_i) and the fitted values ($\hat{y}_i$) from the regression equation. That is:

$$r_i = y_i - \hat{y}_i$$

where

$$\hat{y}_i = \hat{\beta}_0 + \hat{\beta}_1 x_{1i} + \cdots + \hat{\beta}_k x_{ki}$$

The variance of the errors is estimated by

$$s^2 = \sum r_i^2 / (n - k - 1)$$

It is nearly the mean of the squared residuals; the denominator has been reduced by $k + 1$ to account for the loss of degrees of freedom caused by estimating $k + 1$ parameters, and we say s^2 is based on $(n - k - 1)$ degrees of freedom. The least squares procedure ensures that the residual sum of squares $(\sum r_i^2)$ cannot exceed, and is almost certainly less than, the unknown sum of squared errors. The proof that the estimator S^2 is unbiased for σ^2 is given in Appendix 9.

In matrix terms, the column of $\hat{Y}$ is given by

$$\hat{Y} = X\hat{B} = X(X^TX)^{-1}X^TY$$

The matrix which maps the Y onto $\hat{Y}$ is known as the HAT matrix (H). That is

$$H = X(X^TX)^{-1}X^T$$

The column of residuals R is given by:

$$R = Y - \hat{Y} = (I - H)Y$$

Now

$$R = (I - H)Y = (I - H)(XB + E)$$
$$= (I - H)XB + (I - H)E$$
$$= XB - X(X^TX)^{-1}X^TXB + (I - H)E = (I - H)E$$

It follows that the covariance matrix of the residuals is

$$(I - H)\sigma^2$$

The variance of the ith residual is $(1 - h_{ii})\sigma^2$ and is always slightly less than σ^2. This is explained by the loss of degrees of freedom when fitting parameters. If a datum has a relatively strong influence on the parameter estimation, the standard deviation of its associated residual will be slightly less than the average because it tends to pull the surface towards itself. The loss of degrees of freedom also explains the slight correlations between residuals.

Example 5.1
The density of sand-based grout can be increased by introducing, into the voids of sand, a cement which has the same mechanical properties as the granular

material of the sand. This should increase the compressive strength of the grout. Relative density can be defined by

$$\frac{\text{density} - \text{minimum density}}{\text{maximum density} - \text{minimum density}}$$

Measurements of compressive strength on 11 test samples of grout of differing relative densities are given in Table 5.1 (Skipp and Renner, 1963). The data are plotted in Figure 5.1 and the model

$$Y_i = \beta_0 + \beta_1 x_i + E_i$$

Table 5.1 *Relative density (%) and compressive strength (MPa) for 11 test samples of grout (data courtesy of the International Society of Soil Mechanics)*

58	0.74
60	1.06
63	0.85
68	0.81
72	1.41
78	1.34
82	1.41
86	1.20
90	1.62
95	1.30
96	2.01

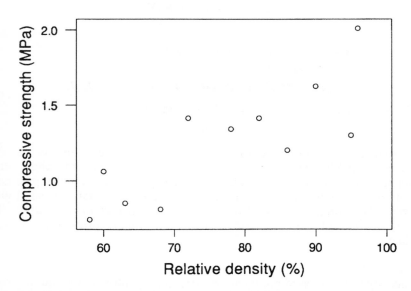

Figure 5.1   Compressive strength against relative density for 11 test samples of grout

where Y is the compressive strength and x is the relative density, seems reasonable over this range of x values. The matrix calculations give

$$\hat{\boldsymbol{B}} = \begin{bmatrix} -0.4535 \\ 0.022097 \end{bmatrix}$$

$$(\boldsymbol{X}^{\mathrm{T}}\boldsymbol{X})^{-1} = \begin{bmatrix} 3.19770 & -0.04030 \\ -0.04030 & 0.00052 \end{bmatrix}$$

$$s^2 = (0.2344)^2$$

and hence the standard deviations of $\hat{\beta}_0$ and $\hat{\beta}_1$ are 0.4192 and 0.00535 respectively. A practical question is how much the compressive strength increases with the relative density. We can answer this by constructing a confidence interval for β_1. For example, the 95% confidence interval is given by

$$\hat{\beta}_1 \pm t_{11-2,\,.025} \,\hat{\text{stdev}}(\hat{\beta}_1)$$

This equals

$$0.022 \pm 0.012$$

and we are fairly confident that β_1 is somewhere between 0.010 and 0.034. If we need a more precise estimate of β_1 we will have to increase the number of tests.

5.2.5 Making predictions

The most common reason for fitting a multiple regression model is to make predictions for the dependent variable when the values of the predictor variables are known. The best point estimate of the dependent variable is its mean, which is given by substituting the values of the predictor variables into the fitted regression equation. We can provide more information by constructing confidence limits for the mean value of the dependent variable, and limits of prediction for a single value taken by it.

Confidence interval for the mean value of Y given x   Let $\boldsymbol{x}_p = (1 \;\; x_{1p} \;\; \dots \;\; x_{kp})^{\mathrm{T}}$. The predicted mean value of Y given that $\boldsymbol{x} = \boldsymbol{x}_p$ is

$$\hat{\mathrm{E}}[Y \,|\, \boldsymbol{x} = \boldsymbol{x}_p] = \boldsymbol{x}_p^{\mathrm{T}}\hat{\boldsymbol{B}}$$

The variance of this quantity is obtained from result (ii) in Exercise (5.3).

$$\mathrm{var}(\boldsymbol{x}_p^{\mathrm{T}}\hat{\boldsymbol{B}}) = \boldsymbol{x}_p^{\mathrm{T}}((\boldsymbol{X}^{\mathrm{T}}\boldsymbol{X})^{-1}\sigma^2)\boldsymbol{x}_p$$

where $(\boldsymbol{X}^{\mathrm{T}}\boldsymbol{X})^{-1}\sigma^2$ is the covariance matrix of $\hat{\boldsymbol{B}}$. Hence a $(1 - \alpha) \times 100\%$ confidence interval for $\mathrm{E}[Y \,|\, \boldsymbol{x} = \boldsymbol{x}_p]$ is

$$\boldsymbol{x}_p^{\mathrm{T}}\hat{\boldsymbol{B}} + t_{n-k-1,\,\alpha/2}\,s\sqrt{(\boldsymbol{x}_p^{\mathrm{T}}(\boldsymbol{X}^{\mathrm{T}}\boldsymbol{X})^{-1}\boldsymbol{x}_p)}$$

Limits of prediction for a single value of Y $(1 - \alpha) \times 100\%$ limits of prediction for a single value of Y when $\boldsymbol{x} = \boldsymbol{x}_p$ are

$$\boldsymbol{x}_p^{\mathrm{T}}\hat{\boldsymbol{B}} \pm t_{n-k-1,\,\alpha/2}\,s\sqrt{(1 + \boldsymbol{x}_p^{\mathrm{T}}(\boldsymbol{X}^{\mathrm{T}}\boldsymbol{X})^{-1}\boldsymbol{x}_p)}$$

For large samples $\pm 2s$ will give approximate, but rather too narrow, 95% limits of prediction. It ignores the uncertainty in the estimation of parameters, and this is why it improves as the sample size increases.

Example 5.1 (continued)
A 95% confidence interval for the mean compressive strength of all test samples when the relative density is 65 is

$$-0.4535 + 0.0221 \times 65 \pm (2.262)(0.234)\sqrt{(1/11 + (65 - 77.09)^2/1912.91)}$$

which gives

$$0.983 \pm 0.217$$

A 95% prediction interval for a single test result when the density is 65 is

$$0.983 \pm 0.573$$

In this application, the confidence interval for the mean strength, when the density is 65, is likely to be more relevant than a prediction interval for a single test result.

5.2.6   Categorical variables

The model can incorporate categorical variables by using x_j as *indicator variables*. Suppose we wish to include three grades of concrete (A, B and C) in a regression. We can code two predictor variables (x_1 and x_2) by:

Grade	x_1	x_2
A	0	0
B	1	0
C	0	1

The coefficient of x_1 is the difference between grade B and grade A, and the coefficient of x_2 corresponds to the difference between C and A. The comparisons are best made against the grade for which there are the most data, assumed to be A in this case. However, the choice of baseline does not affect the sum of squares accounted for by including the categorical variables.

Example 5.2
The data in Table 5.2 are an excerpt from a more extensive table given by Day (1995). Each entry is the average of strength and density measurements on several concrete cubes made from the same batch of cement. The objective is to find whether an accurate prediction of the mean 28-day strength (y) can be made from the data available after 7 days. Variable x_1 is coded as 1 if the concrete is grade 320LWTA and left as 0 otherwise. Similarly, variable x_2 is coded as 1 if the concrete is grade 400. Hence, comparisons are being made with grade 320LWTB. The mean 7-day strength is taken as variable x_3. The data are plotted in Figure 5.2. The fitted regression is:

$$y = 18.863 - 0.0861x_1 + 7.6100x_2 + 0.7269x_3$$

Table 5.2   *Variables measured on concrete cubes (data courtesy of Day, 1995)*

Slump	temp	dens	WTA	400	7-day	28-day
172	25	2058	1	0	40.3	46.2
180	24	2074	1	0	38.1	43.2
180	23	2045	1	0	36.0	42.5
173	24	2085	1	0	40.5	47.8
197	23	2073	0	0	34.9	42.5
170	24	2087	0	0	33.2	42.5
176	24	2069	0	0	34.2	42.0
169	28	2055	0	0	32.5	39.2
182	26	2087	0	0	38.8	44.3
184	25	2412	0	1	44.4	56.7
176	23	2409	0	1	45.2	56.8
174	26	2418	0	1	37.6	52.1
168	25	2406	0	1	42.7	55.3
177	25	2415	0	1	43.8	56.8

From Figure 5.13 of Day (1995)

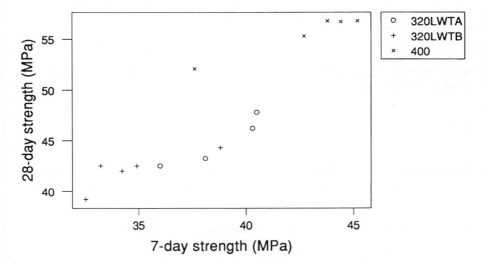

Figure 5.2   28-day strength against 7-day strength for three grades of concrete

and the estimated standard deviation of the errors (s) is 0.9823. We would expect a convincing linear relationship from the plot. This is quantified by various statistics, which also enable us to make further inferences. A useful practical comparison is that between s and the standard deviation of the y if we do not attempt to explain the variability in terms of the x_j. This unconditional standard deviation of the y (s_y) is 6.47, so s is nearly an order of magnitude smaller. It is important to remember that the conditional standard deviation of the y, given the predictor variables, is just the standard deviation of the error (s). A commonly used statistic relating s and s_y is the *coefficient of determination* (R^2), which is the proportion of the variability accounted for by the regression. It is defined by:

$$R^2 = 1 - \sum r_i^2 / \sum (y_i - \bar{y})^2 = (\sum (y_i - \bar{y})^2 - \sum r_i^2) / \sum (y_i - \bar{y})^2$$

It will always be between 0 and 1, and 1 corresponds to an exact fit. This is its limitation. It must increase if an extra predictor variable is added to the regression and will eventually equal 1, if $n - 1$ variables are used as predictors. A modification is the adjusted R^2 which is defined by:

$$R_{adj}^2 = 1 - (\sum r_i^2 / (n - k - 1)) / (\sum (y_i - \bar{y})^2 / (n - 1)) = 1 - s^2 / s_y^2$$

This increases, or decreases, as s becomes smaller or larger. In this example $R^2 = 98.2\%$ and $R_{adj}^2 = 97.7\%$. These are high values which tell us that a large proportion (0.98) of the variance is accounted for by the model. The estimated standard deviations of the coefficients help us assess the importance of the corresponding variables. They are: 0.802, 1.107 and 0.1142 for the coefficients of x_1, x_2 and x_3 respectively. There is strong evidence that the grade 400 concrete has a greater 28-day strength than grade 320LWTB, and that 28-day strength is linearly related to 7-day strength. However, we have no confidence that there is any difference between grades 320LWTA and B, because the coefficient of x_1 is much smaller than its estimated standard deviation. A 95% confidence interval for β_1 is,

$$-0.08 \pm 1.79$$

which extends from -1.87 to $+1.71$. We can get a slight improvement in the fit if we drop this variable. Then

$$y = 17.08 + 7.692x_2 + 0.7199x_3$$

with

$$s = 0.9371$$

We appear to get a still better relationship, $s = 0.8471$, if we include the density (x_4). The estimated coefficients and their standard deviations are given below.

Variable	Estimated coefficient	Standard deviation of coefficient
Intercept	−57.34	40.11
Grade 400	−4.586	6.636
7-day strength	0.70701	0.08120
Density	0.03617	0.01944

The first thing to notice is that the coefficient of grade 400 has changed sign, and is now smaller than its estimated standard deviation. This is because grade 400 contains extra cement which increases the density. The fact that blocks are grade 400 tells us no more about their likely strength than can be inferred from the increased density. There is no requirement that the predictor variables should be uncorrelated, but high correlations between them can lead to numerical problems with the parameter estimation (*ill conditioning*) and possible misinterpretation. In this case, a very large standard deviation is associated with the intercept because it is sensitive to changes in the coefficient of density, which is poorly estimated because grade 400 is included in the analysis. That is, the estimators $\hat{\beta}_0$, $\hat{\beta}_2$ and $\hat{\beta}_4$ are themselves highly correlated. The solution to this quandary is to include either grade 400 or density, but not both, in the regression model.

For example:

$$y = -29.70 + 0.7075x_3 + 0.0228x_4$$

The standard deviation of the errors, s, equals 0.8268 and the standard deviations of the coefficients are: 2.96; 0.0020; and 0.0792 respectively.

The general problem of multicollinearity is discussed in the next sub-section and in Section 5.3.

There is one other point to be made about this example. We are predicting the mean 28-day strength of several blocks, from the same batch, from mean measurements available after 7 days. We would obtain smaller R^2 values if we analysed individual blocks. We would also need to allow for variability within and between batches. One way of doing this would be to use indicator variables for the different batches.

5.2.7 Polynomial relationships

The linear model can be used to fit polynomial relationships. For example, a general quadratic surface is given by

$$Y_i = \beta_0 + \beta_1 x_{1i} + \beta_2 x_{2i} + \beta_3 x_{1i}^2 + \beta_4 x_{1i} x_{2i} + \beta_5 x_{2i}^2 + E_i$$

This is of the general form expressed by equation (5.1) if x_{3i} is defined as x_{1i}^2, and so on.

Example 5.3
Rebeiz *et al.* (1996) give several examples of applications for polyester mortar (PM). Polyester mortar can be used to produce a very thin but durable overlay (5–25 mm) on portland cement concrete, thereby reducing the amount of material needed in a construction. It can also be used as a lining for pipes, and as an adhesive. The data in Table 5.3 are the results of the authors' investigations into the effect of sand and fly ash filler on the tensile strength (y) of PM, which is one of its most important properties. The total amount of filler is restricted to 40%.

If we wish to fit linear and quadratic terms in some predictor variable x, they will be highly correlated if the x values are positive and have a small coefficient of variation. If we mean correct the x_i values (i.e. subtract the mean) the correlation between $(x_i - \bar{x})$ and $(x_i - \bar{x})^2$ will be much less than that between x_i and x_i^2, and zero if the x_i are symmetric about their mean (see Exercise 4). However, we can subtract any other convenient number. In this case we will scale the percentages by

$$x_1 = (\% \text{ sand} - 20)/10$$
$$x_2 = (\% \text{ fly ash} - 20)/10$$

because using the mid-range values reduces the correlations more than subtracting the means, and leads to integer values. The division by 10 avoids inconvenient large numbers when squaring. The correlation between % sand

Table 5.3 *Percentages of sand (column 1) and fly ash filler (column 2), and tensile strength (MPa) of polyester mortar*

0	0	38.9
0	0	37.8
10	0	26.4
10	0	25.8
20	0	23.1
20	0	25.3
30	0	18.2
30	0	17.1
40	0	24.2
40	0	26.4
0	0	39.5
0	0	37.8
0	0	36.7
0	10	31.8
0	10	31.3
0	10	28.0
0	20	24.7
0	20	23.6
0	30	20.9
0	30	17.6
0	30	13.8
0	40	18.2
0	40	17.6
0	40	14.4
0	40	12.7
40	0	22.5
40	0	21.5
40	0	20.4
30	10	23.1
30	10	25.3
20	20	31.3
20	20	26.9
10	30	21.5
10	30	14.9

and (% sand)^2 is 0.964, mean correcting reduces it to 0.726, and the correlation between x_1 and x_1^2 is -0.337. The linear scaling of the predictor variables is completely arbitrary because, if the numerical calculations are sufficiently precise, the estimates of the coefficients compensate for it. The reason for scaling is that high correlations lead to an ill-conditioned problem set-up which is more prone to numerical error. In practice, the need to scale when fitting quadratic surfaces will depend on the algorithm used, and computer precision, as well as the coefficients of variation of the predictor variables. Minitab gives exactly equivalent estimates of the regression if the original percentages of fillers are used as predictor variables, but other software may be less robust.

The fitted surface is:

$$y = 25.3 + 5.44x_1 + 3.48x_2 + 2.24x_1^2 + 4.65x_1 x_2 + 0.77x_2^2$$

with $s = 2.76$, $R^2 = 88.4\%$ and $R_{adj}^2 = 86.4\%$. The standard deviations of the coefficients are 1.35, 1.23, 1.24, 0.42, 0.66 and 0.40 respectively. All the coefficients exceed their standard deviations. A contour plot will give a useful indication of the shape of the surface, but only the region corresponding to less than 40% filler ($x_1 + x_2 \leq 0$) is applicable (Figure 5.3). The tensile strength of PM decreases with an increase in filler content, but other key properties, such as the tensile modulus of elasticity, increase with the amount of filler. Details

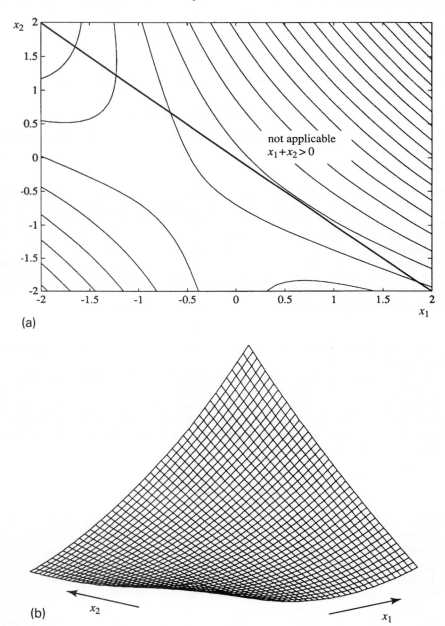

Figure 5.3   Contour plot of tensile strength of polyester mortar mixes

of these experiments are given in Rebeiz *et al.*'s paper. Compromises have to be made, and the authors recommend 20% by weight of sand and 20% by weight of fly ash as the best mix design for PM. The regression gives a 95% confidence interval for the mean tensile strength of all mixes with this design as

$$[22.6, 28.1]$$

and a 95% prediction interval for an individual test as

$$[19.1, 31.6]$$

They considered these limits adequate for the applications they described.

5.2.8 Checking that the assumptions are reasonable

If we have only one predictor variable, we can assess the suitability of the model

$$Y_i = \beta_0 + \beta_1 x_i + E_i$$

from the plot. We can check the assumption of constant variance errors (Assumption A4) is reasonable, by noting whether the points are equally scattered about the line for high and low values of x. Also, any extraordinary points will be apparent. However, if there is more than one predictor variable we need to rely on the residuals. They also allow us to check the normality assumption (A5) and, if the observations can be put into some meaningful order, the assumption of independence (A3). Nevertheless, they do not provide a means for checking whether the errors have a mean of zero (Assumption A1) or whether they are uncorrelated with the x_{ji}. This is because

$$\Sigma \, r_i = 0 \quad \text{and} \quad \Sigma \, r_i(x_{ji} - \bar{x}_j) = 0$$

for any set of data. These identities are a consequence of the normal equations (5.2), see Exercise 5.

We will start by considering Assumption A3. Sometimes consecutive residuals are correlated when they are considered in the same order as the data were obtained. A plot of the residuals against time may indicate that this is occurring. Also, the first autocorrelation of the residuals can be calculated – that is, the correlation between residuals at time t and those at time $t + 1$. This is usually defined as

$$\sum_{t=1}^{n-1} r_t r_{t+1} \Big/ \sum_{t=1}^{n} r_t^2$$

where r_t are the time-ordered residuals (remember that the mean of the residuals is constrained to equal zero). If the assumption that the errors are independent is inappropriate, ordinary least-squares estimators are no longer efficient and the estimators of their variances may be seriously biased. High correlations between residuals are quite common in economic data sets, and the Durbin–Watson statistic provides a formal test for this. If a particular covariance structure can be assumed for the errors, weighted least squares is an answer to both this problem and that of changing variance. Weighted least squares is described in Section 5.5.

It is quite common for the variability of residuals to increase with increasing values of the dependent variable. A plot of the calculated residuals against the fitted values of the dependent variable, calculated from the fitted model, may suggest that assumptions of the form

$$\sigma_{E_i}^2 \propto E[Y_i]$$

or

$$\sigma_{E_i} \propto E[Y_i]$$

would be more appropriate than assuming a constant value. If the assumption of constant variance is inappropriate, the ordinary least-squares estimators will no longer be efficient and the estimators of their variances may be biased. The major practical problem is that limits of prediction for individual values about their mean could be misleading. The residuals can be grouped according to the magnitude of $E[Y_i]$, and a variance structure can then be estimated (see Example 4.11). An alternative is to use a function of Y as the dependent variable. For example, taking logarithms of Y if σ_{E_i} is thought to be proportional to $E[Y_i]$ (Section 5.6).

Most statistical packages will print out the *studentized residuals* as well as the residuals. These are the residuals divided by their estimated standard deviations (see Section 5.2.4) i.e.

$$q_i = r_i / \sqrt{(1 - h_{ii})s^2}$$

where h_{ii} are the diagonal elements of the HAT matrix. If the E_i are independent and $N(0, \sigma^2)$ the studentized residuals should be approximately $N(0, 1)$. Since h_{ii} is a measure of the distance of x_i from $\bar{x}$, q_i will be noticeably larger than the scaled residuals, r_i/s, (also called standardized residuals) at points near the boundary of the region in which the regression is fitted. It is slightly preferable to use the studentized residuals for the following checks, although it is unlikely to make any substantial difference to the conclusions. The general adequacy of the model can be assessed by looking for any patterns in the residuals plotted against individual predictor variables, or in three dimensions over the plane defined by two predictor variables. Such plots may suggest that quadratic terms should be included in the model. Any point at which the studentized residual exceeds about 2 may warrant further investigation. A normal probability plot of the residuals can be used to decide whether the assumption that the errors are normally distributed is reasonable. This is required for the distribution of the $\hat{\beta}_j$ to be normal and for constructing limits of predictions for individual values about their mean. In practice, the $\hat{\beta}_j$ will be approximately normally distributed for most distributions of the errors provided the sample is not too small. This is a consequence of the central limit theorem. In contrast to this, the limits of prediction would change considerably if the errors were to be modelled by an exponential distribution. Furthermore, the least-squares estimators will not necessarily be the minimum-variance linear estimators if the errors are not from a normal distribution. If an assumption of normality is not reasonable, a transformation of the dependent variable may improve matters; but we need to remember that $E[f(Y)]$ is not the same as $f(E[Y])$ (see Section 5.6 for an example).

There is another type of residual which has a very relevant interpretation. The *i*th *prediction error sum of squares* (PRESS) residual is the difference between y_i and the predicted value of y if the datum (x_i^T, y_i) is excluded when estimating the regression relationship. A common notation for this prediction is $\hat{y}_{(i)}$ and the PRESS residual is then

$$r_{(i)} = y_i - \hat{y}_{(i)}$$

and

$$\text{PRESS} = \sum_{i=1}^{n} r_{(i)}^2$$

It is not necessary actually to calculate the n regressions with one datum excluded because of the following identity

$$r_{(i)} = r_i / (1 - h_{ii})$$

If there is a large difference between the residual and PRESS residual, a datum has a high influence on the regression and is not well predicted by a regression fitted to the other points only. As usual, such points will be apparent from a plot of the data if there is only one explanatory variable. There are more variants on the residuals. One which you may come across is the Cook statistics. These are defined by,

$$D_i = q_i^2 (h_{ii} / (1 - h_{ii})) / (k + 1)$$

and are a composite measure of size of residual and distance from centroid of the data; the second term in the product being the ratio of the variances of the *i*th predicted value and *i*th residual. Relatively large values flag outstanding data.

Example 5.3 (continued)
The outstanding residuals correspond to the 31st and 34th observations, whichever criterion is used. Some of the statistics are given below.

i	x_{1i}	x_{2i}	y_i	$\hat{y}_i$	r_i	h_{ii}	$r_{(i)}$	q_i	D_i
31	0	0	31.3	25.345	5.955	0.239	7.825	2.475	0.320
32	0	0	26.9	25.345	1.555	0.239	2.043	0.646	0.022
33	−1	1	21.5	21.741	−0.241	0.158	−0.286	−0.095	0.000
34	−1	1	14.9	21.741	−6.841	0.158	−8.125	2.703	0.229

The next largest Cook statistic, D_6, equals 0.133, and none of the others exceeds 0.100.

Whilst unusual data should be investigated, and may lead to important findings, they must not be excluded from the analysis, unless they are found to be mistakes or are outside the range of values of predictor variables required for the application. Outlying points can be taken as evidence against a normal distribution of errors. If errors have a Laplace distribution (back to back exponentials), the maximum likelihood estimators of the parameters are obtained by minimizing the sum of absolute deviations rather than the sum of squared deviations (Exercise 7). This can be done with a numerical optimization routine. More generally, *M*-estimators minimize

$$\sum f(y_i - x_i' \tilde{B})$$

with respect to $\tilde{\boldsymbol{B}}$, where the function f depends on the maximum likelihood function (hence the M) of the chosen error distribution. Montgomery and Peck (1992) give a more detailed account of these, and other, *robust regression* methods.

5.2.9   How many predictor variables?

A regression model is just an empirical approximation to some underlying physical process, and there is no precise answer to this question. If the predictor variables are significantly correlated among themselves, several models, which may look quite different, can give similar results when making predictions within the domain of values used when fitting them. Extrapolation beyond this domain is inadvisable unless there are good physical reasons for doing so, and even then should not be relied on far beyond it. The following guidelines may be useful.

1.   Keep variables which have an obvious physical relationship with the dependent variable in the model.
2.   Only add further predictor variables if the estimated standard deviation of the errors, s, decreases sufficiently to compensate for the additional complexity. Note that a small reduction in s can be accompanied by wider intervals of prediction, away from the centroid of the predictor variables, because of the increased uncertainty in the estimates of the parameters.
3.   A $(1 - \alpha) \times 100\%$ confidence interval for an individual coefficient is given by:

$$\hat{\beta}_j \pm t_{n-k-1, \alpha/2} \, \hat{\text{stdev}}(\hat{\beta}_j)$$

If the confidence interval does not include zero, we have evidence that the inclusion of x_j leads to more precise predictions than a regression based only on the other $n - k - 2$ predictor variables. However, we may not wish to consider predictor variables in isolation. For example; is a quadratic surface a better fit than a plane? The following result helps us answer such questions. Suppose model 1 is based on p predictor variables and model 2 is based on k predictor variables, where $p < k$. Then

$$\frac{(\text{residual sum of squares for model 1} - \text{residual sum of squares for model 2})/(k - p)}{s^2_{\text{model 2}}} \sim F_{k-p, n-k-1}$$

In Example 4.3 the residual sum of squares for the quadratic surface was 212.99. If only the linear terms are fitted the residual sum of squares increases to 654.03. The calculated value of the test statistic is

$$\frac{(654.03 - 212.99)/3}{(212.99)/(34 - 5 - 1)} = 19.33$$

This far exceeds even the upper 0.1% point of $F_{3,28}$ which is 7.19 and there is no doubt that the quadratic surface is an improvement. This is a foregone conclusion, given the high ratios of coefficients to their standard deviations.

5.3 Multicollinearity

There is no requirement that the x_j be uncorrelated, but physical interpretation of the regression relationship is much easier if they are. This is most unlikely to happen by chance. If we are analysing environmental data, or making a retrospective analysis of some process, we have no control over the values of predictor variables, and they will usually be correlated. In contrast, if we select the values of the predictor variables in advance, when we design an experiment, we can choose them to be uncorrelated (see Chapter 9). Another source of correlation which can be removed is that between x_i and x_i^2 (see Example 5.3). It is certainly advisable to scale the variable if it has a small coefficient of variation. Any constant can be subtracted, but the mean is often convenient and if the x_i are symmetric about their mean the correlation will be zero. When they are not, there will be some other constant (c, say) such that

$$(x_i - c) \quad \text{and} \quad (x_i - c)^2$$

are uncorrelated. Such constants can be found by a quick numerical search, and the advantage of doing this is that $(x_i - c)$ and $(x_i - c)^2$ are uniquely attributable to the linear and quadratic effects respectively. However, I prefer not to scale the predictor variables when fitting a quadratic surface if there is some physical significance to the product, e.g. volume in the case of average yearly rainfall and catchment area. Although the predictor variables do not have to be uncorrelated, it is not possible to include two linearly dependent predictor variables in a regression. For example, if $x_2 = ax_1 + b$

$$
\begin{aligned}
y &= \beta_0 + \beta_1 x_1 + \beta_2 x_2 \\
&= (\beta_0 + \beta_2 b) + (\beta_1 + \beta_2 a) x_1
\end{aligned}
$$

so we can only estimate $(\beta_0 + \beta_2 b)$ and $(\beta_1 + \beta_2 a)$. The matrix $(X^T X)$ is singular if predictor variables are linearly dependent, and Equation (5.2) is ill-conditioned if there are any high correlations. The simplest way of dealing with these is to drop variables from the model. For example, in a trip generation study we might find that the number of vehicles per household is highly correlated with a linear combination of household size, income and distance from city centre. In this case we could leave the number of vehicles per household out of any regression equation which includes the other three variables. However, this may not be such a satisfactory solution in other cases. Suppose we have data on strengths of concrete cubes, the proportions of cement and water used, and the ambient temperature at the time of mixing. If there was a tendency to increase the proportion of water used when it was warmer, the two variables may have a high correlation. We might nevertheless wish to estimate coefficients for both variables. The best strategy would be to run a properly designed experiment, but we may need some immediate indication of the effects of the two variables which we can use in the interim. Two techniques that can be tried if the predictor variables are quite highly correlated are described below. They lead to biased estimators of the regression coefficients, but the reduction in their variance may compensate for this.

Remember that:

$$\text{mean squared error} = \text{variance} + (\text{bias})^2$$

(see Appendix A1).

5.3.1 Ridge regression

The device is to add a small multiple of the identity matrix to the matrix X^TX before inverting it. The ridge estimators $\hat{B}_R$ are given by

$$\hat{B}_R = (X^TX + kI)^{-1}X^TY$$

The *ridge trace* is a plot of the ridge estimates of the coefficients against k. A reasonable guide is that values of k should be between 0 and 0.5, and the purpose of the plot is to select the smallest value of k for which the estimates seem to be stable. The estimators tend to underestimate the absolute values of the coefficients.

5.3.2 Principal component regression

The principle is to replace the original set of predictor variables with a set of their linear combination, which are uncorrelated. These new variables, called *principal components* (PC), are derived in order of magnitude of variance. The rationale of the method is that the first few PCs might contain most of the information available from the data. This may be the case, especially when there is some physical interpretation of the first few PCs, but there is no reason why it must be so. The response could be uniquely associated with the last PC, although this is unlikely to occur in practical applications. Karl Pearson (1901) used the method as a means of fitting planes by orthogonal least squares, and Harold Hotelling used it for analysing correlation structures in the 1930s.

We will now sketch the mathematical derivation of principal components. Suppose x is a random variable with p components:

$$x^T = [X_1, \dots, X_p]$$

The first principal component W_1 is that linear combination of the X_i which has maximum variance, subject to the restriction that its length as a vector equals 1. If

$$W_1 = a_{11}X_1 + \cdots + a_{p1}X_p$$

we require a_{i1} to maximize $\text{var}(W_1)$ subject to $\sum a_{i1}^2 = 1$. Define

$$a_1^T = [a_{11}, \dots, a_{p1}]$$

Then

$$\text{var}(W_1) = \text{var}(a_1^Tx)$$
$$= a_1^TVa_1$$

where V is the variance–covariance matrix of x. The method of Lagrange multipliers (see Exercise 13) can be used for constrained maximization problems, and we then maximize

$$\Psi = a_1^TVa_1 - \lambda(a_1^Ta_1 - 1)$$

where λ is the *Lagrange multiplier*. The derivative is

$$\frac{\partial \Psi}{\partial a_1} = 2Va_1 - 2\lambda a_1$$

and if we set it equal to zero

$$(V - \lambda I)a_1 = 0$$

So, λ is an eigenvalue of V and a_1 is the corresponding eigenvector. Since

$$\text{var}(W_1) = a_1^T Va_1 = a_1^T \lambda Ia_1$$

we should take λ as the largest eigenvalue (they are all positive because V is a covariance matrix). The second principal component is defined by

$$W_2 = a_{12}X_1 + \cdots + a_{p2}X_p$$

where $\text{var}(W_2)$ is a maximum subject to W_2 having length 1 and being uncorrelated with W_1. In matrix notation, these two constraints are:

$$a_2^T a_2 = 1 \quad \text{and} \quad a_1^T a_2 = 0$$

Two Lagrange multipliers are now needed, and a_2 turns out to be the eigenvector corresponding to the second largest eigenvalue. The remaining principal components are defined, and derived, in a similar manner; the pth principal component being uncorrelated with all the others. The full mathematical details can be found in any book on multivariate analysis (e.g. Chatfield and Collins, 1980).

The principal components will depend on the variances of the X_i, and those with a large variance will dominate. We can choose to avoid this by scaling the variables so they all have a unit variance. Then V will be the correlation matrix of the original variables. Principal components are estimated by replacing V with its estimate from a sample ($\hat{V}$).

Example 5.4
The data in Table A16.9 are from a study carried out by the Transportation Centre at Northwestern University, Chicago in 1962. Each datum consists of five variables for a traffic analysis zone: trips per occupied dwelling unit (y); average car ownership; average household size (x_1); socio-economic index (x_2); and urbanization index (x_3).

The sample correlation matrix between x_1, x_2, x_3 is

$$\begin{pmatrix} 1 & -0.518 & -0.572 \\ -0.518 & 1 & 0.021 \\ -0.572 & 0.021 & 1 \end{pmatrix}$$

The fitted regression of y on all three predictor variables is

$$y = 6.80 + 0.361x_1 + 0.0214x_2 - 0.0691x_3$$

with $s = 0.841$ and $R^2 = 61.9\%$. The estimated standard deviations of the constant and other coefficients are: 2.16; 0.448; 0.0090 and 0.0111 respectively.

A PC analysis of the correlation matrix, using Minitab, is:

Eigenvalue	1.782	0.979	0.239
Proportion of variance	0.594	0.326	0.080
Cumulative proportion	0.594	0.920	1.000
Variable	w_1	w_2	w_3
Standardized size (v_1)	0.702	0.003	0.712
Standardized socio-economic (v_2)	−0.479	0.741	0.470
Standardized urbanization (v_3)	−0.527	−0.671	0.522

where $v = (x - \text{mean}(x))/\text{stdev}(x)$. The first PC ($w_1$) will be high for zones with large average household sizes which are less affluent and more rural. The second PC is high for affluent rural zones. A regression of y on w_1 and w_2 is

$$y = 5.37 + 0.426w_1 + 0.876w_2$$

with $s = 0.841$ and $R^2 = 61.2\%$. The estimated standard deviations of the coefficients are 0.11, 0.084 and 0.114 respectively. The ratios of coefficients to their estimated standard deviations are considerably higher than they were in the first regression. However, this does not allow for any sampling error when estimating the PC and we could obtain a similar improvement by dropping the average household size x_1. This does not mean there is no evidence of an association between average household size and average number of trips (the correlation between these two variables is 0.375). It should be interpreted as meaning that knowledge of the household size does not help when making predictions of trips, if the socio-economic and urbanization indices are already in the regression equation. Household size is associated with socio-economic index. Whenever we have several correlated predictor variables, there will be several regression relationships which have a similar performance despite looking rather different.

Principal component regression is particularly useful if the number of potential predictor variables exceeds the sample size. This might be the case if some structure, or scale model of a structure, has strain gauges at many points.

5.4 Recursive calculations

In some circumstances it would be very convenient to update the estimates of parameters in a multiple regression model as new data become available without explicitly inverting the new (X^TX) matrix. Such a facility may be essential in real time control applications. Another application I have recently come across is updating a multiple regression model, which predicts costs of water supply work, in a database which does not include a matrix inversion routine. Young (1974) gives an interesting account of the algorithm that can be traced back to a paper by Carl Gauss in 1821. Plackett (1950) generalized the result for the simultaneous addition of any number of new observations.

The following definitions make the algorithm easier to describe. If we start with n data, the estimates of the parameters, in the notation of Section 5.2, are given by

$$P_n = (X^T X)^{-1} \qquad \hat{B}_n = P_n (X^T Y)$$

Now suppose an additional observation,

$$(x_{n+1}^T, \; y_{n+1})$$

is available. Then:

$$P_{n+1} = P_n - P_n x_{n+1} (1 + x_{n+1}^T P_n x_{n+1})^{-1} x_{n+1}^T P_n$$

and

$$\hat{B}_{n+1} = \hat{B}_n - P_{n+1}[x_{n+1} x_{n+1}^T \hat{B}_n - x_{n+1} y_{n+1}]$$

The proof of these identities is based on a result known as the *matrix inversion lemma*. The increase in the residual sum of squares (RSS) is given by:

$$RSS_{n+1} = RSS_n + (y_{n+1} - x_{n+1}^T \hat{B}_n)^T (1 + x_{n+1}^T P_n x_{n+1})^{-1} (y_{n+1} - x_{n+1}^T \hat{B}_n)$$

If the individual residuals are required, we have to augment the Y and X matrices as:

$$Y_{n+1} = \begin{bmatrix} Y_n \\ y_{n+1} \end{bmatrix} \qquad X_{n+1} = \begin{bmatrix} X_n \\ x_{n+1}^T \end{bmatrix}$$

Then the residuals are given by

$$Y_{n+1} - X_{n+1} \hat{B}_{n+1}$$

The algorithm can be started with guesses of the parameter values ($\hat{B}_0$) and a suitably large P_0, which is proportional to the variances of the estimators. The effect on the initial values diminishes as more observations become available. In some applications we may wish to allow the parameters to change slowly over time. We would then like to give more weight to recent data. There has been considerable interest in this general formulation and Pole *et al.* (1994) is a very useful source for the main results. A relatively simple approach is *exponentially weighted past averages*. The algorithm becomes

$$P_{n+1} =$$
$$(1/(1-\alpha))P_n - (\alpha/(1-\alpha)^2)P_n x_{n+1}(1 + (\alpha/(1-\alpha))x_{n+1}^T P_n x_{n+1})^{-1} x_{n+1}^T P_n$$

$$\hat{B}_{n+1} = \hat{B}_n - \alpha P_{n+1}[x_{n+1} x_{n+1}^T \hat{B}_n - x_{n+1} y_{n+1}]$$

where $0 < \alpha < 1$. The effect of α is to prevent the elements of P from becoming small, so new data continue to affect the estimates. The larger is α, the more the recent observations are emphasized, and values between 0.1 and 0.3 are typically used.

5.5 Generalized least squares

The assumptions that the E_i are independent of each other and have a constant variance can be replaced by any other assumed covariance matrix. Suppose that

$$Y = XB + E$$

where

$$E[E] = 0, \qquad \text{cov}(E) = V\sigma^2$$

and the elements of E are independent of the covariates. This is a generalized linear regression model. Any covariance matrix V can be written in the form:

$$V = Q^2$$

It is straightforward to find Q, because any real symmetric matrix is diagonalizable i.e.

$$V = MDM^{-1}$$

where D is a diagonal matrix of the eigenvalues and M is a matrix of corresponding eigenvectors. Then

$$Q = MD^{1/2}M^{-1}$$

and $D^{1/2}$ is the matrix with square roots of the eigenvalues along the leading diagonal. Now define

$$F = Q^{-1}E$$

then

$$\text{cov}(F) = E[FF^T] = E[Q^{-1}EE^T(Q^{-1})^T]$$

which, since Q is a symmetric matrix of constants

$$\text{cov}(F) = Q^{-1}E[EE^T]Q^{-1}$$
$$= Q^{-1}Q^2Q^{-1}\sigma^2 = I\sigma^2$$

The original model premultiplied by Q^{-1}

$$Q^{-1}Y = Q^{-1}XB + Q^{-1}E$$

is of the form

$$W = UB + F$$

where W contains the dependent variables, U contains the explanatory variables, and the elements of F satisfy the usual assumptions. The practical problem is that V is unlikely to be known, and it is usual to make some assumptions about the form of V. As an example, the standard deviation of the random variations might be assumed to be in proportion to the expected value of the dependent variable. The expected values could be obtained from a preliminary unweighted analysis, and the process can even be iterated.

In some cases there is a theoretical reason for assuming some variance and covariance structure. In principle, generalized least squares is the most efficient way to draw a line through points on an extreme value plot. Johnson *et al.* (1995) give a table of the variances and covariances of order statistics for small samples (up to 10) and references to tables for samples up to size 30. Another example is Mackay *et al.* (1996) who give an analysis of results from a computer model which simulated the transport of contaminants through the water table. The investigation was concerned with the effect of the position of the waste vault on, among other variables, the average times and standard deviation of times taken by particles to reach two waste pits. A simulation consisting of 1000 particles was repeated for different vault positions. The number of particles reaching the pits depended on the position of the vault. The variances of the mean travel times, and their standard deviations, could be assumed to be in inverse proportion to the number of particles.

Despite the versatility of the standard linear model, there are important applications which need a modified analysis. One of the most common is a log-regression, which is an example of an intrinsically linear model.

5.6   Intrinsically linear models

A water company required an equation for predicting the cost of construction of water filtration plants. The costs (monetary units) of 59 past constructions were available, together with the capacity (mega litres per day) and number of filters (one, two or three). The points in a plot of the natural logarithm of cost (y) against the logarithm of volume (x_1) appeared to be evenly scattered about a straight line. A regression of y on x_1 and the logarithm of number of filters x_2 was fitted:

$$y = 2.4596 + 0.7120x_1 + 0.2230x_2$$

with $s = 0.293$. The estimated standard deviations of the coefficients are 0.0708, 0.0544 and 0.1071 for the constant, ln(volume), and ln(number of filters) respectively. The coefficient for the ln(number of filters) is more than twice its estimated standard deviation and this justifies the inclusion of x_2. The assumed model is

$$Y_i = \beta_0 + \beta_1 x_{1i} + \beta_2 x_{2i} + E_i$$

where the E_i satisfy the usual assumptions. The company's main interest in the model, from a planning perspective, is to estimate the average costs of treatment works given a design volume and number of filters. This is given by:

$$\exp(\hat{y} + s^2/2)$$

where $\hat{y}$ is the estimate from the fitted regression equation. The reason for adding $s^2/2$ before taking the exponential becomes clear if we rewrite the model in terms of the original variables as:

$$\text{cost}_i = k(\text{volume}_i)^{\beta_1}(\text{number of filters}_i)^{\beta_2}F_i$$

where $k = \exp(\beta_0)$ and F_i is a log-normal random variable, such that

$$\ln F_i \sim N(0, \sigma^2)$$

We now use the result of Section 2.4.1 (and Exercise 7 at the end of this chapter) to obtain

$$E[F] = \mu_F = \exp(0 + \sigma^2/2) = \exp(\sigma^2/2)$$

It follows that

$$E[\text{cost}_i] = k(\text{volume}_i)^{\beta_1}(\text{number of filters}_i)^{\beta_2} \exp(\sigma^2/2) = \exp(E[Y_i] + \sigma^2/2)$$

This justifies the multiplication by $\exp(s^2/2)$ when estimating mean costs, and we now show that omitting this factor would lead to estimates of median costs.

Since ln F is normally distributed, its median equals its mean of 0, and the median of F is $\exp(0) = 1$. It follows that $\exp(\hat{y})$ estimates the median cost of treatment works with these design parameters. If we take the exponential of a confidence interval for $E[Y \mid x]$ we will obtain a confidence interval for the median cost. In contrast to these relationships, the exponential of a prediction interval for an individual Y given x is a prediction interval for the cost. Let us take a design for 8 mega litres per day with 2 filters as an example.

$$\hat{y} = 2.4496 + 0.7120 \times \ln(8) + 0.2230 \times \ln(2) = 4.0847$$

The estimated mean cost of all such treatment works is

$$\exp(4.0847 + (0.293)^2/2) = 62 \text{ monetary units}$$

which is about 4% higher than the median value. A 95% prediction interval for an individual Y is 4.0847 ± 0.60. So we predict

$$\Pr(3.48 < \ln(\text{cost}) < 4.68) = 0.95$$

and it follows that

$$\Pr(\exp(3.48) < \text{cost} < \exp(4.68)) = 0.95$$

Therefore a 95% prediction interval for the cost of an individual treatment works is between 32 and 108 monetary units.

Example 5.5
The data in Table 5.4 were obtained during 1982–1983 from the River Browney in County Durham, UK, by the Northumbrian Water Authority. They represent peaks flows, rainfall during the two days before the peak, and baseflow on the day before. Baseflow is the contribution to the flow from water which has infiltrated the catchment. It cannot be measured directly, but

Table 5.4 *Rainfall during two days (column 1, mm) baseflow on preceding day (column 2, m^3/s), and peak flow (column 3, m^3/s)*

11.2	0.28	13.16
17.0	0.31	2.52
30.6	0.45	4.58
19.6	0.34	2.99
13.6	0.40	1.01
10.1	0.22	0.86
12.4	0.29	0.97
12.8	0.28	1.08
11.2	0.26	1.30
24.8	0.36	6.75
10.6	0.90	7.00
25.1	2.03	23.90
13.8	2.35	31.55
19.4	4.14	17.27
31.3	1.27	30.97
27.0	2.52	38.82
27.6	0.74	3.21
14.3	0.46	1.42
15.6	0.31	1.58

hydrologists have procedures for estimating it. A regression model for predicting peak flows from given rainfalls and baseflows forms part of a system for estimating seasonal flood risks for up to one month ahead, taking into account the prevailing catchment wetness. Rainfall distribution varies with the seasons, and baseflow is a measure of the catchment wetness which has a major effect on the run-off. Time series methods are used to predict changes in baseflow.

Peak flows have a markedly skewed distribution, and whilst they cannot be negative they do take values close to zero. This suggests that the logarithm of peakflow (y) might be more suitable for the dependent variable in a multiple regression model. The logarithm of peakflow is plotted against rainfall, with baseflow level indicated, in Figure 5.4. The regression of y on rainfall (x_1) and baseflow (x_2) is

$$y = 1.5096 + 0.06328(x_1 - 18.316) + 0.7800(x_2 - 0.94263)$$

$$s = 0.8647 \qquad R^2 = 62.9\% \qquad R^2 \text{ adjusted} = 58.2\%$$

The estimated standard deviations of the coefficients of x_1 and x_2 are 0.0296 and 0.199 respectively. See Exercise 9 for some alternative models.

In general, any relationship between y and x which can be transformed into a relationship which is linear in functions of the unknown parameters is said to be *intrinsically linear* or *linearizable*. The Michaelis–Menten model, which is used to relate the initial velocity (v) of an enzyme reaction to substrate

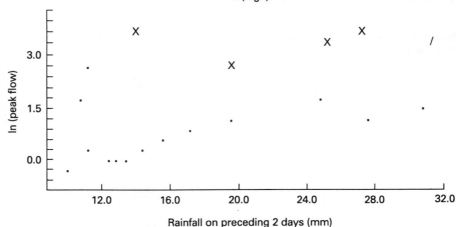

Character TPlot

Logarithm of peak flow against 2 day preceding rainfall.
Baseflow level shown by: . (low); / (above average);
X (high).

'0' < -1.3E-01 < '.' < 0.942632 < '/' < 2.012153 < 'X'

MTB >

Figure 5.4 Logarithm of peak flow against rainfall, with baseflow levels shown, for the River Browney

concentration (c)

$$v = \frac{\theta_1 c}{c + \theta_2}$$

can be expressed as

$$y = \beta_0 + \beta_1 x$$

where $y = 1/v$, $x = 1/c$, $\beta_0 = 1/\theta_1$, $\beta_1 = \theta_2/\theta_1$. There are two snags to this device. The first is that we cannot assume, consistently, $N(0, \sigma^2)$ errors about both the original relationship and the linearized relationship. If we use the usual regression results we assume the errors about the latter are $N(0, \sigma^2)$. This may be reasonable, for example if we take logarithms and the original variable has a constant coefficient of variation, but will not always be so. The second is that $E[f(Y)]$ is not the same as $f(E[Y])$ for non-linear functions. One way around this is to scale the fitted relationship so that the average of the residuals is zero. The alternative is to use the technique of Section 5.7.

5.7   Non-linear regression

A general non-linear model is

$$y_i = f(x_i, \theta) + E_i \qquad \text{for } i = 1, \ldots, n \tag{5.3}$$

where x_i is a vector of predictor variables, θ is a vector of unknown parameters, and the E_i are errors which are defined in the same way as for the multiple regression model. We will linearize the function f, about an initial guess (*starting values*) for the parameters, by the first terms of a Taylor series expansion. We can then apply the solution for the linear problem to obtain improved estimates. The process is repeated until it converges. The initial estimates:

$$\theta_0^T = [\theta_{10}, \ \theta_{20}, \ \ldots, \theta_{p0}]$$

might be based on simple graphical methods or physical arguments,

$$f(x_i, \theta) = f(x_i, \theta_0) + \sum_{j=1}^{p} \left[\frac{\partial f(x_i, \theta)}{\partial \theta_j} \right]_{\theta = \theta_0} (\theta_j - \theta_{j0})$$

We now define:

$$Y = \begin{bmatrix} y_1 - f(x_1, \theta_0) \\ \vdots \\ y_n - f(x_n, \theta_0) \end{bmatrix}$$

$$W = \begin{bmatrix} w_{11} & \cdots & w_{pp} \\ \vdots & & \vdots \\ w_{1n} & \cdots & w_{pn} \end{bmatrix}$$

where

$$w_{ij} = \left[\frac{\partial f(x_i, \boldsymbol{\theta})}{\partial \theta_j} \right]_{\boldsymbol{\theta} = \boldsymbol{\theta}_0}$$

and the vector of differences between our starting values and the hypothetical parameter values as

$$\boldsymbol{H}^{\mathrm{T}} = [\theta_1 - \theta_{10}, \ \theta_2 - \theta_{20}, \ \ldots, \ \theta_p - \theta_{p0}]$$

Then we can approximate the model (5.3) with the linear system

$$Y = WH + E$$

The vector of differences in parameter values is estimated, by using equation (5.2), as

$$\hat{\boldsymbol{H}} = (\boldsymbol{W}^{\mathrm{T}}\boldsymbol{W})^{-1}\boldsymbol{W}^{\mathrm{T}}\boldsymbol{Y}$$

We now replace $\boldsymbol{\theta}_0$ with

$$\boldsymbol{\theta}_0 + \hat{\boldsymbol{H}}$$

and repeat the procedure until the elements of $\hat{\boldsymbol{H}}$ are sufficiently close to 0. The variance of the errors (σ^2) is estimated by the mean of the squared residuals, allowing for the loss of p degrees of freedom. That is, σ^2 is estimated by s^2 which is defined by:

$$s^2 = \sum_{i=1}^{n} (y_i - f(x_i, \hat{\boldsymbol{\theta}}))^2 / (n - p)$$

The estimate of the asymptotic covariance matrix of $\hat{\boldsymbol{\theta}}$ is

$$(\boldsymbol{W}^{\mathrm{T}}\boldsymbol{W})^{-1}s^2$$

where W is the matrix of partial derivatives from the last iteration.

Example 5.6
Myers (1994) fits a growth model of the form

$$y_i = \alpha[1 - \exp(-\beta x_i)] + E_i$$

to subsidence data from mining excavations in West Virginia, USA. The data, from the Mining Engineering Department, Virginia Polytechnic Institute, USA, are given in Table 5.5. The dependent variable (y) is the angle of draw, which is defined as the angle between the perpendicular at the edge of the excavation, and the line joining the edge of excavation to the point beyond which there is no subsidence. The predictor variable (x_i) is the ratio of the width of an excavation to the depth (w/d). As w/d increases, the angle of draw approaches a maximum value of α.

Initial guesses for α and β, of 35 and 1, were based on a plot of the data (Figure 5.5). The estimates converged to 32.46 and 1.511, and the residual sum of squares decreased from 345.2 to 204.6. Hence s, the estimate of σ, was 3.82.

Table 5.5 *Data from 16 mining excavations in West Virginia, USA (data courtesy of the Mining Engineering Department, Virginia Polytechnic Institute and State University)*

Width (ft)	Depth (ft)	Angle of draw (degrees)
610	550	33.6
450	500	22.3
450	520	22.0
430	740	18.7
410	800	20.2
500	230	31.0
500	235	30.0
500	240	32.0
450	600	26.6
450	650	15.1
480	230	30.0
475	1400	13.5
485	615	26.8
474	515	25.0
485	700	20.4
600	750	15.0

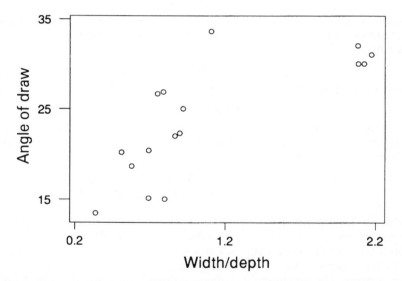

Figure 5.5   Angle of draw against width/depth for mining excavations in West Virginia

5.8   Measurement errors in the predictor variables

We need to distinguish between measurement error, and inherent variability in a population. In Example 5.1 we could measure the density and compressive strength of the test specimens of grout quite accurately. The variation about the regression was due to inherent variability in grout specimens. To be less specific, assume that (X, Y) have a bivariate normal distribution and are measured without error. If a random sample from this distribution is available, either the regression of Y on x or the regression of X on y can be estimated. The choice depends on which variable the investigator wishes to predict. You should note that the assumptions of a random sample are not compatible with choosing the values of x in advance. If the values of x are chosen in advance only the regression of Y on x is appropriate. An alternative to either of the regressions is a correlation analysis (Appendix 1). This is appropriate if a measure of the association between the variables is required rather than an equation for making predictions. In a correlation analysis both X and Y are treated as random variables, whereas the regression of Y on x estimates the conditional distribution of Y for given x, specific values of which happen to have arisen at random.

If there is non-negligible measurement error in both Y and the x_j variables, a modified analysis should be used. An example arises from a comparison of land and aerial survey methods. Let X_i and Y_i represent n measurements to be made by land and aerial survey, respectively. It is assumed that they are subject to independent errors, E_i and H_i, respectively, of zero mean and constant variances σ_E^2 and σ_H^2. Let u_i and v_i represent the unobservable error-free measurements. It is supposed that

$$v_i = \alpha + \beta u_i$$

for some α and β, and one of the aims of the project is to see whether there is any evidence that these parameters differ from 0 and 1, respectively.

Substituting

$$u_i = X_i - E_i \quad \text{and} \quad v_i = Y_i - H_i$$

into the model relating v_i and u_i gives

$$Y_i = \alpha + \beta X_i + (H_i - \beta E_i)$$

Despite first appearances, this is not the standard regression model because the assumption that the errors are independent of the predictor variable is not satisfied:

$$\text{cov}(X_i, (H_i - \beta E_i)) = E[(X_i - u_i)(H_i - \beta E_i)]$$
$$= E[(E_i(H_i - \beta E_i))] = -\beta \sigma_E^2$$

Maximum likelihood solution   The usual approach to this problem is to assume the errors are normally distributed and that the ratio of σ_H^2 to σ_E^2, denoted by λ in the following, is known. In practice, some reasonable value has to be assumed for λ, preferably based on information from replicate measurements. The maximum likelihood estimates of α and β, using a notation S_{xy} for $\Sigma (x - \bar{x})(y - \bar{y})$ etc are:

$$\hat{\beta} = \{(S_{yy} - \lambda S_{xx}) + [(S_{yy} - \lambda S_{xx})^2 + 4\lambda S_{xy}^2]^{1/2}\}/(2S_{xy})$$
$$\hat{\alpha} = \bar{y} - \hat{\beta}\bar{x}$$

These follow from the likelihood function $\mathcal{L}$ which is proportional to

$$\sigma_E^{-1}\sigma_H^{-1} \exp\left\{-\frac{1}{2}\sigma_E^{-2}\sum_{i=1}^{n}(x_i - u_i)^2 - \frac{1}{2}\sigma_H^{-2}\sum_{i=1}^{n}(y_i - (\alpha + \beta u_i))^2\right\}$$

if α, β, σ_E, and u_i are treated as the unknown parameters.

Estimates of σ_E^2 and σ_H^2 are given by

$$\hat{\sigma}_H^2 = \lambda \hat{\sigma}_E^2 = (S_{yy} - \hat{\beta} S_{xy})/(n-2)$$

If $\lambda = 1$, the values of $\hat{\alpha}$ and $\hat{\beta}$ are the slope and intercept of the line such that the sum of squared perpendicular distances from the plotted points to the line is a minimum.

Upper and lower points of the $(1 - \alpha) \times 100\%$ confidence interval for β are given by

$$\lambda^{1/2} \tan(\arctan(\hat{\beta}\lambda^{-1/2}) \pm \tfrac{1}{2}\arcsin(2t_{\alpha/2}\theta))$$

where

$$\theta^2 = \frac{\lambda(S_{xx}S_{yy} - S_{xy}^2)}{(n-2)((S_{yy} - \lambda S_{xx})^2 + 4\lambda S_{xy}^2)}$$

and $t_{\alpha/2}$ is the upper $(\alpha/2) \times 100\%$ point of the t-distribution with $n-2$ degrees of freedom. An approximation to the standard deviation of $\hat{\beta}$ is given by one-quarter of the width of this interval. Since $\hat{\beta}$ is independent of $\bar{x}$

$$\text{var}(\hat{\alpha}) \simeq \hat{\sigma}_H^2/n + \bar{x}^2 \, \text{var}(\hat{\beta}) + \hat{\beta}^2 \hat{\sigma}_E^2/n$$

The data pairs in Table 5.6 are the measured heights (in metres) above sea level of 25 points from the land survey (x) and the aerial survey (y). The points

Table 5.6 *Heights of 25 points determined by land and aerial surveys*

Land survey estimate (x, m)	Aerial survey estimate (x, m)
720.2	732.9
789.5	804.9
749.7	760.5
701.5	712.3
689.2	702.0
800.5	812.8
891.2	902.7
812.8	820.0
780.6	793.6
710.5	720.2
810.4	825.6
995.0	1008.6
890.5	902.4
829.4	845.1
808.7	820.3
781.7	796.1
868.7	885.2
904.2	920.1
780.7	790.2
649.6	660.0
732.1	741.2
770.4	781.2
733.7	745.6
694.9	707.3
620.0	633.4

were equally spaced over a hilly 10 km by 10 km area. Replicate measurements of the height of a single point suggest that errors in the aerial survey measurements have a standard deviation three times that of errors in the land survey. Calculations give

$$\bar{x} = 780.6 \qquad \bar{y} = 793.0$$

$$S_{xx} = 177970 \qquad S_{xy} = 179559 \qquad S_{yy} = 181280$$

λ is assumed to be 9. The formulae of this section lead to the following estimates

$$\hat{\alpha} = 5.379 \qquad \hat{\beta} = 1.00899$$

$$\hat{\sigma}_E = 0.716 \qquad \hat{\sigma}_H = 2.147$$

95% confidence intervals for α and β are

$$[-7.6, 18.3] \quad \text{and} \quad [0.993, 1.025]$$

respectively. There is no evidence of any systematic difference between the results of the two surveys because the confidence intervals for α and β include 0 and 1 respectively. (If a standard regression analysis of Y on x is used, which corresponds to λ tending to infinity, the parameter estimates would be:

$$5.368, 1.00893, 0 \quad \text{and} \quad 2.254$$

respectively. 95% confidence intervals for α and β would be

$$[-3.3, 14.0] \quad \text{and} \quad [0.998, 1.020]$$

respectively.)

A related application is calibrating measuring equipment. In particular, if we need to calibrate a field instrument against a relatively accurate laboratory instrument it is preferable to fit a line with field measurements as y and the accurate measurements as x. If

$$y = \hat{\beta}_0 + \hat{\beta}_1 x$$

is the fitted regression line, we use

$$x = (y - \hat{\beta}_0)/\hat{\beta}_1$$

to infer the accurate measurement from the field reading, although it is not exactly unbiased because $1/\hat{\beta}$ is not an unbiased estimator of $1/\beta$.

5.9   GLIM

5.9.1   Poisson response

In transport engineering applications, it is quite often assumed that traffic accidents at particular sites have Poisson distributions. For example, Al Janahi (1995) investigated the effect of road traffic speed, during free flow conditions, on the frequency of personal injury accidents at nine dual carriageway sites in the Tyne and Wear district of the UK, and ten dual carriageway sites in Bahrain. They assumed that the site means (μ, accidents per 5 year study period) of the Poisson distribution were of the form

$$\mu = kLF^aS^b$$

where L is the length of road at the site (km), F is the traffic flow through the site (10^5 vehicles per dual carriageway year), S is some measure of speed and a, b and k are parameters to be estimated from the data. If we take logarithms we have a linear relationship:

$$\ln(\mu) = \ln(k) + \ln(L) + a \ln F + b \ln S$$

This is an example of a *generalized linear model*. Let Y_i represent the number of accidents at site i. There are two reasons why the standard regression analysis with $\ln(Y_i)$ as the dependent variable is not strictly appropriate. The first reason is that the variance of a Poisson variable is equal to its mean, and the variance of the logarithm of a Poisson variable is, approximately, inversely proportional to the mean (Exercise 10). It follows that the assumption that the errors have a constant variance is not correct. The second reason is that, in general, the variable Y_i may be 0. A common way round this problem is to take the logarithm of 0.5, but there is no mathematical justification for this. Fortunately, the numbers of accidents at specific sites are usually low, and data for monitoring the success of road traffic safety measures are limited. Nevertheless, transport engineers will have to make important decisions on the basis of analyses of whatever data are available. If we are responsible for the analyses we should use the most efficient, and mathematically respectable, method. A maximum likelihood solution involves maximizing

$$\mathcal{L} = \prod_{i=1}^{n} e^{-\mu_i}(\mu_i)^{y_i}/y_i!$$

where $\mu_i = \mathrm{E}[Y_i] = kL_iF_i^aS_i^b$ and n is the number of sites, with respect to k, a and b. Notice that we do not use any adjustment factor, as we did in Section 5.6, because we assume $\ln(\mathrm{E}[Y_i])$, rather than $\mathrm{E}[\ln Y_i]$, is linearly related to the predictor variables. That is:

$$\ln(\mu_i) = \ln(k) + \ln(L_i) + a \ln(F_i) + b \ln(S_i)$$

The Generalized Linear Interactive Modelling (GLIM) system estimates $\ln k$, a and b by maximizing the Poisson likelihood function. The data in Table 5.7 are: the length of the road section (len r); average traffic flow (flow); coefficient of upper spread of speed (cuss); and the number of accidents in the five year period. The coefficient of upper spread is defined by:

$$\frac{85\% \text{ quantile} - \text{median}}{\text{median}}$$

This commonly used measure of the speed distribution on roads is sensitive to differences in speeds of the faster vehicles. The GLIM session commands, and responses, are shown in Figure 5.6. Declaring the logarithm of 'len r' as an offset sets its coefficient to one. The estimates of $\ln(k)$, a and b are -0.5498, 0.9896 and 1.043 respectively. Hence the estimate of k is 0.5771.

A quantity called the *scaled deviance* is shown three lines above the

Table 5.7 *Length of road section (km), average traffic flow (10 vehicles per dual carriageway year), coefficient of upper spread of speed, and number of accidents in 5 year period (1989–93) at nine dual carriageway sites in Tyne and Wear, UK*

Length	Flow	Cuss	Accidents
44.0000	39.63	0.1588	142.000
18.0000	50.00	0.2041	94.000
7.0000	47.67	0.1869	37.000
0.6400	46.91	0.2821	4.000
1.1600	33.33	0.1441	3.000
0.4700	52.51	0.1639	3.000
0.6200	44.78	0.1593	2.000
1.0500	54.21	0.1799	2.000
1.0300	24.45	0.1214	1.000

```
[o] GLIM 3.77 update 2 (copyright)1985 Royal Statistical Society, London
[o]
[i] ? $input 8
[i] File name? g.con
[i] $INP? $
[i]    $units 9    $data lenr flow cuss nacc $read
[i]    44.0000    39.63    0.1588    142.000
[i]    18.0000    50.00    0.2041    94.000
[i]    7.0000    47.67    0.1869    37.000
[i]    0.6400    45.91    0.2821    4.000
[i]    1.1600    33.33    0.1441    3.000
[i]    0.4700    52.51    0.1639    3.000
[i]    0.6200    44.78    0.1593    2.000
[i]    1.0500    54.21    0.1799    2.000
[i]    1.0300    24.45    0.1214    1.000
[i] $calc lnlr = %log(lenr)
[i] $calc lnfl = %log(flow)
[i] $calc lnsp = %log(cuss)
[i] $yvar nacc
[i] $error P
[i] $offset lnlr
[i] $fit lnfl,lnsp
[i] $dis e
[o] scaled deviance = 4.0438 at cycle  3
[o]             d.f. = 6
[o]
[o]            estimate        s.e.       parameter
[o]    1      -0.5498        5.159       1
[o]    2       0.9896        0.9821      LNFL
[o]    3       1.043         0.9157      LNSP
[o]    scale parameter taken as  1.000
[o]
[i] $stop
```

Figure 5.6 GLIM session for Poisson response

parameter estimates. Let n be the number of data. The *deviance* (D) is defined by

$$D = 2 \ln\left(\frac{\text{likelihood of maximal model with } n \text{ parameters}}{\text{likelihood of fitted model with } p \text{ parameters}}\right)$$

The *maximal model* has the same number of parameters as there are data, and must therefore be an exact fit. The *scale factor* for a Poisson model is one (Exercise 9), and if the model were to be correct D would have an approximate chi-square distribution with $n - p$ degrees of freedom. This is a useful means of assessing the adequacy of the model. Large values of D indicate that the model is not realistic, and may be due to influential predictor variables which have not been recorded. Alternatively, a large value of D can arise because the response is more variable than a Poisson variate. This phenomenon is known as *overdispersion*, but it is not exhibited here since the value of D (4.04) is less than the mean of a χ_6^2 distribution (which equals its degrees of freedom).

The estimated standard deviations of the parameter estimates (s.e. in GLIM) are interpreted in the same way as in any multiple regression. Although there is no strong evidence that the measure of speed is related to the number of accidents, we cannot dismiss the possibility. The minimum and maximum values of 'cuss' at the nine sites are 0.12 and 0.28. If the flow is set at the average value of 44 (10^5 vehicles per dual carriageway year), the estimates of μ would be 2.7 and 6.4 respectively. The estimate of the reduction may be imprecise, but it represents a substantial social benefit. If we are prepared to assume accidents under free flow conditions are proportional to the average traffic flow, we could use the product 'len $r \times$ flow' as an offset. Then the estimate of b is almost the same (1.034), but its standard deviation is dramatically reduced to 0.4451. The snag is that, even on a straight road, the number of potential conflicts is not necessarily proportional to the flow. Although individual studies of the relationship between speed and the number of accidents may be inconclusive, substantial conclusions can be drawn from an overview of the many investigations that are carried out (e.g. Ibrahim and Metcalfe, 1993). Whatever the relationship between speed and the number of accidents may be, accidents are much more serious if they occur at high speeds.

5.9.2 Binomial response

Let Y_i be the number of successes in n_i independent trials with probability of success p_i. We now write

$$E[Y_i/n_i] = p_i$$

and assume the p_i are related to explanatory variables, by the relationship

$$\ln(p_i/(1 - p_i)) = \beta_0 + \beta_1 x_{1i} + \cdots + \beta_k x_{ki}$$

The left-hand side is known as the *logit* of p_i. It transforms p_i which is restricted to the range $[0, 1]$ to a range $[-\infty, \infty]$. The inverse relationship is defined by:

$$\ln(p/(1 - p)) = \theta \Leftrightarrow p = e^\theta/(1 + e^\theta)$$

The exact maximum likelihood solution is again preferable to analysing logit ($\hat{p}_i$), where $\hat{p}_i = Y_i/n_i$, using the standard model. This is because $\hat{p}_i$ might equal 0 or 1, and the variance of logit ($\hat{p}_i$) depends on n_i and to a lesser extent p_i. The likelihood function is

$$\mathcal{L} = \prod_{i=1}^{m} p_i^{y_i}(1 - p_i)^{n_i - y_i} \,_{n_i}C_{y_i}$$

where the p_i are defined in terms of the parameters $\beta_0, \ldots, \beta_k$ and the known values of the predictor variables. This has to be maximized with respect to the parameters, and the GLIM system is a convenient means of doing so. An important special case is when all the n_i equal 1, and the response Y_i is a binary variable, possibly representing passing or failing a specification.

We will now look at a practical application. Chlorination is a widely practised means of disinfecting drinking water, but it is not a panacea, and it is most effective when applied to water that has already been treated by other means. The data in Table 5.8 are the results of 29 tests of coliform bacteria (coli) removal at a rapid gravity water filtration plant over a six month period. An operational issue is whether the efficiency depends on the number of coli in the untreated water.

Table 5.8 *Numbers of coli in samples of 100 ml from input stream (column 1), numbers in output stream (column 2), and days after 7th September 1995 (column 3)*

380	32	1
160	30	7
140	20	15
480	22	17
280	20	20
310	18	23
100	80	28
280	10	31
150	6	34
200	6	36
600	15	41
200	18	48
160	28	66
180	18	69
240	21	71
150	30	87
240	130	93
120	110	98
180	60	106
160	24	114
200	100	120
700	200	122
240	160	128
1600	84	136
130	30	147
110	22	155
1000	70	160
300	48	163
130	66	171

Let n be the number of coli in a test sample (100 ml) of water. The results of an analysis using GLIM are shown in Figure 5.7. The *y-variable* is the number of coli removed, the *error* is binomial with denominators n, and the *link* between the proportion removed (p) and the linear combination of predictors is the logit. The fitted formula is:

$$\ln(p/(1-p)) = -3.29 + 0.839 \ln(n)$$

The deviance is 1442 on 27 degrees of freedom, and is much higher than the expected value of 27. One explanation for this overdispersion is that important variables which affect the process, such as ambient temperature, have not been recorded. Another is that the coli tend to cluster together. Furthermore, the assumption of a binomial distribution is only an approximation if the test samples were sampled from the continuous input and output streams. If the conditions for a binomial model are to be satisfied, we must count how many of the specific coli in the test sample are removed by the filter. This is not practical on a full scale filtration plant, with a 100 ml sample, and we have a third contribution to the overdispersion. The displayed standard deviation of the coefficient of $\ln(n)$ is 0.039, but this assumes a scale factor of 1. The deviance is 53 times its expected value and the estimated standard deviation should be multiplied by the square root of this (7.3). An approximate 95% confidence interval for the coefficient of $\ln(n)$ is therefore,

$$[0.27, 1.41]$$

There is evidence that the efficiency improves as the number of coli in the input stream increases, but the relationship has not been identified very precisely. Some predictions are:

Coli in sample of raw water (n)	Percentage removal ($p \times 100\%$)
100	64%
314	82%
1600	95%

We can obtain a substantial decrease in the deviance, to 970 on 25 degrees of freedom, if we fit a quadratic trend. In the following equation t is days from 7 September.

$$\ln(p/(1-p)) = -3.286 + 0.9809 \ln(n) - 0.01404t + 0.0001659(t-79.55)^2$$

If n is set at its average value of 314 we obtain the following fits.

t	Percentage removal ($p \times 100\%$)
1	97%
80	77%
120	72%
160	77%

A possible explanation for the trend is that the efficiency decreases in cold weather.

```
[o] GLIM 3.77 update 2 (copyright)1985 Royal Statistical Society, London
[o]
[i] ? $input 29$data n m t$dinput 8$
[i] File name? table5p8.dat
[i]   380   32    1
        etc ...
[i]   130   66   171
[i] ? $error b n$
[i] ? $calc r=n-m     $yvar r$
[i] ? $calc l=%log(n)$
[i] ? $tab the n mean
[o]             314.5
[i] ? $tab the l mean$
[o]             5.476
[i] ? $tab the t mean$
[o]             79.55
[i] ? $calc y=100*r/n$
1[i] ? $plot y n$
[o]    100.00 |                      Y
[o]     95.00 |    Y Y 2Y      Y                   Y                        Y
[o]     90.00 |      YYY     Y
[o]     85.00 |     3     Y
[o]     80.00 |    Y2
[o]     75.00 |    Y
[o]     70.00 |                      Y
[o]     65.00 |      Y
[o]     60.00 |
[o]     55.00 |
[o]     50.00 |    Y   Y
[o]     45.00 |        Y
[o]     40.00 |
[o]     35.00 |        Y
[o]     30.00 |
[o]     25.00 |
[o]     20.00 |  Y
[o]     15.00 |
[o]     10.00 |  Y
[o]      5.00 |        Y
[o] ---------:---------:---------:---------:---------:---------:---------:
[o]          0.      300.     600.     900.    1200.    1500.    1800.

[i]  $calc lm2=(1-5.476)
[i]  $calc tm2=(t-79.55)**2$calc lmtm=(1-5.476)*(t-79.55)$
[i]  $fit n$dis e$
[o] scaled deviance = 1546.2 at cycle  4
[o]          d.f. =   27
[o]
[o]            estimate        s.e.       parameter
[o]     1      0.9224      0.04375        1
[o]     2      0.001414   0.00007930      N
[o]     scale parameter taken as  1.000
[o]
[i] ? $fit l$dis e$
[o] scaled deviance = 1441.9 at cycle  4
[o]          d.f. =   27
```

Figure 5.7 GLIM session for binomial response

```
[o]
[o]          estimate        s.e.       parameter
[o]    1      -3.290         0.2250      1
[o]    2       0.8390        0.03917     L
[o]    scale parameter taken as  1.000
[o]
[i] ? $fit 1,lm2,t,tm2,lmtm$dis e$
[o] scaled deviance = 950.49 at cycle  4
[o]             d.f. =  23
[o]
[o]          estimate        s.e.       parameter
[o]    1      -3.374         0.4158      1
[o]    2       0.9901        0.07699     L
[o]    3       0.1958        0.06085     LM2
[o]    4      -0.01422       0.0007694   T
[o]    5       0.0001646     0.00001488  TM2
[o]    6      -0.004706      0.001164    LMTM
[o]    scale parameter taken as  1.000
[o]
[i] ? $fit 1,lm2,t,tm2$dis e$
[o] scaled deviance =  967.1 at cycle  4
[o]             d.f. =   24
[o]
[o]          estimate        s.e.       parameter
[o]    1      -2.723         0.3874      1
[o]    2       0.8766        0.07130     L
[o]    3       0.09742       0.05582     LM2
[o]    4      -0.01439       0.0007654   T
[o]    5       0.0001657     0.00001490  TM2
[o]    scale parameter taken as  1.000
[o]
[i] ? $fit 1,t,tm2$dis e$
[o] scaled deviance =  970.2 at cycle  4
[o]             d.f. =   25
[o]
[o]          estimate        s.e.       parameter
[o]    1      -3.286         0.2148      1
[o]    2       0.9809        0.03909     L
[o]    3      -0.01404       0.0007411   T
[o]    4       0.0001659     0.00001495  TM2
[o]    scale parameter taken as  1.000
[o]
[i] ? $fit 1,t$dis e$
[o] scaled deviance = 1100.3 at cycle  4
[o]             d.f. =   26
[o]
[o]          estimate        s.e.       parameter
[o]    1      -3.462         0.2161      1
[o]    2       1.038         0.03919     L
[o]    3      -0.01078       0.0006089   T
[o]    scale parameter taken as  1.000
[o]
[i] ? $stop
```

Figure 5.7 (*continued*)

The significance of the decrease in deviance can be formally tested in a similar way to Section 5.9.1. If there is no trend:

$$\frac{[D(\text{model with no trend}) - D(\text{model with quadratic trend})]/2}{D(\text{model with quadratic trend})/25} \sim F_{2,25}$$

The calculated value of this statistic is

$$[(1442 - 970)/2]/(970/25) = 6.08$$

This exceeds $F_{2,25,0.01} = 5.57$, and there is evidence of a trend. GLIM can be used for a standard multiple regression, in which case the deviance is the sum of squared residuals.

5.10 Summary

1. The standard form of the multiple regression model is

   $$Y_i = \beta_0 + \beta_1 x_{1i} + \cdots + \beta_k x_{ki} + E_i$$

   where the E_i are independent variables from $N(0, \sigma^2)$. The x_j can include squares and cross products of continuous variables, and indicator variables for categories. In general, the x_j are correlated and estimates of β_j depend on which other x_l are included in the model. A particular $\hat{\beta}_j$ is an estimate of the effect of x_j, given that the other x_l variables are included in the model. This should be borne in mind when interpreting the physical significance of a model. If we can choose the x_j values, as in a designed experiment (see Chapter 9), we usually arrange for them to be uncorrelated (orthogonal), or nearly so. This leads to more precise estimators and makes interpretation easier.

2. Generalized least squares (Section 5.5) accommodates E_i which are correlated and have different variances, although they still need to be independent of the x_j.

3. The requirement that the errors are independent of the x_j implies that the latter are measured without error. If measurement error for the x_j is substantial when compared with the variation in Y, which can be measurement error or inherent variability, the model described in Section 5.8 should be used.

4. The standard model is linear in the unknown parameters β_j, and the estimates are the solution of a set of linear equations. It may be possible to transform a model which is non-linear in the parameters to one which is linear in this sense. Assuming normal errors in the transformed model is not equivalent to assuming normal errors in the original model. If $W = f(Y)$ a prediction interval for Y is given by f^{-1} of a prediction interval for W, but $\mu_Y \neq f^{-1}(\mu_W)$ unless f is a linear transform. A Taylor series expansion about μ_Y gives

   $$\mu_W = E[f(Y)] \simeq f(\mu_Y) + f''(\mu_Y)\sigma_Y^2/2$$

5. Models which are non-linear in the parameters can be fitted by the methods described in Section 5.7.

6. Responses that are counts or proportions are explicitly modelled by GLIM (Section 5.9). If all the counts happen to be large, the standard form of the multiple regression with Y defined as the logarithm of the count will be an adequate approximation. Similarly, if all the proportions are well away from 0 and 1 and calculated from like sample sizes the standard form with Y defined as the logit of the proportion will be a reasonable approximation. However, small counts and sample proportions of 0 or 1 often occur in applications.

Exercises

1 (a) Let X_i represent weights of a random sample of bags of cement. Assume

$$X_i \sim N(\mu, \sigma^2)$$

Then we can write

$$X_i = \mu + E_i$$

where $E_i \sim N(0, \sigma^2)$ and E_i satisfy the standard assumptions of a regression model (A1–A5). Show that $\bar{X}$ is the least squares estimator of μ.

(b) The diagram below is of a junction in oil pipelines. There are three meters which measure (Y_i) the flows (θ_i).

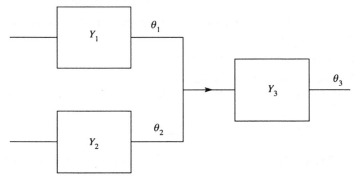

Assume there is no leakage so $\theta_3 = \theta_1 + \theta_2$. Assume that the measurement errors $(Y_i - \theta_i)$ have mean zero, the same variance σ^2, and are independent. Write down an expression for the sum of squares ψ as a function of θ_1 and θ_2. By setting $\partial\psi/\partial\theta_1$ and $\partial\psi/\partial\theta_2$ equal to zero show that the least squares estimators of θ_1 and θ_2 are:

$$\hat{\theta}_1 = (2Y_1 - Y_2 + Y_3)/3$$
$$\hat{\theta}_2 = (2Y_2 - Y_1 + Y_3)/3$$

Deduce that $\hat{\theta}_3 = (2Y_3 + Y_1 + Y_2)/3$. Verify that all these estimators are unbiased and have a variance of $2\sigma^2/3$. Show that the minimized sum of squares is an unbiased estimate of $1\sigma^2$ with one degree of freedom.

In general, the residual sum of squares is an unbiased estimator of $(n-p)\sigma^2$ where n is the number of observations and p is the number of unconstrained parameter estimates.

2 The following general results are required before deriving the formula for the least-squares estimators. Let $\phi(\boldsymbol{v})$ be a scalar function of an $m \times 1$ array $\boldsymbol{v}$, where

$$\boldsymbol{v}^{\mathrm{T}} = [v_1, \ldots, v_m]$$

That is, ϕ is a function of m variables $v_1, \ldots, v_m$. Define the array of partial derivatives

$$\frac{\partial \phi}{\partial \boldsymbol{v}} = \left[\frac{\partial \phi}{\partial v_1}, \ldots, \frac{\partial \phi}{\partial v_m}\right]^{\mathrm{T}}$$

The results given below are consequences of this definition and the usual rules of calculus. Let $\boldsymbol{c}$ be an $m \times 1$ array of constants and M be an $m \times m$ matrix of constants. Then:

(i) $\dfrac{\partial}{\partial \boldsymbol{v}} (\boldsymbol{v}^{\mathrm{T}}\boldsymbol{c}) = \boldsymbol{c}$

(ii) $\dfrac{\partial}{\partial \boldsymbol{v}} (\boldsymbol{v}^{\mathrm{T}}M\boldsymbol{v}) = M\boldsymbol{v} + M^{\mathrm{T}}\boldsymbol{v}$

(iii) A necessary requirement for ϕ to have a maximum or minimum is that $\partial\phi/\partial\boldsymbol{v} = \mathbf{0}$.

Verify these results for the case when $m = 2$.

3 Useful matrix results for statistical analysis.
   (i)   Let $A = (a_{ij})$ be a square $n \times n$ matrix. The *trace* of A is defined by

$$\mathrm{tr}\, A = \sum_{i=1}^{n} a_{ii}$$

If B is another $n \times n$ matrix, it follows from the definition that

$$\mathrm{tr}(A + B) = \mathrm{tr}\, A + \mathrm{tr}\, B$$

Also, if P and Q are $n \times m$ matrices, respectively then

$$\mathrm{tr}(PQ) = \mathrm{tr}(QP)$$

since

$$\sum_{i=1}^{m}\left\{\sum_{a=1}^{n} p_{ia}q_{ai}\right\} = \sum_{i=1}^{n}\left\{\sum_{a=1}^{m} q_{ia}p_{ai}\right\}$$

Verify this result for $m = 1$ and $n = 2$.
   (ii) Let A be a square $n \times n$ matrix and c any $n \times 1$ matrix which is not identically equal to a column of zeros. The scalar expression $\boldsymbol{c}^{\mathrm{T}}A\boldsymbol{c}$ is known as a *quadratic form* in the elements of c. The matrix A is

positive definite if and only if

$$c^T A c > 0$$

and *positive semi-definite* if and only if

$$c^T A c \geq 0$$

For example, let Y_1 and Y_2 be random variables with variances σ_1^2 and σ_2^2 and covariance σ_{12}. The covariance matrix of Y_1 and Y_2 is

$$C = E[(Y - E(Y))(Y - E(Y))^T] = \begin{pmatrix} \sigma_1^2 & \sigma_{12} \\ \sigma_{12} & \sigma_2^2 \end{pmatrix}$$

where $Y^T = (Y_1 Y_2)$. Show that C is positive semi-definite. Also show that

$$\text{var}(c^T Y) = c^T C c$$

This is a generalization of the result for a single random variable. It is true for any number of random variables Y_i in an $n \times 1$ matrix Y. The essential steps in the proof are demonstrated by the 2×1 case considered above.

(iii) If an $n \times n$ matrix A is positive semi-definite and has an inverse, its inverse is also positive semi-definite. To prove this, set $c = A^{-1}v$ in the definition of positive semi-definiteness.

(iv) If Y is an $n \times 1$ matrix of random variables Y_i and A is a constant $n \times n$ matrix, then

$$E[Y^T A Y] = E[Y]^T A E[Y] + tr(AC)$$

where C is the covariance matrix of Y. This is a generalization of the result

$$E[aY_i^2] = a E[Y_i]^2 + a \,\text{var}[Y_i]$$

for a single random variable. Verify the matrix result for $n = 2$; the general proof follows a similar argument.

4 Calculate the correlation between x_i and x_i^2 when x_i are:

$$8 \quad 9 \quad 10 \quad 11 \quad 12$$

Now calculate the correlation between $(x_i - 10)$ and $(x_i - 10)^2$.

5 Consider the model

$$Y_i = \beta_0 + \beta_1 x_i + E_i$$

(i) Write down an expression for $\sum E_i^2$ and hence show that the least squares estimates of β_0 and β_1 are

$$\hat{\beta}_1 = \sum (x_i - \bar{x})(y_i - \bar{y}) / \sum (x_i - \bar{x})^2$$
$$\hat{\beta}_0 = \bar{y} - \hat{\beta}_1 \bar{x}$$

(ii) The fitted line is therefore

$$y = \bar{y} + \hat{\beta}_1 (x - \bar{x})$$

and the residuals are

$$r_i = y_i - \hat{y}_i$$

where $\hat{y}_i = \bar{y} + \hat{\beta}_1(x_i - \bar{x})$. Show that $\sum r_i = 0$ and $\sum r_i(x_i - \bar{x}) = 0$.
Show that $\sum(y_i - \bar{y})^2 = \sum(y_i - \hat{y}_i)^2 + \sum(\hat{y}_i - \bar{y})^2$ i.e. the total (corrected) sum of squares equals the residual sum of squares and a regression sum of squares.

(iii) Show that the formulae in (i) are a special case of the general matrix solution.

6 The following data are the sizes (x_i, thousand properties) and costs of meeting European Community requirements (y_i, coded monetary units) for eight water supply zones.

Size	1.0	2.3	4.5	5.1	6.7	6.8	7.2	9.3
Estimated cost	11	4	41	36	45	87	80	81

Find the least squares estimator of β for a line constrained to pass through the origin:

$$Y_i = \beta x_i + E_i$$

Plot the data and the fitted line. Show that the estimator for β can be obtained as a special case of the general matrix solution.

7 Consider the model

$$Y_i = \beta_0 + \beta_1 x_i + E_i$$

(a) Suppose that the E_i are independent $N(0, \sigma^2)$. Show that the least squares estimators of β_0 and β_1 are the same as the maximum likelihood estimators.

(b) Suppose the E_i have a Laplace distribution with parameter θ. That is, the Laplace distribution, with mean 0, has a pdf

$$f(w) = \frac{1}{2\theta} \exp\left(\frac{-|w|}{\theta}\right)$$

(i) Show that the variance of E_i is $2\theta^2$.
(ii) Show that the ML estimators of β_0 and β_1 are found by minimizing the sum of absolute deviations.

8 The data in Table 5.9 are the costs (in monetary units adjusted to current prices) of schemes to prevent flooding during extreme storms. The number of properties that are affected is also given. More schemes are planned, but the average number of properties protected by each of the prospective schemes is less. A fictitious distribution is given below.

Number of properties at risk	Number of schemes planned
1	28
2	5
3	4
5	2
6	1
8	1

Write a short report giving your estimate of the total cost (current prices) of these schemes and associated 90% limits of prediction (making any approximations you need).

Table 5.9 *Costs of schemes to prevent flooding (monetary units adjusted to current prices). Data provided by Northumbrian Water Ltd.*

Cost of scheme	Number of properties affected
56 770	10
180 769	9
2 133	7
15 267	6
70 644	5
20 708	5
252 351	5
4 300	4
61 234	4
14 609	4
179 567	4
135 796	3
12 060	3
16 501	3
12 101	3
4 291	3
30 275	3
3 031	2
29 746	2
5 420	2
677	1
6 327	1
3 308	1
16 278	1
30 862	1
15 911	1
13 276	1
8 807	1
7 861	1
4 008	1
2 547	1

9 Compare the model fitted in Example 5.5 with a full quadratic model (5 predictors) and a model which includes a quadratic term for baseflow (3 predictors). Can you think of a physical explanation for the signs of the baseflow coefficients in the latter model, for which s equals 0.7338?

10 Let X represent a Poisson variable with mean μ. Use the linear Taylor series approximation to show that:
(i) $\text{var}(\sqrt{X}) \simeq 1/4$
(ii) $\text{var}(\ln X) \simeq 1/\mu$.

11 Suppose $Y_1, \ldots, Y_n$ are independent Poisson random variables. The maximal model estimates n different means for the Poisson distributions by $\hat{\mu}_i = y_i$. Now consider the model which assumes all n distributions have the same mean μ.

(i)   Show that the MLE of μ is $\bar{y}$.
(ii) Show that the deviance

$$D = 2 \sum y_i \ln(y_i/\bar{y})$$

(iii) Show that the limit as $\varepsilon \rightarrow 0$ of

$$\varepsilon \ln \varepsilon \quad \text{is} \quad 0$$

[Hint $\varepsilon^0 = \varepsilon^1 \varepsilon^{-1} = 1$]

12   W_1 and W_2 are independent χ^2 variables with v_2 and v_2 degrees of freedom $(v_1 > v_2)$. Explain why $W_2 - W_2$ cannot have a χ^2 distribution with $v_1 - v_2$ degrees of freedom.

13   Lagrange's method of multipliers
A rectangle has a length x, a width y, and a fixed perimeter p. That is

$$2x + 2y = p$$

Find x and y, in terms of p, which maximize the area, $A = xy$, by the following methods.
(i)   Substitute for y, and solve $dA/dx = 0$.
(ii) Form the function

$$u = xy + \lambda(2x + 2y - p)$$

where λ is the Lagrange multiplier. Now solve the system of equations

$$\frac{\partial u}{\partial x} = 0; \qquad \frac{\partial u}{\partial y} = 0; \qquad 2x + 2y - p = 0$$

Notice that the constraint imposed by the third equation implies that $u = A$ for any value of λ. What is the value of λ at the maximum? Although it is of no practical relevance in this example, the values of the Lagrange multipliers can be useful in other contexts.

14   Suppose $X_1, \ldots, X_n$ are unbiased, and independent, estimates of some parameter θ with variances $\sigma_1^2, \ldots, \sigma_n^2$. Show that the unbiased linear estimator of θ

$$L = a_1 X_1 + \cdots + a_n X_n, \qquad \text{where } \sum a_i = 1$$

has a minimum variance if each a_i is inversely proportional to σ_i^2.

6

Bayesian Methods

6.1 An introduction to Bayesian methods

Imagine we have been asked to advise a city council whether mini-roundabouts are safer than priority controlled (no traffic lights) crossroads. Ibrahim and Metcalfe (1993) described a Bayesian overview of results from six study areas. The number of sites in these areas ranged from 11 to 78. The average reduction in accidents was estimated to be 23%. However, there was evidence that the benefit varied between areas, and the standard deviation of reductions was estimated to be 9%. On the basis of this information, and any more recent reports, we advise the council to install mini-roundabouts in a pilot area, with a review after one year. At this stage we might formalize our assessment of the expected benefit. Suppose the underlying accident rate is reduced from μ/year in the period before the mini-roundabouts to $\theta\mu$/year during the trial period. Our assessment of the likely values for θ can be summarized by some probability distribution, known as the *prior distribution*, with pdf $f(\theta)$. In this case we might choose

$$\theta \sim N(0.8, (0.1)^2)$$

The normal distribution is plausible and convenient. The slight reduction in the expected decrease, from 23% to 20%, and the increase in the standard deviation allows for some scepticism about the results of overviews. For example, studies which demonstrate a significant effect are more likely to have been published.

At the end of the trial period there have been x accidents. We will now revise our assessment of the distribution of θ, using Bayes' theorem. The *posterior distribution* of θ is given by:

$$f(\theta \mid x) = f(\theta, x)/f(x)$$
$$\propto f(x \mid \theta)f(\theta) \tag{6.1}$$

The conditional distribution of x given θ is just the likelihood function for θ, from a different perspective, so the posterior distribution is proportional to the

product of the likelihood and the prior distribution. The constant of proportionality is found as the reciprocal of the integral of the right-hand side. We now
assume that accidents occur as a Poisson process, and that the numbers of
accidents at different sites are independent. Then the total number of accidents
in the pilot area has a Poisson distribution. If we assume the mean before the
trial starts (μ/year) is known, the posterior distribution of θ is proportional to

$$f(\theta \mid x) \propto (e^{-\theta\mu}(\theta\mu)^x/x!)(0.1\sqrt{2\pi})^{-1}e^{-[(\theta-0.8/0.1]^2/2}$$

$$\propto \theta^x \exp(-[(\theta - 0.8)/0.1]^2 - \theta\mu)$$

A more realistic approach is to model our uncertainty about μ by some
distribution. Since μ must be positive, a gamma distribution

$$f(\mu) = c^a\mu^{a-1}e^{-c\mu}/\Gamma(\alpha) \qquad \text{for} \quad 0 < \mu$$

is a reasonable choice. The mean and variance of the distribution of μ are α/c
and α/c^2 respectively, so our knowledge about μ is equivalent to observing α
accidents over c years. We should also allow for any trend, obtained from
national figures, when specifying the distribution of μ. If we assume θ is
independent of μ, equation (6.1) now becomes

$$f(\theta \mid x, \mu) \propto f(x \mid \theta, \mu)f(\theta)$$

In general $f(\phi)$ is related to $f(\phi \mid \mu)$ by:

$$f(\phi) = \int f(\phi, \mu)\,d\mu = \int f(\phi \mid \mu)f(\mu)\,d\mu$$

Applying this result to $f(\theta \mid x, \mu)$ gives

$$f(\theta \mid x) \propto \int f(x \mid \theta, \mu)f(\theta)f(\mu)\,d\mu \qquad (6.2)$$

Example 6.1
The average number of accidents per year at sites in the pilot area, during the 7
years before installing the roundabouts, has been 18.3. During this period the
number of accidents in the UK has decreased by 2% per year. The pilot area
was chosen because it had a relatively poor record for safety at junctions. There
have been 10 accidents in the year since the mini-roundabouts have been
installed. We have been asked to give an updated estimate of θ.

The average of 18.3 accidents per year during the 7 years before installation
is centred $3\frac{1}{2}$ years ago. If mini-roundabouts have no effect we would expect
18.3 to be reduced by $(0.98)^4$ to 16.9 because of the annual decrease. There is
also a possible bias by selection effect. If the pilot area was selected from a
population of physically identical areas, because it happened to have the worst
accident record, a decrease would be expected without applying any treatment.
Wright *et al.* (1988) estimated an expected reduction of 8% for a single site
chosen from a group of junctions in Hertfordshire (UK) because of a bad
record over a 5 year period. However, our pilot area consists of several sites
and this will reduce the bias by selection effect, as will the physical variability
of notionally similar areas.

It seems reasonable to think that the bias by selection effect, β say, has a
uniform distribution over [0.92, 1.00]. Let η be our estimate of the mean in the

after period if mini-roundabouts have no effect and there is no bias by selection. Then η has a gamma distribution with mean 16.9 and standard deviation $\sqrt{(16.9/7)} = (2.42)^{1/2}$, and β has a mean of 0.96 and a standard deviation of $(0.08)/\sqrt{12} = 0.023$. Now

$$\mu = \eta\beta$$

We assume η and β are independent, and then, using the results of Exercise 3

$$\text{mean}(\mu) = \text{mean}(\eta) \times \text{mean}(\beta) = 16.2$$

$$\text{var}(\mu) = (\text{mean } \eta)^2 \, \text{var}(\beta) + (\text{mean } \beta)^2 \, \text{var}(\eta) + \text{var}(\eta)\text{var}(\beta)$$

$$= (16.9)^2(0.023)^2 + (0.96)^2(2.42) + (0.023)(2.42) = 2.53$$

So we will model our belief about the distribution of μ as gamma with a mean of 16.2 and a variance of 2.53. The product of a gamma variable with a uniform variable will not have an exact gamma distribution, but it will be close and the choice of a gamma distribution for η was fairly arbitrary anyway. We already have our prior distribution for θ and the conditional distribution $f(x \mid \theta, \mu)$, which is Poisson with mean $\theta\mu$, so we can now apply equation (6.2) to calculate the posterior distribution of θ. Simpson's rule was used twice; for the integration in equation (6.2) and then to find the scaling factor so that $f(\theta \mid x)$ has an area of 1. Figure 6.1 is a plot of $f(\theta \mid x)$ and the mean of this distribution is approximately 0.76.

This example has demonstrated the distinctive features of a Bayesian analysis. Unknown parameters have probability distributions which represent

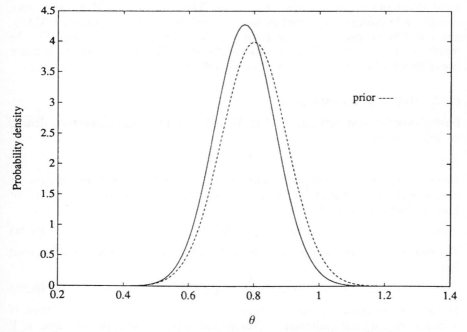

Figure 6.1 Prior and posterior distributions of multiplicative effect of a mini-roundabout

our assessment of their likely values. Our assessments can incorporate subjective judgement, and are often referred to as *beliefs* in this context. We revise our assessments, when we obtain more information, by applying Bayes' theorem. Lee (1989) gives a clear and detailed introduction to Bayesian statistics.

6.2   Asset Management Plans and the Bayes Linear Estimator

6.2.1   Background

Water companies in the UK are required to maintain and periodically revise their asset management plans (AMPs), as a condition of their licences. An AMP embodies company strategy for a 20 year period and estimates the capital investment required to improve, maintain and, where necessary, extend the company's assets. There are many uncertainties associated with such a complex task, for example: the current condition and serviceability of all assets; the scope of work needed to achieve standards; variation in costs for the same nominal work; and trends in demand.

If the AMPs are to be effective they must give a fair reflection of the cost of the work needed, and an indication of the accuracy of the estimate (Metcalfe, 1991, 1994). A single figure for the estimated future cost could be seriously misleading, and contribute to poor decisions. For the purpose of producing an AMP, a company will usually be divided into discrete assets such as trunk mains, treatment works and water towers, and approximately self-contained local distribution areas known as *zones*. The company will already have considerable information about its assets, partly as a result of previous AMPs, and this can be used to set up a prior model. The prior model can then be regularly updated as new information is obtained. The Bayes' linear estimator is a means for doing this (O'Hagan *et al.*, 1992).

6.2.2   Bayes' linear estimator

Basic Bayes' linear estimator   Let X and Y be two random vectors; denote the prior means by

$$m_x = \mathrm{E}[X] \qquad m_y = \mathrm{E}[Y]$$

the prior variance–covariance matrices by V_x and V_y; and the matrix of covariances between X and Y by C. Then the Bayes' linear estimator (BLE) of Y after observing $X = x$ is

$$\hat{Y}(x) = m_y + CV_x^{-1}(x - m_x) \tag{6.3a}$$

The dispersion matrix is the expected value of the squared error, and is given by:

$$D_y(x) = V_y - CV_x^{-1}C^{\mathrm{T}} \tag{6.3b}$$

The BLE is the linear estimator which minimizes the expected squared error. If X and Y have a joint multivariate distribution: $\hat{Y}(x)$ is the expected value of Y given x; the dispersion matrix is the conditional variance of Y; and the BLE is the regression of Y on x, as defined in the population (Appendix 1.10). The derivation of the BLE (see Appendix 10) makes no assumptions about the distribution of X and Y, but if it is reasonably close to normal we can interpret $\hat{Y}(x)$ and $D_y(x)$ as the posterior mean and variance of Y. The BLE can be compared with making predictions from an empirical regression (Section 5.1). In the latter case the CV_x^{-1} of the BLE has been replaced by estimates, $(X^TX)^{-1}X^TY$, from the n data used to fit the regression model. In practice we still have to estimate V_x, C and V_y to use the BLE, but do so in a less formal manner. We can combine information from a variety of sources with engineering judgement. A more sophisticated formulation would allow for uncertainty about these variance–covariance matrices.

In the AMP context, X are the costs for zones for which we have new data. Although we do not have any new data, explicitly, about the costs in zones Y, we know they have similarities with zones X and use the BLE to improve our estimates. However, the basic BLE assumes that we know the values x taken by X and this is not necessarily the case. In general we have an improved, but still imprecise estimate. O'Hagan *et al.* (1992) emphasize that this should be thought of as a posterior estimate of X rather than an observation of X with error and suggest the following modification.

Modified Bayes' linear estimator   Let X be the vector of costs for zones in which further studies have been made. The prior mean and variance–covariance matrix of Y are replaced by posterior values $\mathbf{m}_y^*$ and V_y^* according to:

$$m_y^* = m_y + CV_x^{-1}(m_x^* - m_x) \tag{6.4a}$$

$$V_y^* = V_y - CV_x^{-1}C^T + CV_x^{-1}V_x^*V_x^{-1}C^T \tag{6.4b}$$

where m_x and V_x^* are the posterior mean and variance of X (see Exercise 5). If we have precise estimates, $V_x^* = 0$ and these formulae revert to the basic BLE. In contrast to this, if the posterior information about X is the same as the prior, there is no change in our beliefs about Y. The updating equations 6.2(a) and (b) would not be appropriate if we observed x plus independent error. Then V_x would be increased by the error variance and the new observations would have less effect on $\hat{Y}$. The next example demonstrates the application of the modified BLE to a simplified part of an AMP.

Example 6.2
We have been asked to estimate the cost of mains refurbishment in an area of the local distribution network. Preparatory work has led to the following assumptions.

(i)   The costs are the products of lengths of main and unit cost of refurbishment.
(ii)   The region can be split into five self-contained zones.
(iii) The zones can be split into two strata, rural (zones 1, 2 and 3) and urban (zones 4 and 5).
(iv)   The equations of the cost model are:

$$\text{total cost} = \Sigma \text{ zone cost}$$
$$\text{zone length} = \text{length} \times \text{unit cost}$$

168   *Bayesian methods*

(v)   The prior distribution of lengths and costs is shown in Table 6.1.
(vi)   The errors in lengths are negligible. The unit costs are modelled as

$$C = \mu + \nu_i + \varepsilon_{ij}$$

where μ is an overall average, ν_i is an adjustment for stratum i and ε_{ij} is specific to a zone within stratum i. The standard deviations of the three components of prior uncertainty are given in percentage terms in Table 6.1. They have been estimated from components of variance analysis (Section 2.2) of ratios of scheme out-turn costs to costs predicted using AMP methods. These costings were made retrospectively by engineers who took no part in the schemes (Metcalfe, 1994). The finite population correction is ignored. For example, we do not impose a restriction that the sum of the stratum adjustments equals zero. In a real AMP there will be many strata and zones so this simplification has a negligible effect, particularly when compared with the rather rough nature of the estimates of standard deviations. The estimates of the standard deviations after a detailed study are also given in Table 6.1. They must be smaller than the estimates of prior uncertainty because they include all prior information.

Table 6.1   *(a) Prior data (lengths and unit costs); (b) variance data; (c) update information (for lengths and unit costs)*

a

Zone	Stratum	Length of cleaning (km)	Unit cost (£k/km)	Total cost (£k)
Zone 1	rural	1	1	1
Zone 2	rural	2	1	2
Zone 3	rural	3	1	3
Zone 4	urban	2	2	4
Zone 3	urban	3	2	6

b

Level of uncertainty	Base quantity	Percentage standard deviation		
		Overall %	Stratum %	Individual %
Prior	Length	0	0	0
	unit cost	±30	±80	±50
Detail	Length	0	0	0
	unit cost	±1	±1	±10

c

		Length (km)	Unit cost (£k/km)	Total cost (£k)
Zone 2	rural	2	2	4
Zone 5	urban	3	5	15

It remains to calculate variances and covariances of zone costs. The total error (E) can be split into an error in the overall mean E_0, an error in the stratum adjustment E_s and an individual error (E_I). We assume error components are independent. Then

$$\text{var}(E) = \text{var}(E_0) + \text{var}(E_s) + \text{var}(E_I)$$

Since only the error in unit costs is being modelled, zone cost errors will be the same in relative terms, i.e. let C_i be the cost for zone i

$$\text{var}(C_i) = \text{var}(E) = (0.3)^2 + (0.8)^2 + (0.5)^2 = 0.98$$

Hence the prior standard deviation of a zone is

$$\sigma_E = \sqrt{0.98} = 0.99$$

Now consider another zone in the same stratum. The covariance of costs, C_i and C_j, will be known

$$\text{cov}(C_i, C_j) = \text{E}[(E_0 + E_s + E_{Ii})(E_0 + E_s + E_{Ij})]$$

where E_{Ii} and E_{Ij} are the individual errors. Since

$$\text{E}(E_{Ii}E_{Ij}) = 0$$

the covariance is

$$= \text{E}[E_0^2 + E_s^2] = (0.3)^2 + (0.8)^2 = 0.09 + 0.64 = 0.73$$

The correlation between zones in the same stratum is

$$0.73/(0.99)^2 = 0.74$$

A similar argument can be applied to a zone in a different stratum. The covariance of costs, C_i and C_k, will be

$$\text{cov}(C_i, C_k) = \text{E}[(E_0 + E_{si} + E_{Ii})(E_0 + E_{sk} + E_{Ik})]$$

which, assuming independence of the stratum errors

$$\text{cov}(C_i, C_k) = \text{E}[E_0^2] = (0.3)^2 = =0.09$$

Turning to the update information, the total error variance of a zone is

$$(0.01)^2 + (0.01)^2 + (0.1)^2 = 0.0102$$

and the covariance between zone costs in different strata is 0.0001. As there is only one update zone in each stratum, we do not need to calculate the covariances within strata. We now have enough information to use the modified BLE equations (6.4) to calculate posterior estimates for zones for which update information is not available. We first construct the individual matrix elements.

Prior costs		Prior costs		Update costs	
$m_y = \begin{bmatrix} 1 \\ 3 \\ 4 \end{bmatrix}$	zone 1 zone 3 zone 4	$m_x = \begin{bmatrix} 2 \\ 6 \end{bmatrix}$	zone 2 zone 5	$m_x^* = \begin{bmatrix} 4 \\ 15 \end{bmatrix}$	zone 2 zone 5

The variances and covariances, so far, are in terms of $(\%/100)^2$, and to obtain the variance–covariance matrix they must be multiplied by the estimated mean costs.

The prior covariance between zones where update information will not become available is given by:

$$V_y = \begin{bmatrix} \text{var(zone1)} & \text{cov(zone1, zone3)} & \text{cov(zone1, zone4)} \\ \text{cov(zone3, zone1)} & \text{var(zone3)} & \text{cov(zone3, zone4)} \\ \text{cov(zone4, zone1)} & \text{cov(zone4, zone3)} & \text{var(zone4)} \end{bmatrix}$$

now substituting the numbers, i.e. multiplying the covariance (which is in $(\%)^2$ terms) by the product of the mean of the two zones, gives

$$= \begin{bmatrix} 0.98(1 \times 1) & 0.73(1 \times 3) & 0.09(1 \times 4) \\ 0.73(3 \times 1) & 0.98(3 \times 3) & 0.09(3 \times 4) \\ 0.09(4 \times 1) & 0.09(4 \times 3) & 0.98(4 \times 4) \end{bmatrix} = \begin{bmatrix} 0.98 & 2.19 & 0.36 \\ 2.19 & 8.82 & 1.08 \\ 0.36 & 1.08 & 15.68 \end{bmatrix}$$

The prior covariances between zones where update information will become available are given by:

$$V_x = \begin{bmatrix} \text{var(zone2)} & \text{cov(zone2, zone5)} \\ \text{cov(zone5, zone2)} & \text{var(zone5)} \end{bmatrix}$$

$$= \begin{bmatrix} 0.98(2 \times 2) & 0.09(2 \times 6) \\ 0.09(6 \times 2) & 0.98(6 \times 6) \end{bmatrix} = \begin{bmatrix} 3.92 & 1.08 \\ 1.08 & 35.28 \end{bmatrix}$$

The update covariances between zones where update information will become available are given by:

$$V_x^* = \begin{bmatrix} 0.0102(4 \times 4) & 0.0001(4 \times 15) \\ 0.0001(15 \times 4) & 0.0102(15 \times 15) \end{bmatrix} = \begin{bmatrix} 0.1632 & 0.006 \\ 0.006 & 2.295 \end{bmatrix}$$

Notice that the diagonal elements are the variance of cost for zone i, and the off diagonal elements are the covariance between zone costs.

We now need the matrix of prior covariances between the zones where update information will not become available and the zones where update information will become available:

$$C = \begin{bmatrix} \text{cov(zone1, zone2)} & \text{cov(zone1, zone5)} \\ \text{cov(zone3, zone2)} & \text{cov(zone3, zone5)} \\ \text{cov(zone4, zone2)} & \text{cov(zone4, zone5)} \end{bmatrix}$$

substituting the numbers gives

$$C = \begin{bmatrix} 0.73(1 \times 2) & 0.09(1 \times 6) \\ 0.73(3 \times 2) & 0.09(3 \times 6) \\ 0.09(4 \times 2) & 0.73(4 \times 6) \end{bmatrix} = \begin{bmatrix} 1.46 & 0.54 \\ 4.38 & 1.62 \\ 0.72 & 17.52 \end{bmatrix}$$

Substitution into equation (6.4a) gives

$$m_y^* = [1.78 \ \ 5.33 \ \ 8.55]^T$$

and substitution into equation (6.4b) leads to

$$V^* = \begin{bmatrix} 0.4582 & 0.6247 & 0.0321 \\ 0.6247 & 4.1242 & 0.0962 \\ 0.0321 & 0.0962 & 7.5342 \end{bmatrix}$$

The diagonal terms are the posterior variances for zone1, zone3 and zone4. The standard deviations follow by taking square roots.

$$\sigma(\text{zone1}) = \sqrt{0.4582} = 0.677 \text{ or in \% terms } \quad \frac{(0.677 \times 100)}{1.778} = 38.1\%$$

$$\sigma(\text{zone3}) = \sqrt{4.1242} = 2.031 \text{ or in \% terms } \quad \frac{(2.031 \times 100)}{5.334} = 38.1\%$$

$$\sigma(\text{zone4}) = \sqrt{7.5342} = 2.745 \text{ or in \% terms } \quad \frac{(2.745 \times 100)}{8.55} = 32.1\%$$

The results are summarized in Table 6.2.

The BLE solution is always a compromise between the prior information and the new data. We can see the following.

i.   For Rural zones the update represents a two-fold increase, and we know that the correlation between zones in the same stratum is 0.74, so we would expect the posterior estimates for costs in unsampled zones to go up by a factor rather less than 2. The BLE prediction, for other zones in this stratum, is that prior costs will rise by a factor of 1.78.

ii.   For Urban zones the update represents a 2.5 fold increase, and correlation between zones in the same stratum is again 0.74, so again, we would expect the posterior estimates for unsampled zones to go up by a factor rather less than 2.5. The BLE prediction, for other zones in this stratum, is that prior costs will rise by a factor of 2.15.

Table 6.2   *Summary of analysis using the BLE equations (calculated without using the ABLE software)*

Zone	Prior Zone cost £k	Stdev		Posterior Zone cost £k	Stdev	
Zone 1	1	99%	0.99	1.78	38.1%	0.68
Zone 2	2	99%	1.98	4.00	10%	0.40
Zone 3	3	99%	2.97	5.33	38.1%	2.03
Zone 4	4	99%	3.96	8.55	32.1%	2.74
Zone 5	6	99%	5.94	15.00	10%	1.50
Total	16			34.66		

iii.   Correlation between zones in different strata is only 0.09, so we would not expect updates in one stratum to have much effect on estimates in the other stratum.

iv.   The standard deviation for costs in unsampled zones has fallen from the prior level of 99% to 36%, a figure much closer to the update value of 10%.

v.   Overall the estimated cost for mains cleaning is increased by a factor of 2.17 (from £k 16 to £k 34.66).

6.3   Bayesian Overview

6.3.1   Introduction

When making policy decisions, transport engineers use information from various studies of the success of strategies to improve road safety. We will use results from six studies of the effects of installing mini-roundabouts as an example (Ibrahim and Metcalfe, 1993). Let the effect of replacing priority controlled junctions by mini-roundabouts, within some study area j, be to change the accident rate by a factor θ_j. Values of θ_j less than 1 correspond to improvements. Suppose we have initial estimates (y_j) of the θ_j for J study areas, i.e. $1 \leqslant j \leqslant J$, and the standard deviations of these estimates, σ_j. Although we do not have detailed information which might explain differences in the θ_j, we think it is more realistic to allow for the possibility that they vary between study areas than to assume a common value. A *hierarchical model* which represents this is

$$y_j \sim N(\theta_j, \sigma_j^2) \tag{6.5a}$$

$$\theta_j \sim N(\mu, \tau^2) \tag{6.5b}$$

together with prior distributions for the *hyperparameters* μ and τ. The choice of normal distributions is convenient, rather than essential. An analysis of the variance, between and within studies, leads to point estimates of μ and τ. An approximate confidence interval can be constructed for μ. Revised estimates of the distribution of θ_j conditional on μ, y_j and τ can be made:

$$\theta_j \mid \mu, \tau, y_j \sim N(\hat{\theta}_j, V_j) \tag{6.6a}$$

where

$$\hat{\theta}_j = (y_j/\sigma_j^2 + \mu/\tau^2)/(1/\sigma_j^2 + 1/\tau^2) \tag{6.6b}$$

$$V_j = 1/(1/\sigma_j^2 + 1/\tau^2) \tag{6.6c}$$

(see Exercise 1). However, a full Bayesian analysis has some advantages over this simpler approach. One is that it avoids the difficulties which arise if the estimate of τ^2 is negative. In such cases the sample size is too small for τ^2 to be distinguished from zero. The device of setting $\tau^2 = 0$ implies that the θ_j are equal, and ignores the considerable uncertainty about τ. The full Bayesian analysis also allows for the uncertainty in μ and τ when estimating the distribution of θ_j. This refinement would seem more important if the overview was of studies in different countries, when the θ_j for the home country is directly relevant.

6.3.2 Full Bayesian analysis

The joint posterior distribution Knowledge of μ and τ, as well as θ which is $(\theta_1, \ldots, \theta_k)$, tells us no more about the distribution of y which is $(y_1, \ldots, y_k)$. Therefore

$$f(y \mid \theta, \mu, \tau) = f(y \mid \theta)$$

Therefore the joint posterior distribution for all the parameters, given the data, is:

$$f(\theta, \mu, \tau \mid y) \propto f(\mu, \tau) f(\theta \mid \mu, \tau) f(y \mid \theta) \tag{6.7}$$

All three factors on the right-hand side of equation (6.7) are known, i.e.

$$f(\theta \mid \mu, \tau) = \Pi N(\theta_j \mid \mu, \tau^2)$$

$$f(y \mid \theta) = \Pi N(y_j \mid \theta_j, \sigma_j^2)$$

where $N(\theta_j \mid \mu, \tau^2)$ is a convenient shorthand for the normal pdf of θ_j with mean μ and variance τ^2, and so on, and $f(\mu, \tau)$ is whatever prior distribution we choose to represent our prior knowledge of μ and τ. In principle, this joint posterior distribution is the answer to all our questions. For example, the marginal posterior distribution of θ_j can be found by integrating over the other θ_k, μ and τ. However, as is often the case in Bayesian analysis, direct integration of theoretical solutions may not be practical. Other methods for finding the required distributions, often involving simulation, are a key feature of modern Bayesian statistics. Gelman *et al.* (1995) advise using the following alternative factorization of the joint posterior distribution

$$f(\theta, \mu, \tau \mid y) \propto f(\tau \mid y) f(\mu \mid \tau, y) f(\theta \mid \mu, \tau, y) \tag{6.8}$$

to find the posterior distribution of θ via simulation. The final step is to draw random numbers from $f(\theta \mid \mu, \tau, y)$, which has the normal distribution given in equations (6.6a–c). The posterior conditional distribution of μ, given τ and y, follows from a similar argument. Suppose $\tilde{\mu}$ is a prior estimate of μ with precision ϕ.

$$\mu \mid \tau, y \sim N(\hat{\mu}, V_\mu) \tag{6.9a}$$

where

$$\hat{\mu} = [\Sigma \, y_j / (\sigma_j^2 + \tau^2) + \phi \tilde{\mu}] / [\Sigma \, (1 / \sigma_j^2 + \tau^2) + \phi] \tag{6.9b}$$

and

$$V_\mu^{-1} = \Sigma \, 1 / (\sigma_j^2 + \tau^2) + \phi \tag{6.9c}$$

The posterior distribution of τ follows from a more sophisticated argument (Gelman *et al.*, 1995) which is now summarized. First notice that

$$y_j \mid \mu, \tau \sim N(\mu, \sigma_j^2 + \tau^2)$$

as a consequence of equations (6.5a and b). Also, from the definitions of conditional distributions

$$f(\tau \mid y) = f(\mu, \tau \mid y) / f(\mu \mid \tau, y) \propto f(\mu, \tau) f(y \mid \mu, \tau) / f(\mu \mid \tau, y)$$

Table 6.3 The total number of sites in the group, average lengths of before and after periods (years) and the total numbers of accidents reported in six studies

Study	Sites	Before period		After period	
		Length	Accidents	Length	Accidents
TRRL Leaflet 393	78	3	171	3	117
Lalani	20	1.6	99	1.6	69
Green	54	3.4	405	2.5	192
Hanson and Wilson	11	3.7	49	3.3	63
Avon county Council					
(i) first study	12	1.7	41	1.7	32
(ii) second study	23	2.5	62	2.5	41

The left-hand side does not include μ, so the right-hand side should hold for any value of μ, and in particular for $\hat{\mu}$. Hence

$$f(\tau \mid y) \propto f(\tau) V_\mu^{1/2} \Pi (\sigma_j^2 + \tau^2)^{-1/2} \exp(-(y_j - \hat{\mu})^2 / (2(\sigma_j^2 + \tau^2)))$$

A uniform distribution with some best guess upper bound is a convenient choice for the prior distribution of τ, but the inverse chi-squared family of distributions (χ^{-2}) is often used for variance priors (see Exercise 6.2). We can simulate from $f(\tau \mid y)$ by calculating the numerical cdf and then using the inverse-cdf method.

The reported results from the six studies are given in Table 6.3. We made the following initial estimates, y_j, of the factors, θ_j, following the principles discussed in Example 6.1. The estimates of the variances (σ_j^2) of the y_j are also given.

Study	y_j	(σ_j^2)	$E[\theta_j/y]$ and (standard deviation)^2	
TRRL	0.74	$(0.09)^2$	0.75	$(0.08)^2$
Lalani	0.74	$(0.11)^2$	0.75	$(0.09)^2$
Green	0.68	$(0.18)^2$	0.70	$(0.06)^2$
Hanson and Wilson	1.52	$(0.26)^2$	0.93	$(0.16)^2$
Avon County Council I	0.82	$(0.18)^2$	0.79	$(0.12)^2$
Avon County Council II	0.69	$(0.13)^2$	0.72	$(0.10)^2$

Our prior distribution for μ was uniform over [0.5, 1.1], so μ is 0.8 and θ is 33.3 and the prior distribution for τ was uniform over [0, 0.4]. Using the method of this section the updated estimate of the mean μ, of the distribution of θ_j is 0.77, and the standard deviation of the distribution of μ is 0.08. Revised estimates of the factors θ_j are shown above. The assumption that the y_j are normally distributed is a rather rough approximation.

6.4 Dynamic Linear Models

6.4.1 Definition of the dynamic linear model

The dynamic linear model (DLM) has many engineering applications, including Kalman filtering. The book by Pole *et al.* (1994) is an excellent introduction, and includes the Bayesian Analysis of Time Series (BATS) software. An example concerns a construction company which has a house building division. The sales manager uses the following model to allow for the influence of a general level of sales in the sector and the company's own pricing policy on its sales

$$\text{sales}_t = \text{level}_t + \beta_t \, \text{price}_t + E_t$$

$$\text{level}_t = \text{level}_{t-1} + \Delta \, \text{level}_t$$

$$\beta_t = \beta_{t-1} + \Delta \beta_t$$

The first equation is a linear regression with one predictor variable, and the modification that the slope and intercept can change over time. The E_t, Δ level$_t$, $\Delta \beta_t$ are random error terms which are assumed to be independent over time. The E_t are assumed to be independent of the changes to the level and the slope β. This is an example of the known variance model described by Pole *et al.* (1994):

observation equation: $Y_t = F_t^{\mathrm{T}} \theta_t + v_t$
system equation: $\theta_t = G \theta_{t-1} + w_t$

error distributions: $v_t \sim N(0, V_t)$; $w_t \sim N(0, W_t)$

In these equations: Y_t is the observed dependent variable at time t; F_t is a vector of values of the predictor variables at time t; and θ_t is a vector of unknown time varying parameters, often known as system states, at time t. The errors, v_t and w_t, are assumed not to be correlated over time, and v_t is independent of w_t. The usual development assumes the errors are normally distributed. In the following, D_t represents knowledge at time t. The prior information is:

$$\theta_{t+1} \,|\, D_t \sim N(a_{t+1}, \ R_{t+1})$$

In the example, Y_t is sales, F_t is the price, and θ_t includes the slope and intercept.
   Let the distribution of θ at time $t-1$, based on all our information at that time, be

$$\theta_{t-1} \,|\, D_{t-1} \sim N(m_{t-1}, \ C_{t-1})$$

Then our prior distribution for θ at time t is

$$\theta_t \,|\, D_{t-1} \sim N(a_t, R_t)$$

where $a_t = G m_{t-1}$ and $R_t = G C_{t-1} G^{\mathrm{T}} + W_t$.
   We will derive the posterior distribution for θ_t given y_t, after first finding the one step ahead forecast distribution for Y_t. Since the posterior distribution for θ_t given D_{t-1} and y_t is the updated distribution for θ_t given our information at time t, D_t is the combination of D_{t-1} and y_t.

One step ahead forecast

$$E[Y_t \,|\, D_{t-1}] = E[F_t^{\mathrm{T}} \theta_t + v_t \,|\, D_{t-1}]$$
$$= F_t^{\mathrm{T}} E[\theta_t \,|\, D_{t-1}]$$
$$= F_t^{\mathrm{T}} a_t$$
$$\text{var}[Y_t \,|\, D_{t-1}] = \text{var}[F_t^{\mathrm{T}} \theta_t + v_t \,|\, D_{t-1}]$$
$$= F_t^{\mathrm{T}} R_t F_t + V_t$$

Posterior distribution $\theta_t | D_t$ is equivalent to $\theta_t | D_{t-1}$, y_t, and using Bayes' theorem

$$f(\theta_t | D_{t-1}, y_t) \propto f(y_t | \theta_t) f(\theta_t | D_{t-1})$$

$$\propto \exp[-\tfrac{1}{2}(y_t - F_t^T \theta_t)^2 V_t^{-1}] \times \exp[-\tfrac{1}{2}(\theta_t - a_t)^T R_t^{-1}(\theta_t - a_t)]$$

$$\propto \exp[-\tfrac{1}{2}(\theta_t - m_t)^T C_t^{-1}(\theta_t - m_t)]$$

where m_t and C_t are given by the following algorithm, in which e_t is the error in the one step ahead forecast.

$$e_t = y_t - F_t^T a_t$$
$$Q_t = F_t^T R_t F_t + V_t$$
$$A_t = R_t F_t / Q_t$$
$$m_t = a_t + A_t e_t$$
$$C_t = R_t - A_t A_t^T Q_t$$

Intervention   Our prior distribution for θ at time t would be

$$\theta_t | D_{t-1} \sim N(a_t, R_t)$$

but we now have some external information I_t about θ_t such as a change in taxation law relating to house purchase, or a new competitor in the market. We therefore adjust our prior distribution to

$$\theta_t | D_{t-1}, I_t \sim N(a_t^*, R_t^*)$$

Then a_t^* and R_t^* replace a_t and R_t in the algorithm. We should also increase G for at least one step

$$G^* = (Z_t^{-1} U_t)^T G$$

where Z_t and U_t are obtained from R_t and R_t^* through the relationships

$$R_t^* = U_t^T U_t, \qquad R_t = Z_t^T Z_t$$

6.5   Summary

Bayesian analysis enables us to combine expert judgement with data in a systematic manner. This is an important advantage in many applications. Three examples are: overview studies, also known as *meta-analysis*, of road safety measures; Bayes' linear estimators which can be used to formulate and maintain asset management plans in water supply and other businesses with a distributed network ,e.g. electricity, gas and railways; and the dynamic linear model for sales forecasting when we wish to include personal knowledge of the market. The most general form of the dynamic linear model includes the ARMA models, and regression methods for allowing for any trend and seasonal variation, given in Chapter 7 (Pole *et al.*, 1994).

Exercises

1   Let X and Y be independent unbiased estimators of a parameter θ, with

variances σ_X^2 and σ_Y^2 respectively. Define

$$W = aX + (1 - a)Y$$

and use the result for the variance of a linear combination of random variables (Appendix 1), to show that W has a minimum variance when

$$a = (1/\sigma_X^2)/(1/\sigma_X^2 + 1/\sigma_Y^2)$$

Show that this minimum variance is

$$1/(1/\sigma_X^2 + 1/\sigma_Y^2)$$

In Bayesian statistics the reciprocal of the variance is referred to as the precision. Hence deduce that the precision of W equals the sum of the precision of X and the precision of Y.

2   The variable X has an inverse chi-squared distribution with ν degrees of freedom, denoted

$$X \sim \chi_\nu^{-2}$$

if $1/X \sim \chi_\nu^2$. Show that the pdf is

$$f(x) = x^{-\nu/2-1}\exp(-1/(2x))/(2^{\nu/2}\Gamma(\nu/2)) \qquad \text{for} \quad 0 < x$$

Also show that if $Y = aX$, and $\nu > 4$, then

$$\mu_Y = a/(\nu - 2)$$
$$\sigma_Y^2 = 2a^2/((\nu - 2)^2(\nu - 4))$$
$$\text{mode } (Y) = a/(\nu + 2)$$

3   Recall that $E[X^2] = \mu_X^2 + \sigma_X^2$, and hence show that if X and Y are independent then

$$\text{var}(XY) = \mu_Y^2\sigma_X^2 + \mu_X^2\sigma_Y^2 + \sigma_X^2\sigma_Y^2$$

Derive a similar expression for the product of three independent random variables W, X and Y, by applying the above result to $W(XY)$.

4   Suppose your prior distribution for the probability of success (p) in a binomial distribution is beta with parameters α and β (Chapter 2, Exercise 5). Show that if you observe x successes in n trials your posterior distribution for p is beta with parameters $\alpha + x$ and $\beta + n - x$. Beta distributions are said to be conjagate to the binomial likelihood.

5   Refer to equations 6.4(a) and (b). Suppose x and y are scalars, and take variance of both sides of 6.4(a):

$$V_y^* = Vy + C^2V_x^{-2}(V_x^* + V_x) - 2CV_x^{-1}C$$

Explain this step, extend the argument to matrices, and hence justify equation 6.4(b).

7

Modelling Variability in Time and Space

7.1 Introduction

Many civil engineering applications are dynamic. We perceive time as part of a continuum, but the variables we need to monitor can be discrete, e.g. the number of lorries queueing at a quarry, or continuous, e.g. wind speed. Moreover, we can approximate a continuous time signal by sampling. Providing the sampling frequency is high enough, nothing is lost and the digital signal is available for computer analysis. We can also digitize the variables which we observe over time, known as the *state variables*. For example, it is often convenient to model the capacity of a dam as a finite number of units. If we suppose the state variables vary randomly we have a *random* or *stochastic process*. There are a multitude of models for stochastic processes, and their theory and applications form an active research area. It may be helpful to split them into four categories, depending on whether the state and time are modelled as discrete or continuous, as in Table 7.1.

7.2 Markov chains

7.2.1 Regular chains

Rainfall-runoff models are used for water resource planning and flood predictions. Daily rainfall can be modelled in two stages. First the sequence of wet and dry days, and then the amount of rain on the wet days. A *Markov chain* may be suitable for modelling the occurrences of wet days. The Markov assumption[1] is that the state (wet or dry) on day t + 1 depends only on the state on day t and the probabilities of moving between states. Let

$$p^{(t)} = (p_W^{(t)} p_D^{(t)})$$

[1] Called after the Russian mathematician Andrei Andreevich Markov (1856–1922) who made a systematic study of such sequences.

Table 7.1   *Classification and examples of models for random processes*

Variable	Time Discrete	Continuous
Discrete	A tropical country has a wet and a dry season each year. The volume of water in a reservoir at the end of each dry season is modelled on a scale from 0 to N in increments of 1. This gives $N+1$ levels, ranging from empty to full. An $(N+1)$ by $(N+1)$ matrix of transition probabilities is postulated which takes into account water inputs, water use and evaporation. (This is an example of a *Markov chain*.)	A model for the number of accidents occurring at a new design of road junction any time after it has been built. (This is an example of a *point process*.)
Continuous	A model for monthly inflow to a reservoir.	A continuous-time dynamic model for an offshore oil rig operating in heavy sea conditions. One variable of interest might be the resultant vertical displacement at the drill head.

represent the probabilities that day t is wet and dry. The two components must add to 1. Now define transition probabilities, e.g. p_{WW} is the probability that day $t+1$ is wet given that day t is wet, and gather them into a matrix, M, known as the transition matrix.

$$M = \begin{pmatrix} p_{WW} & p_{WD} \\ p_{DW} & p_{DD} \end{pmatrix}$$

Rows in the transition matrix must add to 1. The probability that day $t+1$ is wet can be expressed in terms of the probabilities that day t is wet and dry, and the transition probabilities: $p_W^{(t+1)} = p_W^{(t)}p_{WW} + p_D^{(t)}p_{DW}$. There is a similar expression for the probability that day t is dry. Both are given in matrix terms by

$$p^{(t+1)} = p^{(t)}M$$

General formulation   There is a finite set of states known as the state space. If there are r of them, they can conveniently be numbered $1, ..., r$. The random process

$$\{X_t\} \qquad \text{for } t = 0, 1, ...$$

is viewed at a discrete set of times, when it is in one of the states. It can change states between these times. The *Markov property* is that the probability that it moves from state i to state j depends on those two states, but not on the history of the process before it arrived in state i. That is:

$$\Pr(X_{t+1} = j \mid X_t = i, \text{ and states taken by } X_{t-1}, ..., X_0) = \Pr(X_{t+1} = j \mid X_t = i)$$

If we assume a *stationary transition mechanism* these probabilities are constant *one-step transition probabilities*, which we will refer to as p_{ij}. The $r \times r$ matrix of transition probabilities, which we will denote by M, is known as the transition matrix. The two-step transition probabilities $p_{ij}^{(2)}$ are given by:

$$p_{ij}^{(2)} = \sum_{k=1}^{r} p_{ik} p_{kj}$$

because we must consider all possible intermediate states. They are the elements of M^2. The more general result

$$p_{ij}^{(n+m)} = \sum_{k=1}^{r} p_{ik}^{(n)} p_{kj}^{(m)}$$

is known as the *Chapman–Kolmogorov equation*. The *probability vector* at time t is the set of probabilities that X_t is in any of the r states:

$$p^{(t)} = (p_1^{(t)} \cdots p_r^{(t)})$$
$$p^{(t+1)} = p^{(t)}M = (p^{(t-1)}M)M = p^{(t-1)}M^2 = \cdots = p^{(0)}M^{t+1} \qquad (7.1)$$

In a *regular chain* some power of the transition matrix has only positive elements. That is, there is some time at which it is possible to be in any of the states regardless of the starting state. All higher powers of the transition matrix must also be positive. In a regular chain the probabilities of being in the various states tend to constant values as t becomes large, irrespective of the initial probabilities. This stationary probability vector, ν, must satisfy

$$\nu = \nu M \qquad (7.2)$$

One way of finding ν is to solve these equations, together with the constraint that the elements sum to one. An alternative is to compute $p^{(t)}$ until it no longer changes.

Example 7.1
The following statistics are calculated from 21 years of records (1965–1986) from a rain gauge at Emley Moor on the River Dearne in South Yorkshire, UK. During the season from March until October there were: 1316 transitions from a wet day to a wet day; 691 transitions from a wet day to a dry day; 686 transitions from a dry day to a wet day; and 1749 transitions from a dry day to a dry day. The estimate of M is therefore

$$M = \begin{matrix} & W & D \\ W & \\ D & \end{matrix}\begin{pmatrix} 0.656 & 0.344 \\ 0.282 & 0.718 \end{pmatrix}$$

If it is raining today, $p^{(0)} = (1\ 0)$, the probability that it is raining in two days' time is given by the first element of $p^{(2)}$. This is 0.527. The stationary probability vector is the solution of:

$$v_1 = 0.656v_1 + 0.282v_2$$
$$v_2 = 0.344v_1 + 0.718v_2$$

which are linearly dependent, giving $v_1 = 0.820v_2$; and

$$v_1 + v_2 = 1$$

Hence v_1 and v_2 are 0.45 and 0.55 respectively. The physical interpretation is that, in the long run, during this season, there is some rain at Emley Moor on 45% of days. The adequacy of this simple model was tested by comparing frequencies of different numbers of consecutive wet days, in the record and in a simulation of the Markov process. There was a good agreement. A Weibull distribution was used to model the amount of rainfall on wet days (Tsang, 1991).

An extension to several sites, n say, can be made by introducing 2^n states. The correspondence might be: state 1 – all sites dry; state 2 – rain at site 1 only; ... ; state 2^n – rain at all sites.

7.2.2 Probability chains

A natural extension of the Markov chain is to let transition probabilities depend on more than just the current state. A *probability chain* is said to be of order q if the probability that $X_{t+1} = j$ depends on the states taken at times t, $t-1, ..., t-q+1$. A probability chain of order 1 is a Markov chain.

The maximum likelihood estimates of the transition probabilities for a chain of order 2 are given by the ratio of the number of ilj sequences to the number of sequences of three states which start il, i.e.

$$\hat{p}_{ilj} = n_{ilj}/n_{il.}$$

where n_{ilj} is the number of ilj sequences, $n_{il.} = \sum_{k=1}^{r} n_{ilk}$, and r is the number of states. This generalizes to any finite order. Tong (1975) described a test which can be used to help decide a suitable order. The test relies on Akaike's Information Criterion (AIC) which is itself based on the concept of entropy. The AIC is defined by:

AIC = -2 ln(maximized likelihood)

$+ 2$(number of independently adjusted parameters used to fit the model)

and the best fit is deemed to be that which minimizes the AIC. Tong's statistic, $T(q)$, is calculated for each order q up to some chosen maximum L

$$T(q) = K_{L+1} - K_L - K_{q+1} + K_q - 2(r^{L+1} - r^L - r^{q+1} + r^q)$$

where

$$K_q = 2 \sum_{\omega} n_\omega \ln n_\omega$$

n_ω is the number of occurrences of a given sequence of q states, and ω ranges over all the possible different sequences. $K_0 = 2(N - L + 1)\ln(N - L + 1)$, where N is the record length. His examples included patterns of wet and dry days in three parts of the world: Hong Kong, Honolulu, and New York. Markov chains gave the best fit for the first two, and an independent sequence ($q = 0$) had the minimum AIC for New York. Chang *et al.* (1984) also considered the use of probability chains for modelling rainfall.

7.2.3 Theory of dams and storage systems

Moran (1954) proposed a Markov chain model for water from dams in countries with clearly defined wet and dry seasons. A schematic representation is shown in Figure 7.1. The dam has a capacity of N units of water, and the volume of water in the dam is assumed to be an integer. A choice of N of about 20 is probably adequate for most purposes. Let D be the usual demand for water during the dry season. We will assume D is a fixed integer, but it could easily be treated as a random variable in simulations. Let W_t be the amount of water in the dam at the end of dry season t. The net inflow during the following wet season is R_t units, and inflows are assumed to be independent with an estimated probability distribution. No water is drawn from this storage dam during the wet season. At the end of the wet season the dam contains:

$$\text{minimum}(N, W_t + R_t)$$

and there may have been an overflow of

$$O_t = W_t + R_t - N$$

During the dry season the water released is:

$$\text{minimum}(D, W_t + R_t)$$

To set up the Markov chain, working in integral multiples of a unit of water, for the probability distribution of W_t we proceed as follows

$$R_t = 0, 1, 2, \ldots \text{ with probabilities } \theta_0, \theta_1, \theta_2, \ldots \text{ respectively}$$

$$W_t = 0, 1, 2, \ldots, N - D \text{ with probabilities } p_0^{(t)}, \ldots, p_{N-D}^{(t)}$$

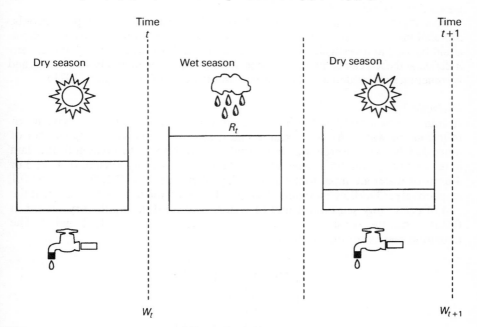

Figure 7.1 Moran's Markov chain model for dam storage

If we suppose $N > 2D$, it follows that

$$p_0^{(t+1)} = p_0^{(t)}(\theta_0 + \theta_1 + \dots + \theta_D) + p_1^{(t)}(\theta_0 + \dots + \theta_{D-1}) + \dots + p_D^{(t)} \theta_0$$

$$p_1^{(t+1)} = p_0^{(t)} \theta_{D+1} + \dots \qquad\qquad\qquad + p_D^{(t)} \theta_1 + p_{D+1}^{(t)} \theta_0$$

$$\vdots$$

$$p_{N-D}^{(t+1)} = p_0^{(t)}(\theta_N + \dots) + p_1^{(t)}(\theta_{N-1} + \dots) + \dots + P_{N-D}(\theta_D + \dots)$$

In matrix terms

$$p^{(t+1)} = p^{(t)} \begin{bmatrix} (\theta_0 + \theta_1 + \dots + \theta_D) & & \dots & (\theta_N + \dots) \\ (\theta_0 + \theta_1 + \dots + \theta_{D-1}) & & \dots & (\theta_{N-1} + \dots) \\ \vdots & & & \\ \theta_0 & & & \\ 0 & & & \\ \vdots & & & \\ 0 & & \dots & (\theta_D + \dots) \end{bmatrix}$$

It is intuitively reasonable to expect $p^{(t)}$ to tend to some limiting distribution, for any initial condition given by $p^{(0)}$. For example, an empty dam would be represented by

$$(1, 0, 0, \dots, 0)$$

and we can obtain the limit of $p^{(t)}$ by iterating the last equation. Alternatively we can solve equation (7.2).

In a typical application, the θ_i and D will be estimated from past records. Simulations which apply similar models for several interconnected storage dams could be used to investigate different operational rules for water supply. If there are d dams, and each can be in any of s states, the Markov chain will have s^d states and is sometimes referred to as a multivariate Markov chain (e.g. Pegram, 1971).

Example 7.2
A country has a rainy and a dry season in each year. A dam has a capacity of four units of water. We will assume that it always contains a whole number of units. Let W_t be a discrete random variable representing the content of the dam at the end of the tth dry season. During the preceding rainy season, 0, 1, 2, 3 or 4 units of water are input to the dam with probabilities 0.4, 0.2, 0.2, 0.1 and 0.1 respectively. Overflow is regarded as lost. Provided the dam is not empty at the end of each rainy season one unit of water is released during the dry season. It follows that the possible values which W_t can take are 0, 1, 2 and 3. The transition matrix for this problem is

	0	1	2	3
0	0.6	0.2	0.1	0.1
1	0.4	0.2	0.2	0.2
2	0.0	0.4	0.2	0.4
3	0.0	0.0	0.4	0.6

If the dam started as empty $p^{(0)} = (1, 0, 0, 0)$

$$p^{(1)} = (0.6, 0.2, 0.1, 0.1)$$
$$p^{(2)} = (0.44, 0.20, 0.16, 0.20)$$
$$p^{(3)} = (0.34, 0.19, 0.20, 0.27)$$

The fixed probability vector is the solution of

$$(v_1, v_2, v_3, v_4) \begin{pmatrix} 0.6 & 0.2 & 0.1 & 0.1 \\ 0.4 & 0.2 & 0.2 & 0.2 \\ 0.0 & 0.4 & 0.2 & 0.4 \\ 0.0 & 0.0 & 0.4 & 0.6 \end{pmatrix} = (v_1, v_2, v_3, v_4)$$

together with the requirement that

$$v_1 + v_2 + v_3 + v_4 = 1$$

It is (0.17 0.17 0.26 0.40), and in the long run the dam will be empty at the end of 17% of dry seasons.

7.2.4 Absorbing Markov chains

A Markov chain is *irreducible* if it is possible to move between any pair of states, though not necessarily in one step. A regular chain is irreducible, but the converse is not necessarily true, because there may be periodic states. A state i has a *period d* if when the chain starts in i, subsequent occupations of i can only occur at times which are multiples of d, i.e. d, $2d$, $3d$, ... A regular chain with no periodic states is called *aperiodic*. An aperiodic irreducible chain is regular. If an irreducible chain has periodic states, there is still a probability vector v satisfying equation (7.2), but $p^{(t)}$ does not tend to the limit v as t tends to infinity.

A state in a Markov chain is an *absorbing state* if it is impossible to leave it. A Markov chain is absorbing if it has at least one absorbing state. An absorbing chain is reducible, because the process must eventually end in an absorbing state. We will now find expressions for the probabilities of ending up in each of the absorbing states for each starting state, and the expected number of steps until absorption.

Canonical form of the transition matrix Assume there are r states of which s are non-absorbing and $(r - s)$ are absorbing. The transition matrix can be written in a partitioned form, known as the *canonical form*, as

$$M = \begin{pmatrix} I & O \\ R & Q \end{pmatrix}$$

where I is the $(r - s) \times (r - s)$ identity matrix; O is the $(r - s) \times s$ zero matrix; R is an $s \times (r - s)$ matrix; and Q is an $s \times s$ matrix. Powers of this transition matrix will be of the form

$$M^t = \begin{pmatrix} I & O \\ * & Q^t \end{pmatrix}$$

where $*$ is some $s \times (r - s)$ matrix. The elements of Q^t are the t-step transition probabilities between the non-absorbing states. Since $Q^t \to 0$

$$I + Q + Q^2 + Q^3 + \cdots = (I - Q)^{-1}$$

This result is analogous to summing a geometric progression, and can easily be verified by pre-multiplying both sides by $(I - Q)$. The *fundamental matrix* is defined as:

$$N = (I - Q)^{-1}$$

Number of steps until absorption　Assume the process starts in the non-absorbing state i. Consider a non-absorbing state j, and define

$$u_{ij}^{(s)} = \begin{cases} 1 & \text{if the process is in state } j \text{ at end of step } s \\ 0 & \text{otherwise} \end{cases}$$

Then

$$E[u_{ij}^{(s)}] = 1 \times p_{ij}^{(s)} + 0 \times (1 - p_{ij}^{(s)}) = p_{ij}^{(s)}$$

Define $w_{ij}^{(t)}$ as the total number of times the process has been in state j, starting from i, at the end of t steps. Then

$$w_{ij}^{(t)} = u_{ij}^{(0)} + u_{ij}^{(1)} + \cdots + u_{ij}^{(t)}$$

For example, suppose a chain goes: state 4; state 3; state 4; state 4; absorbing state. Then $u_{44}^{(0)} = 1$; $u_{44}^{(1)} = 0$; $u_{44}^{(2)} = 1$; $u_{44}^{(3)} = 1$; $u_{44}^{(4)} = 0$; and $w_{44}^{(4)} = 3$. Taking expectation gives

$$E[w_{ij}^{(t)}] = E[u_{ij}^{(0)}] + E[u_{ij}^{(1)}] + \cdots + E[u_{ij}^{(t)}] = p_{ij}^{(0)} + p_{ij}^{(1)} + \cdots p_{ij}^{(t)}$$

which is the entry in the ith row and jth column of $I + Q + \cdots + Q^t$. It follows that

$$n_{ij} = \lim_{t \to \infty} (E[w_{ij}^{(t)}])$$

so the entries of N are the mean number of times in each non-absorbing state for each possible non-absorbing starting state. Adding the components of each row of N will give the mean number of steps before being absorbed for each possible non-absorbing starting state.

Mean first passage time for irreducible chain　In the dam example (Example 7.2), one performance indicator is the expected time until it is first empty. This can be found by changing 0 to an absorbing state, and calculating the mean number of steps until absorption for the newly constructed chain.

Example 7.2 continued
The modified transition matrix with the empty state made absorbing is:

	0	1	2	3
0	1	0	0	0
1	0.4	0.2	0.2	0.2
2	0	0.4	0.2	0.4
3	0	0	0.4	0.6

Therefore

$$Q = \begin{pmatrix} 0.2 & 0.2 & 0.2 \\ 0.4 & 0.2 & 0.4 \\ 0 & 0.4 & 0.6 \end{pmatrix}$$

and

$$N = \begin{pmatrix} 2.5 & 2.5 & 3.75 \\ 2.5 & 5.0 & 6.25 \\ 2.5 & 5.0 & 8.75 \end{pmatrix}$$

If the dam contains the maximum three units at the end of one dry season, the mean number of years until it is empty is 16.25.

Probabilities of ending in each absorbing state   Let b_{ij} be the probability that an absorbing chain will be absorbed in state j if it starts in the non-absorbing state i. We can go directly to j, or go via any of the non-absorbing states, i.e.

$$b_{ij} = p_{ij} + \Sigma \, p_{ik} b_{kj}$$

where the summation is over all the non-absorbing states. In matrix terms

$$B = R + QB$$
$$B - QB = R$$
$$B(I - Q) = R$$
$$B = (I - Q)^{-1} R = NR$$

Example 7.3
Wave height data, which are needed for the safe design of offshore structures, can be recorded by satellite. A civil engineer has requested observations for a particular site. The satellite can be in four states

  0   failed

  1   badly positioned

  2   slight deviation from correct position

  3   correctly positioned.

We take 0 and 3 as absorbing states. Corrective action can be taken from an earth station. The transition matrix between attempts at corrective action is

$$\begin{array}{c c} & \begin{array}{c c c c} 0 & \;\;3 & \;\;1 & \;\;2 \end{array} \\ \begin{array}{c} 0 \\ 3 \\ 1 \\ 2 \end{array} & \begin{pmatrix} 1 & 0 & 0 & 0 \\ 0 & 1 & 0 & 0 \\ 0.2 & 0 & 0.1 & 0.7 \\ 0.2 & 0.7 & 0 & 0.1 \end{pmatrix} \end{array}$$

in canonical form. Therefore

$$Q = \begin{pmatrix} 0.1 & 0.7 \\ 0 & 0.1 \end{pmatrix} \qquad R = \begin{pmatrix} 0.2 & 0 \\ 0.2 & 0.7 \end{pmatrix}$$

and it follows that

$$N = \begin{pmatrix} 1.11 & 0.86 \\ 0 & 1.11 \end{pmatrix}$$

$$B = NR = \begin{pmatrix} 0.40 & 0.60 \\ 0.22 & 0.78 \end{pmatrix}$$

If the satellite starts out badly positioned, there is a 60% chance of correcting it.

7.3   Time series

7.3.1   Introduction

We now consider a continuous state variable measured at discrete time points. The discrete time points may arise because the variable has been summed over the sampling interval, e.g. rainfall per five minutes, and inflow to a reservoir per month. Alternatively, an underlying continuous signal has been sampled at a discrete set of times. This is known as analogue-to-digital conversion (A/D) conversion. The digital signal can then be analysed by computer. Provided sampling is rapid enough, no information is lost. A/D converters can now provide sampling rates of millions per second, which is far in excess of rates needed for analysing wind and wave forces, or river flows. The sequence of measurements

$$\{x_t\} \qquad \text{for} \quad t = 1, 2, \ldots, n$$

is known as a time series. The purposes of time series analysis include: control of water resources; prediction of floods; and investigation of the dynamic behaviour of structures under wind or wave loading.

We will denote a corresponding probability model for the time series by

$$\{X_t\} \qquad \text{for integer } t$$

The mean of this *random* (*stochastic*) *process* is defined by

$$\mu_t = E[X_t]$$

The expectation is taken over the *ensemble*, that is, it is the average of all possible time series that could be generated by the probability mechanism (Figure 7.2). In principle it can vary at every time point, but this flexibility would not be helpful. The snag is that we usually only have one time series, i.e. one realization of the stochastic process. This is certainly the case for environmental series. We have to make simplifying assumptions,such as a linear trend or seasonal variation in the mean, if we are to progress. The first steps in the analysis are to plot the series, and identify and remove any trend or seasonal patterns. We can then fit a stationary model, as defined below.

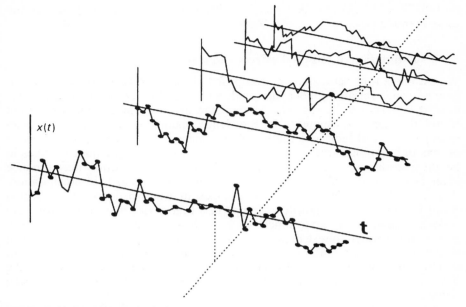

$x(t)$

t

Figure 7.2   Ensemble averaging

7.3.2   Moments of a random process, stationarity and ergodicity

A random process is *stationary in the mean* if the mean does not change over time i.e.

$$E[X_t] = \mu \qquad \text{for all } t$$

It is *second-order stationary* if the covariance structure depends only on the *time lag* between variables (k), and not on absolute time (t). That is

$$E[(X_t - \mu)(X_{t+k} - \mu)] = \gamma(k)$$

where the *autocovariance function* $(acvf)$, $\gamma(k)$, does not depend on t. The lag k can be negative, and $\gamma(-k) = \gamma(k)$. The variance of the process is given by $\gamma(0)$. The *autocorrelation function* (acf) is defined by

$$\rho(k) = \gamma(k)/\gamma(0)$$

and like any correlations these lie between -1 and 1 (see Exercise 4). A plot of $\rho(k)$ against k is called the *correlogram*. The process is *stationary* if all its moments, e.g. skewness, are independent of time.

A stationary random process is *ergodic in the mean* if the time average over one realization equals the ensemble average μ. It is *ergodic* if all time averages equal the corresponding ensemble averages. For hydrological, or other environmental series, we assume underlying ergodic processes. An example of a process that need not be ergodic is the vibration at the driver's seat in earth moving equipment. There would be systematic differences between long records, depending on the type of soil, or soft rock, being moved. There might also be systematic differences between excavators of the same nominal type.

7.3.3   Estimation of seasonal effects and trends

We assume a time series may be composed of: an underlying trend; deterministic seasonal variation about the trend; and a random component. However, even if a series can be represented as a sum of such components it does not follow that they correspond to independently operating causal systems. As an example, we will look at some possible explanations for trends and seasonal effects in hydrological time series. A deposit of silt could produce changes in height measurements made at a specific location. Urbanization may produce changes in run-off and even, if it is on a large scale, changes in precipitation. Agricultural practices can produce changes in watershed conditions which result in changes in stream-flow. Jumps in a time series may result from any sudden change in the environment such as closure of a new dam, starting to pump groundwater, forest fires and so on. Annual cycles are often present in stream-flow, precipitation, evaporation, groundwater level, soil moisture deficit and other hydrological data. Variation within the week and within the day may be present in water use data, such as industrial, domestic or irrigation demands.

The objectives are: to estimate any trend and seasonal effects; to remove these effects so the series can reasonably be considered a realization of a stationary random process; and to identify a plausible stationary process for the random component. We reverse this order to make forecasts, or run simulations. In this section we cover three methods for achieving the first two objectives with monthly data. They can easily be adapted to other sampling intervals, or be modified in other ways.

Centred moving average   This method does not make a separate adjustment for seasonal variation in standard deviations, but if they are proportional to the means, i.e. a constant coefficient of variation, the deseasonalized series will have a constant standard deviation. Hydrological time series often have approximately constant coefficients of variation.

(i)   Calculate a centred 12-month moving average:

$$Sm(x_t) = (\tfrac{1}{2} x_{t-6} + x_{t-5} + \cdots + x_{t+5} + \tfrac{1}{2} x_{t+6})/12$$

Thus, if the data run from January 1909 to December 1980, as they do in Example 7.4, the first 12-month average will be centred on July 1909 and the last will be centred on June 1980. The reasoning behind the formula given is that the average

$$(x_{t-6} + x_{t-5} + \cdots + x_{t+5})/12$$

corresponds to the average value at time $t - \tfrac{1}{2}$, and the average

$$(x_{t-5} + x_{t-4} + \cdots + x_{t+6})/12$$

corresponds to time $t + \tfrac{1}{2}$. The formula given is the average of these two averages and corresponds to time t.

(ii)   Calculate monthly ratios for t equal 7 up to $n - 6$:

$$x_t / Sm(x_t)$$

(iii)   If there are 72 years of data there will now be 71 ratios for each month. Average these 71 ratios to obtain an unadjusted index for each month.

(iv)   Average the 12 unadjusted indices, and divide each by this average to obtain an adjusted index for each month. The average of these adjusted indices is exactly 1, so if they are applied to a series of constant values the total will be unchanged.

(v)   The data are deseasonalized by dividing by the appropriate, adjusted, indices, e.g. the deseasonalized January 1909 datum is the original datum divided by the January index.

(vi)   Plot the deseasonalized data, and consider fitting a trend.

The method described gives multiplicative seasonal indices. An alternative is to assume additive seasonal effects: calculate differences $x_t - Sm(x_t)$; calculate average differences for each month and adjust these so they average 0; and deseasonalize by subtracting the additive effects.

Fitting a trend to deseasonalized data   Trend curves can be fitted by least squares, but the standard statistical tests should not be relied on because the errors are likely to be correlated. We should be wary of extrapolating trends, although it may be reasonable to argue that a linear trend changes relatively slowly when making short-term forecasts. In such cases it might be preferable to fit a local trend to later values instead of an overall trend. Possible trend curves include the following.

(i)   Polynomials such as

$$x_t = \beta_0 + \beta_1 t + \beta_2 t^2$$

The simplest case is a straight line.

(ii)   Modified exponential

$$x_t = a - br^t \qquad \text{where} \quad 0 < r < 1$$

If x_t is replaced by $\ln x_t$ this is known as the Gompertz curve.

(iii)   Logistic

$$x_t = a / (1 + be^{-ct})$$

Both (ii) and (iii) tend towards finite upper limits, and might be suitable for sales of an innovative product, e.g. JCBs when they were first marketed. The least squares equations for the parameters in (ii) and (iii) have to be solved iteratively. The estimated trend can be removed by subtracting the trend value from the observations. Forecasts can be made by extrapolating trend curves into the future and adjusting for seasonal effects. However, there is often no physical basis for choosing a trend curve and several plausible curves may lead to widely different forecasts. In these cases we have to present a range of scenarios.

Standardization method If we use this method we estimate a mean and standard deviation for each month, conditional on any trend having been removed. It requires many years of data for the 24 estimates to have acceptable precision, but a compromise is to smooth them, for example, by fitting harmonic cycles. The procedure involves the following steps.

(i)   Calculate the means for each calendar year and plot them. If appropriate, fit a trend curve and then subtract the trend from the data.
(ii)   For each month, calculate the sample mean and standard deviation.
(iii) The data are deseasonalized by subtracting the monthly means and dividing by the monthly standard deviations.

Harmonic cycles The centred moving average method estimates 12 indices, and the standardization method estimates 24 parameters. If it is reasonable to suppose that monthly variation is part of a harmonic cycle, we can fit the three parameter model

$$X_t = \beta_0 + \beta_1 \cos(2\pi t/12) + \beta_2 \sin(2\pi t/12) + D_t$$

where D_t represents random variation with zero mean, and the frequency $2\pi/12$ corresponds to 1 cycle per year. This can be done with a standard multiple regression routine. The predictor variables are calculated as the cosine and sine of $(2\pi t/12)$ as t goes from 1 to n and the residuals are the deseasonalized time series. A trend can be incorporated by adding, for example, $\beta_3 t$. A more complex seasonal pattern could be allowed for by adding more harmonics. Two cycles per year has a frequency of $4\pi/12$ and we would add $\beta_3 \cos(4\pi t/12)$ and $\beta_4 \sin(4\pi t/12)$ to the model. However, we should be cautious about adding extra harmonics simply because at least one of the pair of coefficients appears to be marginally significant. The D_t are likely to be positively correlated, and the standard deviations of coefficients, and hence P-values, given by a standard regression routine will then be smaller than they should be. The slight lack of efficiency of the standard regression, compared with a weighted least squares, is unimportant if we have a reasonable length series.

Although the method does not allow for any seasonal variation in standard deviation, it can still be useful for hydrological time series if the logarithm of streamflow is analysed. If the original series has a constant coefficient of variation throughout the year, the series of logarithms will have a constant standard deviation. Another advantage of working with logarithms is that forecast and simulated values will always correspond to positive streamflow.

Example 7.4
Monthly effective inflows ($m^3 s^{-1}$) to the Font reservoir in Northumberland (UK Grid Reference NZ049938) from January 1909 until December 1980 have been made available by Northumbrian Water and are given in Table A16.10. The data are the actual inflows from rivers during each month, less an allowance for evaporation from the surface of the reservoir. The purpose of the reservoir is to supply local customers with at least 12 000 m^3 day^{-1}, which cannot be provided from alternative sources, and to maintain a minimum of flow of 2270 m^3 day^{-1} in the river downstream. If the reservoir is full, any excess inflow spills over and is wasted. If the level of water in the reservoir

falls below the lowest draw-off pipe, it fails to fulfil the requirements. A typical problem for the water supply industry is to operate all its sources of water supply so as to minimize losses and the time over which minimum requirements are not met.

The 72-year time series for the Font reservoir is long enough to make useful comparisons between different operating schemes. However, the overall water supply system involves many inter-linked reservoirs and other components, for which the available records are much shorter. One approach is to investigate different models for the Font series and select the most appropriate to fit to the shorter length records. These fitted models could then be used to generate synthetic series for 72 years. Any correlation between inflows to the different reservoirs also has to be estimated and used in the generation procedure.

In this example we restrict our attention to the series for the Font reservoir, investigate whether there is any evidence of a trend and compare some of the deseasonalizing procedures. The yearly means are plotted in Figure 7.3. The fitted regression line is

yearly average effective inflow $= 0.52052 - 0.0013065$ (year $- 1908$)

Using the usual results for regression analysis, the probability that the estimated coefficient exceeds 0.00131 in absolute value, if there is no trend, is 0.07 (the correlation between the residuals, i.e. estimated yearly means after the trend has been removed, is negligible). A possible explanation for a negative trend is that the area under pine forest has been increasing. Whilst we have no conclusive evidence of a trend, there is no doubt that there is seasonal variation in the inflows. This can be seen clearly in a plot of the first six years of the series shown in Figure 7.4. The monthly means and standard deviations are given in Table 7.2 correct to two decimal places. The estimated standard deviations tend to increase with the means but not in proportion to them. The

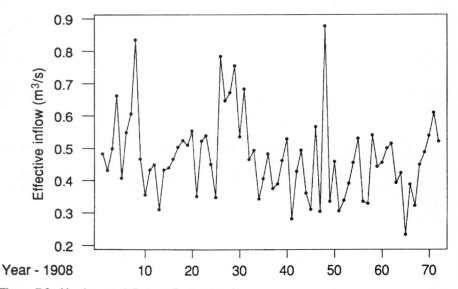

Figure 7.3 Yearly mean inflows to Font reservoir

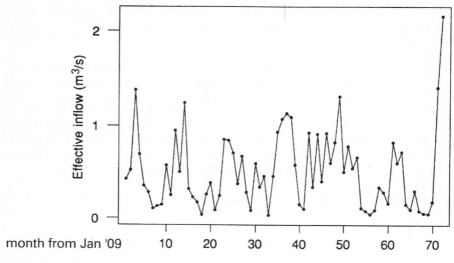

Figure 7.4   Monthly effective inflows to Font reservoir

Table 7.2   *Monthly variation in monthly inflows to the Font reservoir*

Month	Original data			Data after trend removed	
	Mean	Stdev	CV (%)	Mean	Stdev
January	0.76	0.40	53	0.2910	0.4044
February	0.70	0.39	56	0.2218	0.3896
March	0.68	0.50	74	0.2050	0.5031
April	0.40	0.28	70	−0.0739	0.2865
May	0.28	0.20	71	−0.1960	0.2100
June	0.21	0.21	100	−0.2597	0.2144
July	0.19	0.17	89	−0.2819	0.1737
August	0.31	0.38	123	−0.1668	0.3838
September	0.29	0.25	86	−0.1867	0.2552
October	0.44	0.33	75	−0.0296	0.3356
November	0.69	0.40	58	0.2189	0.3991
December	0.73	0.41	56	0.2580	0.4037

net result is that the coefficients of variation (standard deviation/mean) appear to decrease as the means increase.

The 12-month centred moving average method gives the following indices for the months of January to December: 1.73; 1.53; 1.44; 0.88; 0.67; 0.49; 0.46; 0.70; 0.69; 1.01; 1.52; 1.62.

If a harmonic wave is fitted, the equation is

$$\text{monthly inflow} = 0.473 + 0.265 \cos\left(\frac{2\pi t}{12}\right) + 0.129 \sin\left(\frac{2\pi t}{12}\right)$$

where t is the number of the month, starting at 1 for January 1909. If t is added as another predictor variable

$$\text{monthly inflow} = 0.520 - 0.000109t + 0.265 \cos\left(\frac{2\pi t}{12}\right) + 0.128 \sin\left(\frac{2\pi t}{12}\right)$$

This time series is long enough to make reasonably precise estimates of the monthly means and standard deviations. Since the coefficient of variation does not appear to be even approximately constant, I would choose to deseasonalize the data by the standardization method. First remove the trend. The trend fitted to years can be scaled for monthly increments (year 1 is taken as centred at $t = 6.5$ where t is the defined number) to give

$$\text{monthly trend value} = 0.51992 - 0.00010888t$$

The monthly means and standard deviations of the time series with the trend subtracted are given in Table 7.2.

7.3.4   Estimator of the acvf and acf

We will now assume that the time series $\{x_t\}$ represents a realization of a stationary random process, that is, any trend and seasonal patterns have been identified and removed. The mean (μ) is estimated by

$$\bar{x} = \sum_{t=1}^{n} x_t/n$$

The acvf ($\gamma(k)$) is estimated by

$$c(k) = \sum_{t=1}^{n-k} (x_t - \bar{x})(x_{t+k} - \bar{x})/n$$

The devisor n is commonly used because it guarantees a positive spectrum (see Chapter 8). The expected value (Exercise 5) of $c(k)$ is

$$E[c(k)] \simeq \gamma(k)(1 - k/n)$$

and the bias is negligible if k is small in comparison to n. The acf ($\rho(k)$) is estimated by

$$r(k) = c(k)/c(0)$$

A plot of $r(k)$ against k is known as the sample correlogram. It is usually only plotted for positive k because it is symmetric about 0. If $\{X_t\}$ is just a sequence of independent, identically distributed, random variables, known as *discrete white noise (DWN)* then

$$E[r(k)] \simeq -1/n$$
$$\text{var}(r(k)) \simeq 1/n$$

and $r(k)$ is approximately normally distributed for large n (see for example, Kendall and Ord, 1990, for a proof of this result). It is helpful to add a 5% significance pair of lines at

$$-1/n \pm 2/\sqrt{n}$$

to the correlogram. Even if there is no autocorrelation in the process we would expect 1 in 20 $r(k)$ to lie outside these limits, at arbitrary lags. If there is autocorrelation in the process, consecutive values of the estimator $r(k)$ can be highly correlated.

Example 7.4 (continued)
The sample correlogram for the deseasonalized inputs to the Font Reservoir, the slight trend having been removed, is shown in Figure 7.5. We would expect some short-term autocorrelation, and the autocorrelation coefficients at lags 1 and 2 are clearly outside the 1 in 20 limits. The cluster of positive autocorrelations at lags corresponding to about 2 years suggests that there may be some persistence in seasonal patterns, although the absence of anything similar at 1 year is slightly odd. We now show that the lag 23 autocorrelation is statistically significant at an approximate 0.1% level. If we have a realization from a DWN process

$$\frac{r(23) - (-1/864)}{1/\sqrt{(864)}} \sim N(0, 1)$$

The observed value of this statistic is found by substituting 0.109 for $r(23)$ to get 3.24. The P-value is $\Pr(|Z| > 3.24)$, which is 0.001. However, the more

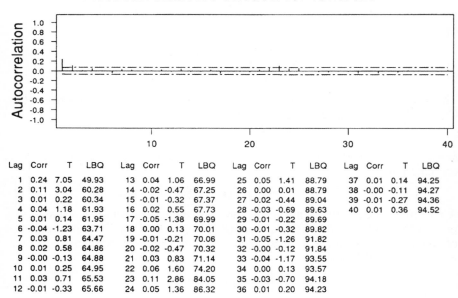

Autocorrelation Function for fontdtds

Lag	Corr	T	LBQ	Lag	Corr	T	LBQ	Lag	Corr	T	LBQ	Lag	Corr	T	LBQ
1	0.24	7.05	49.93	13	0.04	1.06	66.99	25	0.05	1.41	88.79	37	0.01	0.14	94.25
2	0.11	3.04	60.28	14	-0.02	-0.47	67.25	26	0.00	0.01	88.79	38	-0.00	-0.11	94.27
3	0.01	0.22	60.34	15	-0.01	-0.32	67.37	27	-0.02	-0.44	89.04	39	-0.01	-0.27	94.36
4	0.04	1.18	61.93	16	0.02	0.55	67.73	28	-0.03	-0.69	89.63	40	0.01	0.36	94.52
5	0.01	0.14	61.95	17	-0.05	-1.38	69.99	29	-0.01	-0.22	89.69				
6	-0.04	-1.23	63.71	18	0.00	0.13	70.01	30	-0.01	-0.32	89.82				
7	0.03	0.81	64.47	19	-0.01	-0.21	70.06	31	-0.05	-1.26	91.82				
8	0.02	0.58	64.86	20	-0.02	-0.47	70.32	32	-0.00	-0.12	91.84				
9	-0.00	-0.13	64.88	21	0.03	0.83	71.14	33	-0.04	-1.17	93.55				
10	0.01	0.25	64.95	22	0.06	1.60	74.20	34	0.00	0.13	93.57				
11	0.03	0.71	65.53	23	0.11	2.86	84.05	35	-0.03	-0.70	94.18				
12	-0.01	-0.33	65.66	24	0.05	1.36	86.32	36	0.01	0.20	94.23				

Figure 7.5 Correlogram of deseasonalized, detrended, inputs. The t-value is the ratio of the correlogram coefficient to its approximate standard deviation if the population value is zero. The Ljung–Box Q (LBQ) statistic is a cumulative sum of squared coefficients and has an approximate chi-square distribution with k degrees of freedom if the population correlations are zero. (k is the lag. The appropriate degrees of freedom are $k - p$ where p is the number of parameters in the fitted ARIMA model when it is used with residuals from an ARIMA model)

accurate Minitals *t*-value (Figure 7.5) has a considerably higher *P*-value. The log 23 autocorrelation has no practical significance for short term forecasts, but persistence does affect the long term performance of reservoirs (see Section 7.3.10).

7.3.5   ARMA models

In this section, $\{E_t\}$ will denote a sequence of DWN with zero mean, which is the basis for more complex processes. The variance of E_t will be written σ_E^2, and the autocovariances are all zero.

Moving average processes   A process $\{X_t\}$ defined by

$$X_t = \mu + E_t + \beta_1 E_{t-1} + \cdots + \beta_q E_{t-q}$$

is a *moving average process* of order q $(\mathrm{MA}(q))$[2]. Since $\mathrm{E}[E_t] = 0$ for any time, $\mathrm{E}[X_t] = \mu$. The autocovariance is given by

$$
\begin{aligned}
\gamma(k) &= \mathrm{E}\left[(X_t - \mu)(X_{t+k} - \mu)\right] \\
&= \mathrm{E}\left[(E_t + \ldots + \beta_q E_{t-q})(E_{t+k} + \ldots + \beta_q E_{t+k-q})\right] \\
&= \sum_{i=0}^{q-k} \beta_i \beta_{i+k} \qquad \text{for } k \leq q, \qquad \text{and } \beta_0 = 1
\end{aligned}
$$

If k exceeds q, then $\gamma(k) = 0$. The process is stationary for any finite value of q.

Now suppose we have a time series $\{x_t\}$ from a stationary process. If the correlogram, after lag q, appears to be sampling variation, about $-1/n$, a $\mathrm{MA}(q)$ model would be suitable. It cannot be fitted by multiple regression because we do not know the E_t. The MOM estimates given by equating the sample and theoretical correlograms are inefficient, although they would provide good starting values for the following iterative least squares estimation (LSE) procedure, described for q equal to 2.

(i)   Estimate μ by $\bar{x}$.
(ii)   Guess, or use MOM estimates for, β_1 and β_2. Denote these estimates by $\hat{\beta}_1$ and $\hat{\beta}_2$, and write e_t for estimates of the values taken by E_t.
(iii)   Take e_1 and e_2 equal to zero. Calculate

$$e_3 = x_3 - \bar{x} - \hat{\beta}_1 e_2 - \hat{\beta}_2 e_1$$

and hence

$$e_4 = x_4 - \bar{x} - \hat{\beta}_1 e_3 - \hat{\beta}_2 e_2$$

and so on down to

$$\vdots$$

$$e_n = x_n - \bar{x} - \hat{\beta}_1 e_{n-1} - \hat{\beta}_2 e_{n-2}$$

Then calculate the error sum of squares

$$\psi = \sum e_t^2$$

---

[2] In some books, and Minitab, the MA process is defined with a minus sign for the β_k.

(iv)   Use a numerical optimization routine to minimize ψ with respect to $\hat{\beta}_1$ and $\hat{\beta}_2$. The LSE are the values of $\hat{\beta}_1$ and $\hat{\beta}_2$ at this minimum.
(v)   Estimate σ_E^2 by $\psi/(n-3)$.
(vi)   Use a histogram of the e_t to choose a plausible distribution for the E_t.

A refinement would be to treat e_1 and e_2 as additional parameters to be estimated.

Autoregressive processes   A process $\{X_t\}$ is autoregressive of order p (AR(p)) if:

$$(X_t - \mu) = \alpha_1(X_{t-1} - \mu) + \cdots + \alpha_p(X_{t-p} - \mu) + E_t$$

The mean μ is estimated by $\bar{x}$. It will make the algebra simpler, and nothing will be lost, if we assume the process has a mean of 0. We will start by deriving the acvf for an important special case. This is the AR(1) process, which is also known as a Markov process because it has the Markov property

$$X_t = \alpha X_{t-1} + E_t$$

Repeated substitution gives

$$X_t = \alpha(\alpha X_{t-2} + E_{t-1}) + E_t$$
$$= \alpha^2(X_{t-3} + E_{t-2}) + \alpha E_{t-1} + E_t$$

and so on until

$$X_t = E_t + \alpha E_{t-1} + \alpha^2 E_{t-2} + \cdots + \alpha^t E_0$$

The variance of X_t is given by

$$\text{var}(X_t) = \sigma_E^2(1 + \alpha^2 + \alpha^4 + \cdots + \alpha^{2t})$$

If we use the standard result for the sum of a geometric progression,[3] we obtain

$$\sigma_X^2 = \sigma_E^2 \frac{1 - \alpha^{2(t+1)}}{1 - \alpha^2}$$

If α is greater than, or equal to, 1 the variance will increase over time and the process is not stationary. If α is less than 1, α^{2t} will rapidly tend to zero as t increases. It is usual to assume t is large enough for α^{2t} to be negligible, and then the process is stationary

$$\gamma(k) = \text{E}[X_t X_{t+k}]$$
$$= \text{E}[(E_t + \alpha E_{t-1} + \alpha^2 E_{t-2} + \ldots)(E_{t+k} + \alpha E_{t+k-1} + \alpha^2 E_{t+k-2} + \ldots)]$$
$$= \text{E}\left[\left(\sum \alpha^i E_{t-i}\right)\left(\sum \alpha^j E_{t+k-j}\right)\right]$$

---

[3] If $S_n = 1 + r + r^2 + \cdots + r^{n-1}$, then $S_n - rS_n = 1 - r^n$ and $r^n \to 0$ as $n \to \infty$ if $|r| < 1$.

where the summations are formally from $i = 0$ to $i = \infty$. Since the E_t is DWN the only non-zero expectations are when $t - i = t + k - j$, and

$$\gamma(k) = \sigma_E^2 \sum a^i a^{k+i}$$
$$= a^k \sigma_E^2 \sum a^{2i}$$
$$= a^k \sigma_E^2 / (1 - a^2)$$

It follows that the acf is

$$\rho(k) = a^k \qquad \text{for} \quad |a| < 1, \, k = 0, \pm 1, \dots$$

If a is positive the correlogram decays exponentially. This can be quite realistic for some hydrological time series. If a is negative the correlogram oscillates between positive and negative when decaying. This could be appropriate if there is over-correction of some water treatment process after each sample is taken.

A more convenient way of deriving the acf for higher order AR processes is to assume they are stationary, and then set up, and solve, the Yule–Walker equations. The conditions for stationarity will emerge when we solve the equations. We will demonstrate the technique by deriving the acf, and variance, of an AR(2) process

$$X_t = a_1 X_{t-1} + a_2 X_{t-2} + E_t$$

Multiply both sides of the defining equations by X_{t-k}, take expectation, and divide by σ_X^2. This gives

$$\rho(k) = a_1 \rho(k-1) + a_2 \rho(k-2) \qquad \text{for} \quad 0 < k$$

since E_t is independent of the earlier variable X_{t-k}. This is a linear difference equation. It can be solved by trying a solution of the form

$$\rho(k) = \theta^k$$

For the AR(2) process we only have to solve the quadratic equation

$$\theta^2 - a_1 \theta - a_2 = 0$$

The standard solution of this equation is

$$\theta = (a_1 \pm \sqrt{(a_1^2 + 4a_2)})/2$$

Since $\rho(k) = \theta^k$, the process is stationary if, and only if, $|\theta| < 1$. This is satisfied if:

$$a_1 + a_2 < 1, \qquad a_1 - a_2 > -1, \qquad a_2 > -1$$

See Exercise 6 (after Chatfield, 1989). The general solution is

$$\rho(k) = A_1 \theta_1^k + A_2 \theta_2^k$$

where θ_1 and θ_2 are the two roots, which may be complex, of the equation and A_1, A_2 are arbitrary constants. We now need two conditions to determine the constants. These are $\rho(0) = 1$; and, from the Yule–Walker equation with $k = 1$

$$\rho(1) = a_1 \rho(0) + a_2 \rho(-1)$$
$$= a_1 + a_2 \rho(1)$$

If θ_1 and θ_2 are complex the acf is a damped harmonic. The variance of $\{X_t\}$

can be found by taking the variance of both sides of the defining equation and using the expression for $\rho(k)$. It is not possible to give $\rho(k)$ explicitly in terms of the parameters of the model if p exceeds 4. Galois $(1802-23)^4$ proved that it is not possible to construct a formula for solving a general quintic equation in terms of its coefficients.

Although it is harder to find the acf for an AR process than for a MA process, fitting the models is relatively easy. The LSE can be found by multiple regression, which has the advantage that we can add other predictor variables. The LSE are not the same as the ML estimators (see Exercise 7), even if the E_t have a normal distribution. Gardner *et al.* (1980) give an algorithm for the MLE of ARMA processes, and Jones (1980) applies ML to fitting ARMA with missing observations.

ARMA models   A process $\{X_t\}$ of the form:

$$X_t = \alpha_1 X_{t-1} + \cdots + \alpha_p X_{t-p} + E_t + \beta_1 E_{t-1} + \cdots + \beta_q E_{t-q}$$

is ARMA(p, q) with a zero mean. The rationale for introducing it is that it may be possible to model a series with fewer parameters than with an AR or MA on its own. It is often written more concisely by introducing the backward shift operator B, defined by:

$$BX_t = X_{t-1}$$
$$B^2 X_t = B(BX_t) = BX_{t-1} = X_{t-2}$$

and generally

$$B^m X_t = X_{t-m}$$

Then

$$\phi(B)X_t = \theta(B)E_t$$

where $\phi(B)$ and $\theta(B)$ are polynomials in B of order p and q respectively. Formal use of B reduces the algebra involved in derivations. For example, for an AR(1) process

$$(1 - \alpha B)X_t = E_t$$
$$X_t = (1 - \alpha B)^{-1} E_t$$

which is expanded as a Taylor series

$$= (1 + \alpha B + (\alpha B)^2 + \cdots)E_t$$
$$= E_t + \alpha E_{t-1} + \alpha^2 E_{t-2} + \cdots$$

Model order   ARMA models are simple approximations to complex physical processes. We need to choose a suitable order for the process, but should remember that there is no correct answer. The residuals, after fitting the chosen ARMA model, should appear to be uncorrelated and have a minimum estimated standard deviation, or near minimum, amongst the models tried. The chosen model should also be as simple as possible. The correlogram is a useful guide to a suitable order for MA and AR models. If it appears to be close to

<hr>

[4] Galois wrote out his remarkable proof of this result shortly before his tragic death in a duel.

zero after lag q, a MA(q) model is indicated. An apparent exponential decay suggests an AR(1) model might suffice. A damped harmonic is characteristic of an AR(2) process. Another guide is the *partial correlation function (pacf)*. This is a plot of $\hat{\alpha}_p$ against p when fitting an AR(p) model. If the pacf is close to zero above some particular value of p, this is the indicated order.

Example 7.4 (continued)
The correlogram of the deseasonalized inputs (with the trend removed) to the Font Reservoir (Figure 7.5) indicates that an MA(2) model should be a reasonable approximation. I used Minitab to fit:

$$X_t = -0.00029 + E_t + 0.2329E_{t-1} + 0.1116E_{t-2}$$

The standard deviation of the E_t is estimated as 0.963. The estimated standard deviation of the estimators of the coefficients β_1 and β_2 is 0.034. A histogram of the residuals (Figure 7.6) is not well approximated by a normal distribution, and we need to find a more realistic distribution for simulations.

7.3.6 Forecasting and simulation

A point forecast is obtained by assuming future errors, $E_{t+1}\ldots$, equal their expected values of 0. In a simulation we use random numbers from some realistic distribution for the E_{t+1}. Simulation can also be used to give limits of prediction for forecasts. The limits do not then depend on an assumed normal distribution of the E_t.

Example 7.4 (concluded)
The last two residuals from the fitted MA(2) model are 0.62071 and 0.24040 for $t = 863$ and $t = 864$ respectively. The forecast for the next time step,

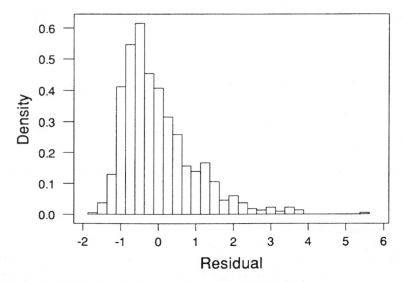

Figure 7.6   Histogram of residuals after fitting MA(2) model to Font data

$t = 865$, is given by

$$\hat{x}_{865} = -0.00029 + 0.00000 + 0.2329 \times 0.24040 + 0.1116 \times 0.62071$$
$$= 0.1250$$

The forecast for $t = 866$ is:

$$\hat{x}_{866} = -0.00029 + 0.00000 + 0.2329 \times 0.000000 + 0.1116 \times 0.24040$$
$$= 0.0265$$

The forecast for $t = 867$, and for higher lead times, equals the mean value of -0.00029. We now have to reseasonalize these forecasts. The times 865, 866 and 867 correspond to January, February and March respectively. The reseasonalized forecasts are therefore:

$$0.1250 \times 0.4044 + 0.2910 = 0.3416$$
$$0.0265 \times 0.3896 + 0.2218 = 0.2321$$
$$-0.0003 \times 0.5031 + 0.2050 = 0.2048$$

Finally the trend has to be added back:

$$0.3416 + 0.51992 - 0.00010888 \times 865 = 0.7673$$
$$0.2321 + 0.51992 - 0.00010888 \times 866 = 0.6577$$
$$0.2048 + 0.51992 - 0.00010888 \times 867 = 0.6303$$

In a simulation E_{865}, E_{866}, E_{867}, ... would be replaced by random numbers from a suitable distribution for the residuals rather than by their expected values of zero. The minimum residual is -1.6526, and a Weibull distribution, with $\alpha = 1.887$ and $\beta = 1.924$, is a reasonable fit to the residuals with 1.7 added. A simulation will nearly always be done on a computer, but to demonstrate the method we will carry out three iterations. Three random numbers from the Weibull distribution are: 0.5746, 0.6722 and 2.2146. Hence:

$$e_{865} = -1.1254, \quad e_{866} = -1.0278, \quad e_{867} = 0.5146$$

$$x_{865} = -0.00029 - 1.1254 + 0.2329 \times 0.24040 + 0.116 \times 0.62071$$
$$= -1.0004$$

$$x_{866} = -0.00029 - 1.0278 + 0.2329 \times (-1.1254) + 0.1116 \times 0.24040$$
$$= -1.2634$$

$$x_{867} = -0.00029 + 0.5146 + 0.2329 \times (-1.0278) + 0.1116 \times (-1.1254)$$
$$= 0.1493$$

These simulated values of the stationary series would be reseasonalized and have the trend added back.

An AR(1) model would give fairly similar results. The estimate of α is 0.2403, and the standard deviation of the E_t is estimated as 0.965. Forecasts for the stationary series are: 0.1018, 0.0243 and 0.0056 for $t = 865$, 866 and 867 respectively.

7.3.7 Integrated and seasonal ARMA processes

Another way of fitting polynomial trends is to *difference* the data. The difference operator is written as ∇, and defined by

$$\nabla X_t = X_t - X_{t-1}$$

Notice that $\nabla = 1 - B$. First order *differencing* will remove a linear trend, but the structure of the differenced process is different from that obtained by fitting a line and subtracting the trend (see Exercise 8). Second order differencing will remove a quadratic trend:

$$\nabla^2 X_t = \nabla(\nabla X_t) = \nabla(X_t - X_{t-1}) = X_t - 2X_{t-1} + X_{t-2}$$

If the dth difference of $\{X_t\}$ is ARMA(p, q) we say that X_t is ARIMA(p, d, q). The I stands for 'integrated' (summing in the discrete time context), which is the inverse operator to differencing (see Exercise 9). The stationary ARMA(p, q) model fitted to the differenced data has to be summed to provide a model for the original time series. A seasonal pattern in the mean, of period s, can also be removed by differencing. A general seasonal ARIMA model, sometimes referred to as a SARIMA model, ARIMApdq PDQs is of the form

$$\phi(B)\phi_s(B^s)(1 - B)^d(1 - B^s)^D X_t = \theta(B)\theta_s(B^s)E_t$$

where $\phi(B)$ and $\theta(B)$ are polynomials in B of degrees p and q respectively and $\phi_s(B^s)$ and $\theta_s(B^s)$ are polynomials in B^s of degrees P and Q respectively, where s is the seasonal period. The operator B^s is defined by

$$B^s X_t = X_{t-s}$$

D is almost always 0 or 1 and for monthly variables s is 12. Then

$$(1 - B^{12})X_t = X_t - X_{t-12}$$

i.e. the difference between a monthly variable and its value in the same month of the previous year. If d equals 1, $(1 - B)(1 - B^{12})X_t$ are the differences of the sequence $\{(X_t - X_{t-12})\}$, i.e.

$$\{(X_t - X_{t-1}) - (X_{t-12} - X_{t-13})\}$$

Notice that, whatever the parameter values, a SARIMA model expresses X_t as a linear combination of past values of X, E_t and past errors. Reasonable choices of P and Q can be made from the correlogram of differenced values by looking at lags which are multiples of s. In particular, suppose we have a systematic monthly pattern with DWN, $\{E_t\}$, added. Differencing, with $D = 12$, will give a sequence ... $(E_t - E_{t-12})$, $(E_{t+1} - E_{t-11})$ ... and Q will be 1.

Example 7.5
After looking at the correlogram of the deseasonalized detrended inputs to the Font Reservoir (Figure 7.4) we might try an

$$\text{ARIMA}(0, 0, 2)(0, 0, 2)12$$

to model the positive correlations near lag 24. The fitted model is

$$X_t = -0.00029 + E_t + 0.2332E_{t-1} + 0.1108E_{t-2}$$
$$-0.0300E_{t-12} + 0.0184E_{t-24}$$

and the estimated standard deviation of E_t is 0.964. This is no improvement on the MA(2) model. The standard deviations of the estimators of the coefficients of β_{t-12} and β_{t-24} are 0.034.

Differencing does not have any effect on the seasonal standard deviations, so these models are more likely to be suitable for the logarithms of river flow than the original data. SARIMA models are often used by hydrologists, but so are the simple ARMA models, trend and seasonal effects having been accounted for in a preliminary analysis. If the model is to be used for simulation the second approach should be used because differencing is unstable (Exercise 10). If the objective is to develop a forecasting scheme, a choice between models can be made by fitting them to the first 80% of the data and then comparing forecasting errors for the remaining 20%.

Another way of modelling seasonal variation is to allow the parameters of an ARMA model to change with the time of the year. These are known as periodic ARMA (PARMA) models (Lin and Lee, 1994, for example).

7.3.8   Gaussian and non-Gaussian processes

If the sequence $\{E_t\}$ in an ARMA model has a normal distribution, usually called a Gaussian distribution in the context of stochastic processes, the joint distribution of

$$(X_{t_1}, \ldots, X_{t_m})$$

will be multivariate Gaussian for any set of t_i. A Gaussian process will, like any multivariate normal distribution, be uniquely defined by its mean and covariance structure, i.e. acvf. Furthermore, the marginal distribution of the $\{X_t\}$ will have a Gaussian distribution. However, it does not follow that we can obtain a Gaussian process by taking some transform of $\{X_t\}$. An ARMA process is a weighted average of the E_t, and this averaged variable will be closer to being normally distributed than is E_t. So even if the marginal distribution of $\{X_t\}$ is close to Gaussian, the distribution of $\{E_t\}$ need not be. The joint distribution of (X_t, X_{t-1}) is crucially dependent on the distribution of E_t, so the former will not then be Gaussian. The advantages of a Gaussian process are that the LS estimates of the parameters are relatively efficient, and that prediction intervals for forecasts, and confidence intervals for parameters, are easy to construct using standard formulae. The prediction intervals are sensitive to the distribution of E_t. If we do manage to transform to a near Gaussian process we must remember, when predicting the mean, that the mean of a transformed variable does not equal the transform of the mean of the original variable. No such adjustment is needed for simulations.

A limitation of Gaussian ARMA processes is that they are time reversible. River flow series often rise rather quickly and then recede more slowly. Such series have a definite direction, and are realizations of time-irreversible processes. These can be modelled, to some extent, by using a highly skewed distribution, such as the exponential, for the E_t. The degree of time irreversibility imparted depends on the parameters of the process. For example, α in an AR(1) process needs to be quite high to see much effect (Figure 7.7). Another ARMA-based approach, which involves stretching the series, is described by Atan and Metcalfe (1994). Point process (Section 7.4) models

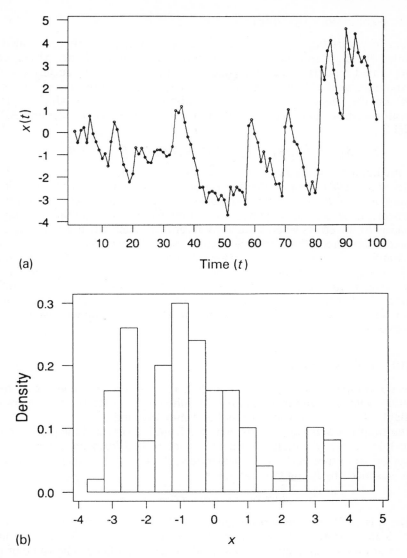

Figure 7.7 Realization of AR(1) process with $\alpha = 0.9$ and exponential noise

are probably more suitable for modelling series with pronounced time asymmetry.

Explanatory variables, such as rainfall when analysing river flow series, can also account for some asymmetry. However, there are so many other variables involved, e.g. the spatial distribution of rainfall and the catchment wetness, that use of a rainfall series from a single site does not always improve the fit very much. When rainfall does improve the fit substantially (e.g. Miller *et al.*, 1981), it will have to be modelled for simulations, or predicted for forecasts. Their model is given below.

7.3.9   Explanatory variables

Explanatory variables can be included in an autoregressive model, which can be fitted by least squares using a multiple regression routine. Miller *et al.* (1981) fitted models, which included rainfall, to daily average flows, during the summer, at the 70th Street monitoring station in Milwaukee, Wisconsin, USA, on the Menomonee River. The following model, in which Y_t is the logarithm of flow and R_t is rainfall on day t was a better fit than any ARMA model.

$$Y_t = \beta_0 + \beta_1 Y_{t-1} + \beta_2 Y_{t-2} + \beta_3 R_t^2 + \beta_4 R_t + \beta_5 R_{t-1} + \beta_6 R_{t-2} + E_t$$

The predictor variables in the regression are: y_{t-1}, y_{t-2}, r_t^2, r_t, r_{t-1} and r_{t-2}. Miller *et al.* (1981) noticed that the variance of the residuals increased with the amount of rainfall. A weighted estimation procedure had little effect on the fit. However, it would be important to allow for the change in the standard deviation of the errors when simulating river flows, and when calculating limits of prediction. Predictions of flow require rainfall forecasts for the same day, and simulations rely on an ancillary model for generating rainfall. Preliminary results using similar models for river flow at Eynsham on the River Thames, UK, have been quite encouraging. Evaporation was used as a predictor variable, and it appeared to incorporate the seasonal variation. Including an interaction between rainfall and some measure of catchment wetness may improve the model.

7.3.10   Multi-site models

A water company asked a colleague (P.S.P. Cowpertwait) and me to model monthly water demand in 11 centres. Monthly supply data from 13 years were available. The model is used to simulate future demands.

The supply has usually met the demand from industry and the public, but we were told of a few months during which there had been hosepipe bans. We began by analysing each centre separately. A rough adjustment for hosepipe bans was made by fitting the regression

$$\text{supply}_t = a + bt + ct^2 - dx_t + e_t$$

where x_t is 1 if there was a ban during month t and 0 otherwise. The estimates of a, b and c were of no particular interest because they define a close fitting curve through the data rather than a trend to be extrapolated. The value of d for the largest centre, a major city, was 15 Ml/day (about 7%). Therefore, the demand series, for this city, was taken to be the supply series with 15 added for months during which there had been a hosepipe ban. We then deseasonalized the demand series using the 12 month centred moving average, fitted a linear trend to the deseasonalized series, and fitted an AR(1) model to the residuals from the trend line. The results for the major city are given in Table 7.3.

At this stage we had 11 series of residuals $\{\varepsilon_{j,t}\}$; one for each centre (j). Each series could reasonably be considered a realization of DWN as the autocorrelations were close to zero. However, we expected some synchronous correlation between centres. The sample covariance matrix V is given in Table 7.4. The corresponding correlations are surprisingly low, the highest being 0.16

Table 7.3 *Fitted demand model for the large city*

January	1.008
February	1.046
March	1.040
April	0.999
May	1.000
June	0.995
July	1.005
August	0.974
September	0.980
October	0.987
November	0.979
December	0.989

$$\text{Demand}_t = 226 - 0.057t + e_t$$
$$e_t = 0.60\,e_{t-1} + \varepsilon_t$$

Table 7.4 *Covariance matrix of residual errors from demand model*

	A	B	C	D	E	F	G	H	I	J	K
A	4.244	0.147	0.125	0.661	0.063	0.643	0.042	0.245	0.196	0.219	0.091
B	0.147	0.791	0.078	0.187	0.044	0.226	0.057	0.035	−0.012	0.012	0.027
C	0.125	0.078	0.739	0.322	0.085	0.107	−0.020	0.151	0.057	0.030	0.026
D	0.661	0.187	0.322	3.913	0.188	0.995	0.021	0.660	0.098	0.131	0.025
E	0.063	0.044	0.085	0.188	0.714	0.204	0.002	0.117	−0.012	0.025	0.009
F	0.643	0.226	0.107	0.995	0.204	7.101	−0.067	0.651	0.191	0.051	0.100
G	0.042	0.057	−0.020	0.021	0.002	−0.067	0.348	−0.014	0.018	0.027	0.023
H	0.245	0.035	0.151	0.660	0.117	0.651	−0.014	2.970	0.036	0.011	−0.017
I	0.196	−0.012	0.057	0.098	−0.012	0.191	0.018	0.036	0.498	0.022	0.042
J	0.219	0.012	0.030	0.131	0.025	0.051	0.027	0.011	0.022	0.578	0.076
K	0.091	0.027	0.026	0.025	0.009	0.100	0.023	−0.017	0.042	0.076	0.601

between A and D. We simulate the spatial covariance by drawing 11 independent random numbers from a unit normal distribution, and applying the following result from Section 5.5. Since V is a covariance matrix there will be a matrix Q such that

$$V = Q^2.$$

Let $z_t = (z_{1,t}, \ldots, z_{11,t})^T$ be the vector of independent standard normal random numbers. The correlated $\varepsilon_{j,t}$ are related to z_t by

$$\varepsilon_t = Qz_t$$

where $\varepsilon_t = (\varepsilon_{1,t}, \ldots, \varepsilon_{11,t})^T$.

7.3.11 Fractional differencing

In the late 1940s, H.E. Hurst, a hydrologist who was working on the design capacity of reservoirs, noticed a tendency for deficits, or surpluses, of inflows to persist (Hurst, 1951). He quantified this by investigating a statistic known as the rescaled adjusted range. This is calculated for a time series $\{x_t\}$, of length m, as follows. First compute the mean $\bar{x}$ and standard deviation s of the series.

Then calculate the adjusted partial sums:

$$S_k = \sum_{t=1}^{k} x_t - k\bar{x} \qquad \text{for } k = 1, \dots, m$$

Notice that $S(m)$ must equal zero, and that large deviations from 0 are indicative of persistence. The rescaled adjusted range,

$$R_m = \{\max(S_1, \dots, S_m) - \min(S_1, \dots, S_m)\}/s$$

is the difference between the largest surplus and the greatest deficit. If we have a long time series of length n, we can calculate R_m for values of m from, for example, 20 upwards to n in steps of 10. When m is less than n, we can calculate $n - m$ values for R_m by starting at different points in the series. Hurst plotted $\ln(R_m)$ against $\ln(m)$ for many long time series. He noticed that lines fitted through the points were usually steeper for geophysical series, such as streamflow, than for realizations of independent Gaussian variables (Gaussian DWN). The average value of the slope (H) of these lines for the geophysical time series was 0.73, significantly higher than the average slope of 0.5 for the independent sequences. The linear logarithmic relationship is equivalent to,

$$R_m \propto m^H$$

where H is known as the Hurst exponent. The discrepancy could not be resolved by using short memory processes, such as an AR(1), rather than DWN for the comparison. An AR(1) process has an acf which decays exponentially, i.e. as α^{-k}, whereas the acf for the geophysical series had slower, hyperbolic, decay, i.e. proportional to $1/k$. Short memory models for inflow, which ignore the persistence feature, can lead to serious overestimation of reservoir performance (Klemeš *et al.*, 1981).

Example 7.5
A plot of $\ln(R_m)$ against $\ln(m)$ for the deseasonalized, detrended, inflows to the Font Reservoir is shown in Figure 7.8. There is marked sampling variability, but no plausible line through the plot has a slope greater than 0.5. This may be because the Font reservoir is in a small catchment by comparison with major overseas reservoirs. Beran (1994) provides convincing evidence of persistence for flows in the River Nile. Hosking simulated persistence by generalizing ARIMA models to have fractional values of the differencing parameter, d, (e.g. Hosking, 1984). The formal definition follows from the binomial expansion

$$\nabla^d X_t = (1 - B)^d X_t = (1 - dB + [d(d-1)/2!]B^2 - [d(d-1)(d-2)/3!]B^3 + \cdots)X_t$$
$$= X_t - dX_{t-1} + [d(d-1)/2]X_{t-2} - [d(d-1)(d-2)/6]X_{t-3} + \cdots$$

If $\{\nabla^d X_t\}$ is DWN we have an ARIMA $(0, d, 0)$ process, which is stationary if $-\frac{1}{2} < d < \frac{1}{2}$, and persistent if $0 < d$, The acf tends towards

$$\rho(k) = [\Gamma(1 - d)/\Gamma(d)]k^{2d-1}$$

for large k. The partial autocorrelations are given by $d/(k-d)$ for $k = 1, \dots$. This is an important result which is used when simulating the process. If the $\{\nabla^d X_t\}$ is ARMA(p, q) the process is ARIMA (p, d, q). The properties of this

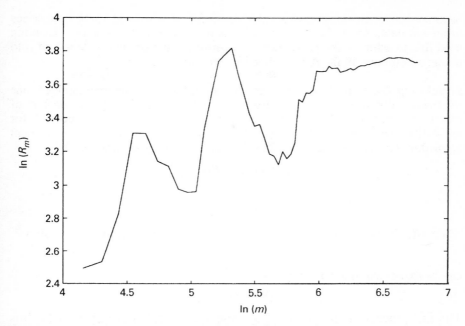

Figure 7.8 Logarithm of rescaled range against $\ln(m)$ for inflows to Font reservoir

process at high lags are the same as those of an ARIMA $(0, d, 0)$. So we can control the short term behaviour by p and q, and the persistence by d. The theoretical relationship

$$d = H - \tfrac{1}{2}$$

can be used to estimate d. Hosking (1984) points out that the corresponding estimator is biased towards 0.2 and is less efficient than approximate MLE, although the mean squared error of the latter, in the example given in the paper, is only reduced by a factor of 0.8. Whether or not we follow Hosking's advice, we need to take fractional differences of a series before estimating the ARMA parameters p and q.

Fractional differences of a time series Let $\{x_t\}$ represent a time series from a stationary process. The fractional differenced series $\{y_t\}$ is calculated from $\{x_t\}$ by the formula:

$$y_t = \sum_{j=0}^{t+M-1} \psi_j x_{t-j}$$

where $\psi_j = (-1)^j d(d-1) \cdots (d-j+1)/j!$
  The infinite sum has been truncated at y_{-M+1}, but this still requires M pre-sample values. These fictitious data can be generated by backcasting, that is: reversing the order of $\{x_t\}$; fitting an ARMA(M) process; predicting M steps ahead; and reversing the order back again so that the k-ahead prediction

becomes y_{-k+1}. Hosking used a value of 30 for M when analysing long series of annual data. An ARMA(p, q) model is now fitted to the y_t series. Hosking used ML to estimate the parameters and perturbed the original estimate of d to obtain a minimum of the AIC (defined in Section 7.2.2).

Simulating fractionally differenced processes   One of the main applications of fractionally differenced processes in civil engineering is the design of reservoir capacities. Long computer generated time series are needed for reliability assessment. Hosking (1984) gives the following convenient algorithm. He decomposes the ARIMA(p, d, q) zero mean process $\{Y_t\}$

$$\phi(B)\nabla^d Y_t = \theta(B)E_t$$

into

$$\phi(B)Y_t = \theta(B)X_t$$

when $\{X_t\}$ is ARIMA$(0, d, 0)$. This follows from noting that the relationship

$$\nabla^d X_t = E_t$$

can be expressed in inverse form

$$(\nabla^d)^{-1}E_t = X_t$$

The $\{x_t\}$ process can be simulated by using the algorithm in Appendix 11. It is simplified somewhat because

$$\phi_{tt} = d/(t - d)$$

Then the $\{y_t\}$ process is generated from

$$y_t = \alpha_1 y_{t-1} + \cdots + \alpha_p y_{t-p} + x_t + \beta_1 x_{t-1} + \cdots + \beta_q x_{t-q}$$

Walden (1994) describes an application of fractional differencing to the analysis of geophysical borehole data.

7.3.12   Non-linear time series

Even with strange distributions of residuals, ARMA models do not always give a satisfactory fit to environmental time series. We may find a more plausible representation amongst the many non-linear time series models that have been described in the engineering and mathematical literature (see Tong, 1990 and Thompson and Stewart, 1986 for example).

Threshold models   In a threshold model the values of the parameters depend on past values of the process. Tong and Lim (1980) fitted a threshold model to the logarithm of daily flow of the Kanna River with daily rainfall as a predictor variable.

The Kanna River is a river with a small catchment area (under 1000 km^2) in Japan. Seasonal variations of Japanese rivers are quite regular due to the rather well-defined rainy season there, and the ground soil is rarely dry. Tong and Lim thought it reasonable to expect that most of the cyclical variation of the riverflow data can be explained by that of the rainfall data if the transformation from the latter to the former is adequately modelled. They fitted this model using the first 280 days of the record.

$$X_t = \begin{cases} \begin{aligned} & 0.0185 + 0.9992X_{t-1} + 0.0065R_{t-1} - 0.1519X_{t-2} - 0.0017R_{t-2} \\ & + 0.1236X_{t-3} - 0.0004R_{t-3} - 0.0295X_{t-4} - 0.0014R_{t-4} \\ & + 0.0065X_{t-5} + E_t^{(1)} \qquad\qquad \text{if } R_{t-1} \leqslant 4.6000 \end{aligned} \\ \begin{aligned} & 0.1281 + 0.5044X_{t-1} + 0.0146R_{t-1} + 0.2767X_{t-2} + 0.0014R_{t-2} + E_t^{(2)} \\ & \qquad\qquad\qquad\qquad\qquad\qquad\qquad\quad \text{if } R_{t-1} > 4.6000 \end{aligned} \end{cases}$$

where var $E_t^{(1)} = 0.0012$, var $E_t^{(2)} = 0.0173$, X_t is logarithm base 10 of flow and R_t is rainfall (mm). They then made one step ahead predictions for 86 days and found that the root mean square error was 18% smaller than when they used an ordinary regression. Longer term predictions would rely on rainfall forecasts.

Threshold models do not have to include additional explanatory variables. The simplest first-order threshold autoregressive model is

$$X_t = \begin{cases} \alpha_1 X_{t-1} + E_t & \text{if } X_{t-1} < d \\ \alpha_2 X_{t-1} + E_t & \text{if } X_{t-1} \geqslant d \end{cases}$$

These models can be fitted by least squares, for a range of threshold values, and the threshold value giving the best fit is retained.

Bilinear models A first-order bilinear model is of the form

$$X_t = (\alpha + \beta E_t)X_{t-1} + E_t$$

or variants of this such as βE_{t-1} replacing βE_t, where E_t is DWN. They can be used to model series that show sudden bursts of large amplitude oscillations at irregular intervals, and you are asked to investigate one of these in Exercise 12. ARMA models, with normal errors, have the moment property, i.e. moments of all orders exist. If the coefficients are subject to random perturbations, high order moments will no longer be finite. We have encountered this concept before. The generalized Pareto distribution of Section 4.5.1 is an example of a distribution which need not have finite high order moments, and the Cauchy distribution (Exercise 8 in Chapter 2) has no defined moments. However, a lack of moments does not prevent a stochastic process from being stationary. Stationarity only requires that the multivariate distribution of any subset of variables $(X_1, X_2, \ldots, X_p)$ is the same as that of $(X_{1+k}, X_{2+k}, \ldots, X_{p+k})$. Either form of the first-order bilinear model is stationary if

$$\alpha^2 + \beta^2 \sigma_E^2 < 1$$

in which case the first two moments, at least, are finite.

Tong (1990) gives a comprehensive account of non-linear models, and their intriguing properties, from a statistical perspective. You are asked to consider least squares estimation of the parameters of a first-order bilinear model in Exercise 11.

7.4   Stochastic models based on point processes

Although we hope to have some sensible physical interpretation of the coefficients in time series models of river flow, they are simple approximations based on wide scale aggregation. A more refined approach is to combine rainfall models with physically based rainfall-runoff models such as the Système Hydrologique Européen (SHE) (Bathurst *et al.*, 1993). There is also the advantage that rainfall records are far more comprehensive than river flow data. The processes modelled by the SHE are as follows.

(i)   Interception of rainfall on vegetation canopy.
(ii)   Evaporation of intercepted rainfall, ground surface water and channel water.
(iii)   Transpiration of water drawn from the root zone.
(iv)   Snowpack development.
(v)   Snowmelt.
(vi)   One-dimensional flow in the unsaturated zone.
(vii)   Two-dimensional flow in the saturated zone.
(viii) Two-dimensional overland flow.
(ix)   One-dimensional channel flow.
(x)   Saturated zone/channel interaction, including an allowance for an unsaturated zone below the channel.
(xi)   Saturated zone/surface water interaction.

The sizes of the grid squares depend on the catchment area. Suyanto *et al.* (1995) use 1 km squares for the Brue catchment, in south-west England, which has an area of 135 km^2. A realistic model for rainfall is essential if we are to obtain the full benefit of the SHE system. One application is real-time runoff forecasting as part of a flood warning system (O'Connell *et al.*, 1996). This work was part of a UK research programme (Hydrological Radar Experiment). The rainfall model described below was developed as part of an earlier UK research programme, the Urban Pollution Management Research Programme. The synthetic rainfall generator was linked to hydraulic models of storm water sewer systems.

A point process is a continuous time model for a stochastic process in which events occur at points in time. The simplest example is a Poisson process, which is characterized by the Markov property. That is, the probability of an occurrence in the next interval, of arbitrary length, is independent of the time of the last occurrence and of the entire history of the process. Poisson processes are the basis of many useful models, such as the Markov processes used for the reliability analyses in Section 2.4.5 and the Neyman-Scott rectangular pulses (NSRP) rainfall model described in Section 1.1.2.

7.4.1   A procedure for fitting the NSRP model

Storm origins occur as a Poisson process with rate λ hr^{-1}. The average number of rain cells per storm is ν. The number of cells, minus one, has a Poisson distribution. The waiting times from storm origins to cell origins have an exponential distribution with mean β^{-1} hr. The cell durations and intensities have exponential distributions with means η^{-1} hr and ξ^{-1} mm hr^{-1} respectively. Therefore, there are five parameters to be estimated for a single site NSRP model. These parameters can be estimated from rainfall depths aggregated over fixed time intervals. Let $Y_i^{(h)}$ be the rainfall depth in the ith time interval of length h. Rodriguez-Iturbe *et al.* (1987) derived the following expressions for population moments (see also Cox and Isham, 1980).

$$E(Y_i^{(h)}) = \lambda v h / (\eta \xi)$$

$$\mathrm{var}(Y_i^{(h)}) = \frac{\lambda(v^2 - 1)[\beta^3 A_1(h) - \eta^3 B_1(h)]}{\beta \xi^2 \eta^3 (\beta^2 - \eta^2)} + \frac{4\lambda v A_1(h)}{\xi^2 \eta^3}$$

$$\mathrm{cov}(Y_i^{(h)}, Y_{i+k}^{(h)}) = \frac{\lambda(v^2 - 1)[\beta^3 A_2(h, k) - \eta^3 B_2(h, k)]}{\beta \xi^2 \eta^3 (\beta^2 - \eta^2)} + \frac{4\lambda v A_2(h, k)}{\xi^2 \eta^3}$$

where

$$A_1(h) = \eta h - 1 + e^{-\eta h}$$
$$B_1(h) = \beta h - 1 + e^{-\beta h}$$
$$A_2(h, k) = \tfrac{1}{2}(1 - e^{-\eta h})^2 e^{-\eta h(k-1)}$$
$$B_2(h, k) = \tfrac{1}{2}(1 - e^{-\beta h})^2 e^{-\beta h(k-1)}$$

Rainfall records at hourly intervals are ideal for fitting the model. A straightforward approach is to equate the sample mean, one and six hour variances, and one and six hour lag one covariances with the population expressions. The five non-linear equations can be solved numerically. Cowpertwait *et al.* (1996) fitted the model to a 20 year record of hourly rainfall at Manston, on the southeast coast of England. The sample variances were taken over the whole 11 or 12 year period, about the overall mean for the appropriate month. The sample autocovariances were the average of the 11 or 12 autocovariances, calculated within each year but taken about the overall mean for the appropriate month to avoid bias (e.g. see Trenberth, 1984). The formulae − in which $Y_{ijk}^{(h)}$ is the j-th h-hourly total in year i for month k, $n_k^{(h)}$ is the number of h-hourly totals in month k and n is the number of years of record − are:

for the mean

$$\hat{\mu}_k(h) = \sum_{i=1}^{n} \sum_{j=1}^{n_k^{(h)}} Y_{i,j,k}^{(h)} \bigg/ \{n_k^{(h)} n\}$$

for the variance

$$\hat{\gamma}_k(h) = \sum_{i=1}^{n} \sum_{j=1}^{n_k^{(h)}} \{Y_{i,j,k}^{(h)} - \hat{\mu}_k(h)\}^2 \bigg/ \{n_k^{(h)} n\}$$

and for the lag 1 covariance, which is all that was used,

$$\hat{\gamma}_k(h, 1) = \sum_{i=1}^{n} \sum_{j=1}^{n_k^{(h)}-1} \{Y_{i,j,k}^{(h)} - \hat{\mu}_k(h)\}\{Y_{i,j+1,k}^{(h)} - \hat{\mu}_k(h)\} \bigg/ \{(n_k^{(h)} - 1)n\}$$

Cowpertwait *et al.* (1996) noticed that this method of fitting the parameters could not be relied upon to give a realistic distribution of lengths of dry periods. The impact of a storm sewer overflow discharge on a receiving watercourse will be worse if the storm follows a dry period, when low river flows offer reduced dilution and pollutant concentrations are high within the sewers. We modified the fitting procedure by incorporating an expression for the probability that an interval of arbitrary length is dry (Cowpertwait, 1991).

$$\Pr\{Y_i^{(h)} = 0\} = \exp\left(-\lambda h + \lambda \beta^{-1}(v-1)^{-1}\{1 - \exp[1 - v + (v-1)e^{-\beta h}]\}\right.$$

$$\left. - \lambda \int_0^\infty [1 - p_h(t)]dt\right)$$

where

$$p_h(t) = \{e^{-\beta(t+h)} + 1 - (\eta e^{-\beta t} - \beta e^{-\eta t})/(\eta - \beta)\}$$
$$\times \exp\{-(v-1)\beta(e^{-\beta t} - e^{-\eta t})/(\eta - \beta) - (v-1)e^{-\beta t} + (v-1)e^{-\beta(t+h)}\}$$

The transition probabilities, $\Pr\{Y_{i+1}^{(h)} > 0 \mid Y_i^{(h)} > 0\}$ and $\Pr\{Y_{i+1}^{(h)} = 0 \mid Y_i^{(h)} = 0\}$, denoted as $\phi_{\text{ww}}(h)$ and $\phi_{\text{DD}}(h)$, respectively, can be expressed in terms of $\phi(h)$ and $\phi(2h)$ where $\phi(h)$ is $\Pr\{Y_i^h = 0\}$:

$$\phi_{\text{DD}}(h) = \phi(2h)/\phi(h)$$
$$\phi(h) = \phi_{\text{DD}}(h)\phi(h) + \{1 - \phi_{\text{ww}}(h)\}\{1 - \phi(h)\}$$

so that

$$\phi_{\text{ww}}(h) = \{1 - 2\phi(h) + \phi(2h)\}/\{1 - \phi(h)\}$$

Modified fitting procedure   There are five parameters to be estimated, and equating five, suitably chosen, functions of the parameters $f_l(\lambda, \beta, \eta, v, \xi)$ with the corresponding sample values $\hat{f}_l$ is one way to get estimates. However, we thought it would be preferable to fit a larger set of statistics approximately. So we minimised an expression of the form

$$S = \sum_{l=1}^m w_l(1 - f_l/\hat{f}_l)^2$$

where λ, β, η, $\xi > 0$, $v > 1$ and $\hat{f}_l > 0$. The w_l allow greater weight to be given to those sample functions, which are thought to be most important for a particular application. We gave the mean a weight of 100, and all other terms weights of 1. Some sample moments are given in Table 7.5. The parameter estimates obtained by fitting: $\hat{\mu}(1)$; $\hat{\gamma}$ for 1, 3, 6, 12 and 24; $\hat{\phi}_{\text{ww}}$ for 1, 3, 6, 12 and 24; $\hat{\phi}_{\text{DD}}(24)$; and $\hat{\phi}(24)$ are given in Table 7.6.

Further developments   Cowpertwait (1994) has extended the model to allow for different types of rain cells. The special case of two cell types: heavy cells with a short expected duration, and light cells with a longer expected duration is consistent with observational studies of rainfall fields. The two-cell model gives a better fit to extreme values than the single cell model. A further major

Table 7.5 *Sample moments for the Manston data*

Month	$\hat{\mu}(1)$ (mm)	$\hat{\gamma}(1)$ (mm²)	$\hat{\rho}(1,1)$	$\hat{\gamma}(6)$ (mm)²	$\hat{\rho}(6,1)$	$\hat{\gamma}(24)$ (mm)²	$\hat{\rho}(24,1)$	$\hat{\phi}(24)$
Jan	0.063	0.082	0.54	1.2	0.23	7.5	0.08	0.52
Feb	0.052	0.068	0.49	1.0	0.33	6.5	0.25	0.58
Mar	0.054	0.071	0.54	1.2	0.27	7.4	0.19	0.57
Apr	0.055	0.082	0.55	1.4	0.27	7.9	0.13	0.60
May	0.056	0.120	0.43	1.7	0.19	8.1	0.11	0.60
June	0.062	0.174	0.39	2.3	0.30	14.0	0.27	0.68
July	0.061	0.204	0.46	3.1	0.34	21.0	0.09	0.70
Aug	0.067	0.386	0.33	4.5	0.32	36.0	0.07	0.71
Sep	0.093	0.530	0.47	9.1	0.51	61.0	0.25	0.64
Oct	0.080	0.208	0.46	3.1	0.37	21.0	0.32	0.62
Nov	0.088	0.153	0.55	2.4	0.34	17.0	0.15	0.54
Dec	0.066	0.097	0.54	1.6	0.27	9.5	0.07	0.56

Table 7.6 *Parameter estimates when using the transition probabilities in the fitting procedure*

Month	λ (h⁻¹)	β (h⁻¹)	η (h⁻¹)	ν	ξ (h mm⁻¹)
January	0.0242	0.455	1.20	4.24	1.39
July	0.0117	0.151	0.687	2.00	0.558

development is a multi-site version (Cowpertwait, 1995), in which rain cells are modelled by discs. The disc centres are points in a spatial Poisson process. In practice, the number of sites at which rainfall needs to be simulated will equal the number of inputs to the hydraulic model of the sewer system. A less elegant, but easier, approach is to disaggregate a single site model in some ad hoc manner. Some other important advances are described in the paper by Wheater *et al.* (1997).

Rainfall field models A rainfall field model gives a continuous space and time description of storms. Field models are particularly suitable for use with radar data. Waymire *et al.* (1984) developed a comprehensive field model. Mellor (1996) proposed a model which incorporates more of the deterministic features that are evident in frontal rainfall. Procedures for fitting this model are given in Mellor and O'Connell (1996) and Mellor and Metcalfe (1996).

Neyman-Scott streamflow model Cowpertwait and O'Connell (1993) introduced a Neyman-Scott shot noise model with a plausible physical structure for the generation of synthetic streamflow data. The model has two components: a rainfall event process which is represented by a NSRP model, and a linear response process, which consists of two linear reservoirs in parallel. A proportion of each pulse is routed through a fast linear reservoir to represent surface runoff, and the remainder through a slow linear reservoir to represent baseflow; the latter providing a means of preserving persistence at the monthly level. This is shown in Figure 7.9. A mathematical description is given below. If $X_{t-u}(t)$ denotes the streamflow response at time t due to a cell with origin at

(a)

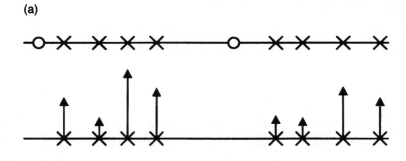

(b)

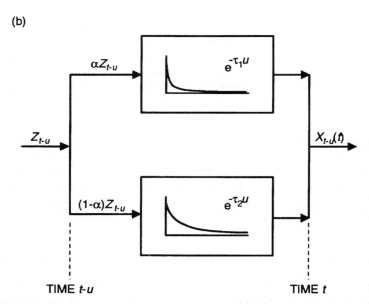

TIME *t-u*   TIME *t*

Figure 7.9 Neyman–Scott shot noise model for streamflow: (a) storm origins (O) and cell origins (X), (b) schematic of the proportional-input linear system model used in the NSSN model, (after Cowpertwait and O'Connell, 1992)

time $t - u$, then:

$$X_{t-u}(t) = \alpha Z_{t-u}e^{-\tau_1 u} + (1 - \alpha)Z_{t-u}e^{-\tau_2 u}$$

where Z_{t-u} is the instantaneous pulse occurring at the cell origin.

Furthermore, it is assumed that the total flow $Y(t)$ at time t is the sum of all the surface flow and baseflow contributions due to all cells with origins preceding t, i.e.

$$Y(t) = \int_{u=0}^{\infty} X_{t-u}(t)\mathrm{d}N(t - u)$$

where

$$dN(t - u) = 1 \text{ if there is a cell origin at time } t - u$$
$$= 0 \text{ otherwise}$$

i.e.

$$Y(t) = \int_{u=0}^{\infty} \{\alpha Z_{t-u} e^{-\tau_1 u} + (1 - \alpha) Z_{t-u} e^{-\tau_2 u}\} dN(t - u)$$

Cowpertwait and O'Connell (1993) obtained expressions for moments of the process in terms of the parameters (reproduced in Appendix 12), and used these to fit the model to data from the River Coquet in Northumberland.

7.5 Spatial processes

In this section we deal with variables that are distributed over an area, or within a volume, rather than through time. Examples range from the mineral content of ores (e.g. Royle, 1977; Sinclair, 1974 and 1978), to depth, and hydraulic properties, of sand in a study of contaminant dispersion (Mackay *et al.*, 1996). Sampling usually involves drilling cores, and we often need to estimate variables at other locations from rather small numbers of cores. The technique that is commonly used is known as *kriging*, after Danie Krige (e.g. Krige, 1951). Ordinary kriging is a weighted average of sample values, in which the weights are determined from the covariance structure, so as to give a minimum variance unbiased estimator. The covariance structure is generally described by the *variogram*.

7.5.1 Variogram

Let $g(x)$ be a random function of position x in some spatial medium. The covariance function, $\gamma(h)$, is the covariance of random variables separated by the vector h. That is

$$\gamma(h) = \text{cov}[g(x)g(x + h)]$$

We will assume the medium is *isotropic*, the same physical properties in all directions, in as much as $\gamma(h)$ only depends on the distance between the two points ($|h| = h$). The simplest way to deal with anisotropic media is to scale the distances. Clark (1979) gives an example of porphyry molybdenum with a range of influence of 70 m vertically through the deposit and a range of influence of 350 m in all horizontal directions. We could change all horizontal directions to multiples of 5 m instead of 1 m. We will also assume that the random process is stationary. In principle, we might estimate a planar trend, for example, by a regression

$$g = \beta_0 + \beta_1 x + \beta_2 y$$

where $x^T = (x, y)$, and then subtract it. In practice, we are unlikely to have sufficient data to justify this approach. The alternative is to split the area into sub-areas within which we are prepared to assume stationarity. The *variogram*

$Y(h)$ is defined by

$$Y(h) = \tfrac{1}{2} E[(g(x) - g(x + h))^2]$$
$$= \tfrac{1}{2} E[(g(x))^2] + \tfrac{1}{2} E[(g(x + h))^2] - E[(g(x)g(x + h)]$$

which, assuming second-order stationarity

$$Y = E[(g(x))^2] - E[(g(x)g(x + h)]$$

The hypothesis of second-order stationarity is often replaced by a less restrictive requirement, called the *intrinsic hypothesis*. This states that, even if the variance of $g(x)$ is not finite, the variance of the increments of $g(x)$ is finite and the increments are second-order stationary and satisfy the following properties:

$$E\{g(x + h) - g(x)\} = 0$$
$$\mathrm{Var}\{g(x + h) - g(x)\} = 2Y(h)$$

7.5.2   Variogram models

If the random process has a finite variance (σ^2), the variogram must tend towards this upper limit known as the *sill*. Variogram models with this property are known as *transition models*. Models which do not reach a plateau are used when there is a trend or drift to be modelled. The *range* of the variogram is the distance at which it reaches the sill, or 95% of the sill value if the sill is an asymptote. The spherical model is, perhaps, the most used.

Spherical model

$$Y(h) = \begin{cases} \sigma^2(1.5h/a - 0.5h^3/a^3) & \text{if } h \leqslant a \\ \sigma^2 \end{cases}$$

Exponential model   The exponential model is assumed in many applications

$$Y(h) = \sigma^2(1 - \exp(-3h/a))$$

Gaussian model   This is used to model slowly varying phenomena

$$Y(\mathrm{h}) = \sigma^2(1 - \exp(-3h^2/a^2))$$

Nugget effect   The *nugget effect* complements the Gaussian model. Gold can occur in nuggets, and the concentration can vary dramatically within a short distance. This is modelled by a discontinuity at the origin, so an exponential model with a nugget effect (b) is

$$Y(h) = b + (\sigma^2 - b)(1 - \exp(-3h/a)) \qquad \text{for } h \neq 0$$

It follows from the definition that $Y(0) = 0$.

7.5.3   Sample variogram

Suppose we have a sample of measurements of the field variable at n locations, $\{g_i\}$. Let h_{ij} be the distance between location i and location j, and let $n(h)$ be the number of pairs (i, j) which are a distance h apart. If sampling points have been distributed systematically over a grid there will be many pairs for most distances h. A sample variogram can be calculated from:

$$v(h) = \Sigma \ (g_i - g_j)^2 / (2n(h))$$

where the summation is over (i, j) such that $h_{ij} = h$. A different estimate of the sill is the sample variance

$$s^2 = \Sigma \ (g_i - \bar{g})^2 / (n - 1)$$

Clark (1979) recommends calculating sample variograms in orthogonal directions, e.g. N-S and E-W, to check if an assumption of an isotropic medium is reasonable. She also points out that $v(h)$ is rarely calculated for h greater than half the total sampled extent. The sample variogram can be used to choose between variogram models and, together with s^2, to choose sensible parameter values. It is also possible just to smooth the points, without assuming any standard form. The alternative to estimating a variogram model from sample data, is to rely on the literature for a suitable form and its parameter values (Mackay *et al.*, 1996). Bayesian estimation is a compromise.

7.5.4 Kriging

Kriging We have measurements of a field variable g at n points $\{x_i\}$ and wish to estimate the value at a point x_0. We will do this by taking a linear combination of the n measurements

$$\hat{g}(x_0) = \sum_{i=1}^{n} w_i g(x_i)$$

We assume g is locally stationary over the region in which the measurements have been made. If $\hat{g}(x_0)$ is to be unbiased the weights must add to one. A requirement that the error of the estimator should have a minimum variance leads to an expression for the weights in terms of the covariance structure, which is itself estimated as the sample variogram. The variance of the estimator error is

$$\psi = \text{var}(\hat{g}(x_0) - g(x_0)) = \text{var}(\hat{g}(x_0)) + \text{var}(g(x_0)) - 2 \ \text{cov}(\hat{g}(x_0)g(x_0))$$

We will now write

$$\text{cov}_{ij} \quad \text{for} \quad \text{cov}(g(x_i)g(x_j))$$

with the understanding that cov_{ii} is the variance of the random function σ^2, and use the result for the variance of a linear combination of variables to obtain

$$\psi = \sum_{i=1}^{n} \sum_{j=1}^{n} w_i w_j \, \text{cov}_{ij} + \sigma^2 - 2 \sum_{i=1}^{n} w_i \, \text{cov}_{i0}$$

We now apply Lagrange's method of multipliers (see Exercise 13 at the end of Chapter 5) and find the w_i which minimize

$$\psi + 2\lambda \left(\sum_{i=1}^{n} w_i - 1 \right)$$

This is done by setting the $n + 1$ partial derivatives, with respect to the w_i and λ, equal to zero

$$\sum_{j=1}^{n} w_j \, \text{cov}_{ij} + \lambda = \text{cov}_{i0} \qquad \text{for } i = 1, \ldots, n$$

$$\sum_{i=1}^{n} w_i \qquad = 1$$

This is the *ordinary kriging system*, which can be expressed in matrix terms as:

$$\begin{bmatrix} \text{cov}_{11} & \cdots & \text{cov}_{1n} & 1 \\ \vdots & & \vdots & \vdots \\ \text{cov}_{n1} & \cdots & \text{cov}_{nn} & 1 \\ 1 & \cdots & 1 & 0 \end{bmatrix} \begin{bmatrix} w_1 \\ \vdots \\ w_n \\ \lambda \end{bmatrix} = \begin{bmatrix} \text{cov}_{10} \\ \vdots \\ \text{cov}_{n0} \\ 1 \end{bmatrix}$$

The minimized error variance can be calculated by substituting into the expression for ψ, but it is quicker to use the expression

$$\psi = \sigma^2 - (\,\Sigma\; w_i \, \text{cov}_{i0} + \lambda\,)$$

The ordinary kriging system can be expressed as easily in terms of the variogram, since

$$\text{cov}_{ij} = \sigma^2 - Y(h)$$

where h is the distance between x_i and x_j.

Co-kriging   In some cases we may have measurements of some secondary variable $h(x)$, which is correlated with $g(x)$, at many more points. Isaaks and Srivastava (1989) give an example of all core samples being assayed for one particular mineral, while only those core samples providing a high assay value are assayed for a second mineral. A contrasting application is the estimation of rainfall from rain-gauge and radar data (Azimi-Zonooz *et al.*, 1989). The estimator is now of the form

$$\hat{g}(x_0) = \sum_{i=1}^{n} w_i \, g(x_i) + \sum_{j=1}^{m} a_j h(x_j)$$

with two constraints for unbiasedness:

$$\sum_{i=1}^{n} w_i = 1, \qquad \sum_{j=1}^{m} a_j = 0$$

A similar development to that which led to the ordinary kriging system, with two Lagrange multipliers λ_1 and λ_2, gives the co-kriging system.

$$\sum_{i=1}^{n} w_i \, \text{cov}(g_i\, g_j) + \sum_{i=1}^{m} a_i \, \text{cov}(g_j\, h_i) + \lambda_1 = \text{cov}(g_j\, g_0) \qquad \text{for } j = 1, \ldots, n$$

$$\sum_{i=1}^{n} w_i \, \text{cov}(g_i\, h_j) + \sum_{i=1}^{m} a_i \, \text{cov}(h_i\, h_j) + \lambda_2 = \text{cov}(h_j\, g_0) \qquad \text{for } j = 1, \ldots, n$$

$$\sum_{i=1}^{n} w_i = 1$$

$$\sum_{i=1}^{m} a_i = 0$$

This requires the estimation of spatial cross-covariances as well as the spatial covariances for the two variables. It may be more practical to use regression methods to estimate the primary variable at sites where only the secondary variable is known (Phillips *et al.*, 1992).

Calculations   It is possible to obtain a negative kriging weight for a site, if another site lies between it and the point at which the estimate is required. This is known as a *screening* effect. The performance of a kriging system can be assessed by cross-validation. Each site, in turn, is removed from the data set. The value of the variable at the missing site is estimated by kriging, and the error is the difference between this estimate and the known value. This is the spatial equivalent of PRESS residuals. Deutsch and Journel(1992) have published a set of Fortran routines for geostatistical analysis.

Example 7.6
Burton *et al.* (1997) collected the data from clay samples (given in Table A16.11) as part of their study of the variability of landslide parameters. The UK national grid numbers, cohesion (10 kPa) depth (m) and slope (degrees) are given for 250 points (Figure 7.10(a)). The maximum depth that could be measured was 1.2 m, and asterisks represent depths greater than this. The N-S and E-W (approximate directions) variograms for cohesion are shown in Figure 7.10(b) and (c). The variance of the 250 cohesion data is 2.02. The sample variograms are reasonably consistent with an assumption of a circular variogram with a range of 30 m, a sill of 2.02, and no nugget effect. To demonstrate the calculations involved in kriging we will use only the nearest four sample points to a prediction point in the lower right corner of the field. It is convenient to use coordinates which have been scaled by subtracting the minimum values of x and y; specifically:

prediction point	(190, 10)
sample point 1	(203, 0, 4.1)
sample point 2	(180, 5, 5.4)
sample point 3	(206, 10, 3.8)
sample point 4	(184, 15, 4.8)

where the triples are (x, y, cohesion). We first calculate the distances between the points, e.g. the distance (h) between the prediction point and point 1 is $[(203 - 190)^2 + (0 - 10)^2]^{1/2}$ which equals 16.40. The covariance for these two points will be:

$$2.02 - 2.02(1.5 \times (16.40/30) - 0.5 \times (16.40/30)^3) = 0.53$$

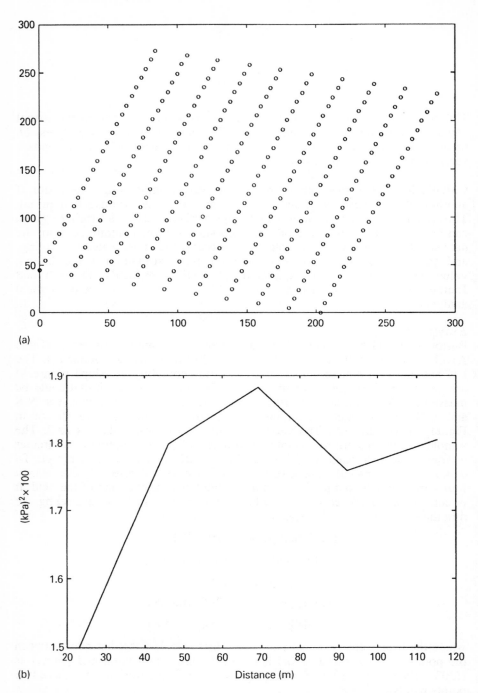

Figure 7.10 (a) Locations of sampling points and variograms for cohesion of clay in landslip study (b) N–S (c) E–W

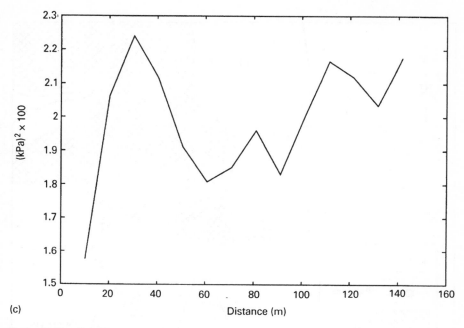

(c)

Figure 7.10 *Continued*

The ordinary kriging system is obtained from similar calculations for the remaining pairings

$$
\begin{bmatrix}
2.02 & 0.13 & 1.01 & 0.11 & 1 \\
0.13 & 2.02 & 0.04 & 0.98 & 1 \\
1.01 & 0.04 & 2.02 & 0.17 & 1 \\
0.11 & 0.98 & 0.17 & 2.02 & 1 \\
1 & 1 & 1 & 1 & 0
\end{bmatrix}
\begin{bmatrix} w_1 \\ w_2 \\ w_3 \\ w_4 \\ \lambda \end{bmatrix}
=
\begin{bmatrix} 0.53 \\ 0.94 \\ 0.56 \\ 1.25 \\ 1 \end{bmatrix}
$$

The solution for the weights is:

$$w_1 = 0.14 \quad w_2 = 0.21 \quad w_3 = 0.16 \quad w_4 = 0.49$$

and the kriged value at the prediction point is

$$0.14 \times 4.1 + 0.21 \times 5.4 + 0.16 \times 3.8 + 0.49 \times 4.8 = 4.67$$

It is worth checking that the larger weights do correspond to points nearer the prediction point.

7.5.5   Simulating random fields

Nuclear power generation produces hazardous waste, and burial of low level material in land sites has been proposed. Mackay and Cooper (1996) report a

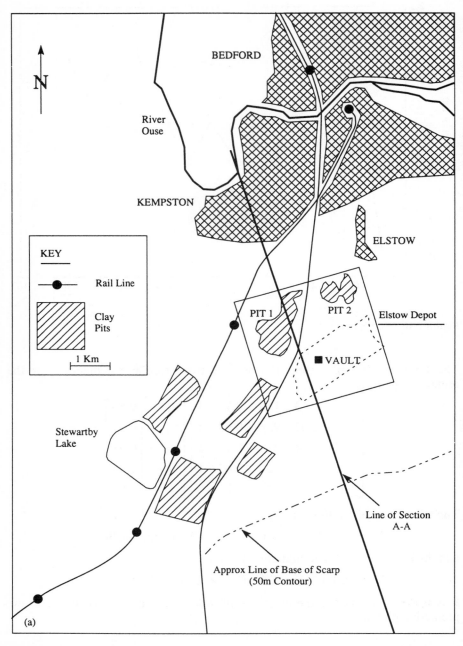

Figure 7.11 Hypothetical burial site for radioactive material in Bedfordshire, UK. (a) The area modelled in detail is the square containing the vault, pit 1 and pit 2. (b) Schematic lithological section along the line A–A

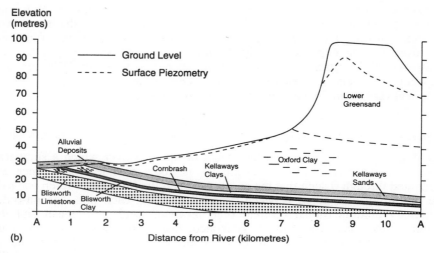

Figure 7.11   *Continued*

risk assessment for a hypothetical vault buried at a site in Bedfordshire, UK. The location of the site and a line section are shown in Figure 7.11. The vault is buried in the Oxford Clay Layer but particles will eventually seep out and migrate both upwards and downwards. If they reach the Kellaways Sands they will move much more rapidly than through the clay and are likely to reach the clay pits to the North of the vault. A quasi three-dimensional flow model of the regional hydrogeology of the Kellaways Sands and its semi-confining layers was devised. Three primary escape zones to the biosphere were modelled, the two clay pits (Pit 1 and Pit 2) and the ground surface above the vault itself (Near Vault). A fourth possible destination for the particles was the Blisworth Limestone.

The hydraulic properties of the Sands were not accurately known, so realizations from a plausible stochastic model were generated on a square grid consisting of 2500 cells of sides 50 m. Before constructing a stochastic model for this purpose, it was necessary to sample the region and measure the hydraulic properties of the sampled cores. These properties vary throughout the medium, in a loosely structured way with apparent random variation superimposed. That is, the properties are correlated in space and differences in measurements at two points are smaller, on average, the closer the points become. The variogram is a convenient device for modelling the correlation structure.

The Turning Bands Method, described by Mantoglou and Wilson (1982), assumes that the field to be simulated is second-order stationary and isotropic with a known covariance function. It is also assumed that values of the field at each point are normally distributed with zero mean. This is not too restrictive because it is often reasonable to assume that some transformation of the field data are normally distributed. The procedure will be described for two dimensions although the ideas easily extend to three, when the detailed mathematics, somewhat surprisingly, becomes easier. Instead of simulating a two-dimensional field directly, we perform simulations along several lines using a one-dimensional covariance function corresponding to the given two-dimensional one. The correspondence between the two covariance functions has usually been expressed in the frequency domain (see Chapter 8), but Dietrich (1995) describes an alternative approach (see also Dietrich and Newsam, 1993, and Tompson *et al.*, 1989). One of the simpler results is for the spherical variogram. The corresponding linear correlation function is

$$\rho(l) = 1 - 3\pi l(2a^2 - l^2)/(8a^3) \qquad \text{for } l \leqslant a$$
$$\rho(l) = 1 + [3 \arcsin(a/l)(l^2 - 2a^2)l - 3al(l^2 - a^2)^{1/2}]/(4a^3) \qquad \text{otherwise.}$$

Let A in Figure 7.12 represent the extent of the area over which a simulated field is required. An arbitrary origin, 0, is taken outside A and lines are generated through 0 such that the angles between the lines and a reference axis are uniformly distributed between 0 and 2π. Along each line, second-order stationary unidimensional discrete time processes with zero mean and the required covariance function are generated. Denote the process on line i by $z_i(\xi_i)$. The value of the field function f at any point, R say, is proportional to the sum of the values of these processes evaluated at the points where the orthogonal projection from R meets the lines. That is

$$f(\boldsymbol{x}_R) = \frac{1}{\sqrt{L}} \sum_{i=1}^{L} z_i(\boldsymbol{x}_R.\boldsymbol{\hat{u}}_i)$$

where $\boldsymbol{x}_R$ is the vector OR and $\boldsymbol{\hat{u}}_i$ is a unit vector in the direction of the line i. The factor of $1/\sqrt{L}$ is used to keep the variance finite as L tends to infinity and it enables asymptotic results to be proved. In practice, L can be about 16 and it is preferable, and theoretically justifiable, to space the lines evenly. Hosking has shown that long term dependence in the field can be simulated by using fractionally differenced ARIMA models for the univariate processes. Spherical models of the variograms were adopted for this study, and their parameters were estimated by Kriging. Two sample realizations of the logarithm (base ten) of hydraulic conductivity in the Kellaway Sands, obtained from the Turning Bands Method, are shown in Figure 7.12. The variogram parameters were 0.0 for the nugget variance, 0.2 for the sill and 2000 m for the range. The mean logarithm of hydraulic conductivity was taken as -0.6 $(\log(\text{m/s}))$ for the simulation experiments but is higher than would normally be expected. The variogram parameters used for the thickness of the Kellaway Sands were 0.0 for the nugget variance, 0.6 m^2 for the sill and 2000 m for the range. The mean thickness was taken as 4.0 m. Contaminant transport was modelled using a particle tracking methodology which included advection and diffusion. The particle trajectories to the pits were dominated by advection, whereas diffusion accounted for particles reaching the ground surface through the Oxford Clay. Estimators of variogram parameters made from limited field data have a large variance, and it is essential to investigate the effects of changing assumed values of the model parameters in any simulation study.

So far, experiments have concentrated on the effects of changing the conductivity variogram parameters, with those for the thickness being unchanged. Simulations have been performed for low, medium and high estimates of the sill, and the range, with the nugget variance assumed to be

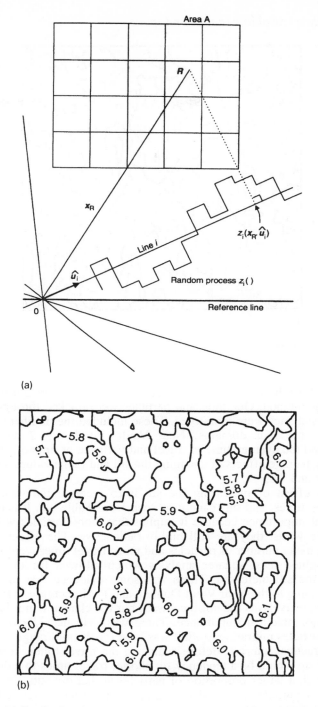

(a)

(b)

Figure 7.12 (a) Turning bands method, (after Mantaglou and Wilson 1982), and (b) and (c) typical realizations

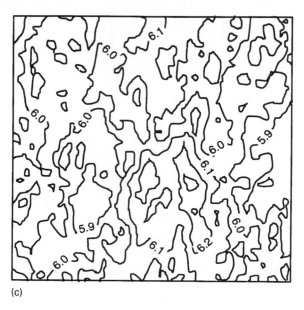

(c)

Figure 7.12   *Continued*

zero. For each of these nine scenarios, 28 random fields were generated by the turning bands method. One thousand particles were tracked through each field and various statistics were monitored. These included the numbers of particles migrating to the four domains, Near Vault, Pit 1, Pit 2 and Blisworth Limestone and the average time of travel and variance of time of travel to each domain. The average effects of changing the sill and range values are presented in Tables 7.7 and 7.8.

The main effect of increasing the sill is to increase the variance of the particle travel times, and this cannot be seen from Table 7.8. It is of considerable practical importance because it implies that more particles will escape to the biosphere whilst they are still highly radioactive. The effects of sill and range on the mean number of particles and mean travel times were investigated using a weighted least squares analysis which allowed for the increase in variance as the sill increased. There was some evidence that the number of particles escaping to the Near Vault Domain and Pit 1 decreased ($P < 0.05$) and increased respectively as the sill increased, but the change is not large enough to be of major practical importance. There was evidence ($P < 0.01$) that the number of particles migrating to Pit 2 was least for the mid range value but this is of little practical interest. Before analysing the travel times, it was thought likely that there would be a tendency for shorter travel times to a domain to be associated with a larger proportion of particles at that domain. For this reason the number of particles is included as a covariate in the regression model for each domain

$$Y_{ijk} = \mu + \alpha_i + \beta_j + B_k + \gamma n_{ijk} + E_{ijk}$$

In this model:

Y_{ijk} is the mean travel time to the domain of interest for realization k with sill i and range j, where i and j range from 1 to 3 and k ranges from 1 to 29;

n_{ijk} is the number of particles arriving;

Table 7.7   *Mean number of particles at domains*

	Sill 0.10	0.20	0.66	Range	Average over sill
Near Vault	616	604	554	1000	591
Pit 1	179	204	249		211
Pit 2	66	62	75		68
Blisworth	139	130	122		130
Near Vault	616	611	596	2000	608
Pit 1	151	163	161		161
Pit 2	89	96	134		106
Blisworth	136	131	109		125
Near Vault	614	595	527	4000	579
Pit 1	176	214	317		236
Pit 2	73	60	37		57
Blisworth	137	130	119		129
	Average over range				*Overall average*
Near Vault	615	603	559		593
Pit 1	171	194	242		203
Pit 2	76	73	82		77
Blisworth	137	130	117		128

Table 7.8   *Mean times of travel to domains (years)*

	Sill 0.10	0.20	0.66	Range	Average over sill
Near Vault	3380	3370	3270		3340
Pit 1	4670	4500	4690	1000	4620
Pit 2	5240	5420	6640		5770
Near Vault	3350	3380	3300		3340
Pit 1	4450	4540	4570	2000	4520
Pit 2	4870	5260	4870		5010
Near Vault	3360	3330	3200		3300
Pit 1	4360	4550	3430	4000	4120
Pit 2	5170	5050	6130		5450
	Average over range				*Overall average*
Near Vault	3360	3360	3260		3330
Pit 1	4490	4530	4230		4420
Pit 2	5090	5240	5890		5410

(Travel times to the Blisworth limestone are longer.)

α_i and β_j are the sill and range effects where $\sum \alpha_i = 0$ and $\sum \beta_j = 0$ and μ is an overall mean

B_k is the block effect which was included because the same sequence of random numbers was used with each combination of sill and range values.

E_{ijk} are the random variations which were assumed to be independent of the other parameters in the model and each other. Their variance was assumed to increase in proportion to the sill, and the variances of the residuals were consistent with this. The ANOVA, in order of adding variables, for the Near Vault Domain is given in Table 7.9. It can be seen that there is very strong evidence to support the suggestion that shorter travel times to a domain are associated with a larger proportion of particles arriving at it. The results are qualitatively similar for the other domains. The report was only concerned with developing the geostatistical methodology and did not attempt to provide a 'Probabilistic Risk Assessment'. In particular, the possibility of faulting, which could have drastic effects, has not been allowed for.

7.5.6 Gibbs' distributions

The Gibbs distribution is a probability distribution for a spatial process with a specified conditional probability distribution. Gibbs distributions are called after Josiah Willard Gibbs (1839–1903), Professor of Mathematical Physics at Yale from 1871, who is renowned for his work in thermodynamics. The main statistical applications of Gibbs distributions are related to image processing. The Gibbs distribution is used to generalize the ideas of Markov processes to spatial processes so that the probability distribution for the grey scale value of a pixel is assumed to depend only on its nearest neighbours. Troutman and Karlinger (1992, 1994) draw on an analogy between the energy in a physical system and the degree of sinuosity in a channel network. They start by approximating the drainage network by a lattice of points. The probability of a particular drainage network (s) is proportional to

$$\exp(-\beta H(s))$$

where β is a parameter to be estimated and $H(s)$ is defined to be the difference between total distance to the outlet through the drainage network and total distance along shortest paths to the outlet.

Table 7.9 *Analysis of variance of travel times to Near Vault Domain*

Source of variation	Degrees of freedom	Corrected sum squares	Mean square	F-ratio
Blocks	28	109	3.9	0.7
Number of particles	1	220	220.0	42.6
Sill	2	10	5.0	1.0
Range	2	18	9.0	1.7
Residual	227	1171	5.2	
Total	260			

They allowed β to depend on drainage area and average channel slope and estimated the relationship for 50 drainage networks in the vicinity of Willow Creek in Montana. The choice of the Hamiltonian energy function $H(s)$, or the form of conditional probabilities in the image processing context, is not completely arbitrary. Some consistency conditions are necessary (Israel, 1990).

7.6 Summary

1. Markov chains are discrete state space, discrete time models for random processes with the Markov property, i.e. future states depend on the present state but not on past ones.

2. Times series are usually thought of as being measurements of a continuous variable at, or summed between, discrete time points. Trends and seasonal patterns should be identified and allowed for before any more detailed analysis is attempted. Three useful methods are the centred moving average, standardization and harmonic cycles. In the first, the series is deseasonalized (also known as seasonally adjusted) before any trend is identified. In the second, any trend is identified and removed before the seasonal variation is investigated. Harmonic cycles can be fitted at the same time as a trend. Results that rely on the extrapolation of trends should be treated with caution. Differencing is not suitable for long simulations. A stochastic process is second order stationary if its mean and covariance structure do not change over time. Standard ARMA models are stationary. Once a time series has been deseasonalized and had any trend removed a suitable ARMA model can be fitted. The residuals should appear to be uncorrelated, and have a smaller standard deviation than the deseasonalized, detrended, series. Although ARMA models are useful for many purposes, there are important hydrological applications which need physically based models. This has led to the development of sophisticated rainfall models to provide realistic inputs.

3. Point process models are a flexible means of modelling time series which show marked asymmetry. Although these models are often built up from Poisson processes, the Markov property is not preserved.

4. Geostatistical methods, such as kriging, were originally developed to model spatial variation when changes over time were negligible. Models that combine spatial and temporal variation include spatial disaggregation, multisite models, and field models.

Exercises

1 A dealer can order mini-computers at the end of one week for delivery at the beginning of the next. If he has 0 or 1 computer in stock at the end of a week he places an order for 2, otherwise no order is placed. Let D_k be the demand for computers during the kth week. Assume that the demand in

any week is independent of the demand in previous weeks, and that

$$\Pr\{D_k = 0\} = 0.4$$
$$\Pr\{D_k = 1\} = 0.2$$
$$\Pr\{D_k = 2\} = 0.2$$
$$\Pr\{D_k = 3\} = 0.2$$

Note that demand can exceed the stock. The dealer starts with three mini-computers in stock.

Let X_k be the number of computers in stock at the end of the kth week. The sequence $\{X_k\}$ is a Markov chain.
(i)   Write down the one-step transition matrix.
(ii) Find the probability that he will have no computers at the end of the third week.
(iii) Find the stationary probability vector and deduce that the long run proportion of times that he has no computers in stock at the end of the week is 1/3.

2. A space-probe can be in exactly one of the following four states: on course (1), slightly off course (2), badly off course (3) and aborted (4). If it is off course course-correction signals are sent out every second. Let X_k be the state of the probe after the kth signal has been received. Assume that the sequence $\{X_k\}$ is a Markov chain with one step transition matrix

$$\begin{array}{c c c c c} & 1 & 2 & 3 & 4 \\ 1 & \begin{pmatrix} 1 & 0 & 0 & 0 \\ 2 & 0.4 & 0.3 & 0.2 & 0.1 \\ 3 & 0.1 & 0.5 & 0.2 & 0.2 \\ 4 & 0 & 0 & 0 & 1 \end{pmatrix} \end{array}$$

If it starts slightly off course, show that the probability that it will eventually be on course is 0.74.

3. The following data are the average flows (m³/s) through the spillway of a dam on the Island of Losden over the past five years. There are two seasons on Losden, the dry season which is quite wet, and the wet season which is very wet. During the past five years there have been extensive building programmes in many parts of the catchment area of the dam.

17   148   86   285   97   213   72   375   4   301

(i)   Plot the data and comment.
(ii) Deseasonalize the data using the centred moving average method, assuming (a) additive seasonal effects; and (b) multiplicative seasonal effects. Estimate the seasonal effects and plot both sets of deseasonalized data on the same graph as the original data.
(iii) Draw a straight line through whichever deseasonalized set looks the more appropriate.
(iv) Predict the average flows for next year.

Notice that if we deseasonalize and then estimate, and remove, the trend, we forecast by adding the trend back and finally reseasonalizing.

4. Show that $|\rho(k)| \leq 1$, by expanding the left-hand side of the inequality

$$\text{var}(X_t + bX_{t+k}) \geq 0$$

and substituting 1 and -1 for b.

5. Complete the following derivation of the approximate expected value of the sample acvf, defined as

$$c(k) = \sum_{t=1}^{n-k} (X_t - \bar{X})(X_{t+k} - \bar{X})/n$$

where the upper case is used to emphasize random variables.

(i) Write $\bar{X}$ as $\mu + (\bar{X} - \mu)$ and multiply out to obtain a sum of four terms.

(ii) Assume $\bar{X} \approx \sum_{t=1}^{n-k} X_t/(n-k) \approx \sum_{t=1}^{n-k} X_{t+k}/(n-k)$

(iii) Hence obtain

$$E[c(k)] \approx (n-k)\gamma(k)/n - (n-k)\text{var}(\bar{X})/n$$

(iv) Remember that

$$\text{var}(X_1 + \cdots + X_n) = \text{var}(X_1) + \cdots + \text{var}(X_n) + 2\,\text{cov}(X_1 X_2)$$
$$+ \cdots + 2\,\text{cov}(X_1 X_n) + 2\,\text{cov}(X_{n-1} X_n)$$

and hence deduce

$$\text{var}(\bar{X}) = \gamma(0)/n + 2\sum_{i=1}^{n-1} (n-1)\gamma(i)/n^2$$

(v) Since $\sum \gamma(i)$ is bounded as n tends to infinity for a stationary process the result

$$E[c(k)] \approx \gamma(k)(1 - k/n)$$

follows.

6. Consider the AR(2) process

$$X_t = a_1 X_{t-1} + a_2 X_{t-2} + E_t$$

This is stationary if, and only if, θ defined by

$$\theta = a_1 \pm \sqrt{(a_1^2 + 4a_2)}/2$$

is such that $|\theta| < 1$. Show that if θ satisfies this condition, and θ is real, then

$$a_1 + a_2 < 1 \qquad \text{and} \qquad a_1 - a_2 > -1$$

Show that if θ satisfies this condition, and θ is complex, then

$$a_2 > -1$$

7. Consider two steps of a zero mean AR(1) process, i.e. the random variable (X_1, X_2) where:

$$X_{t+1} = aX_t + E_{t+1}$$

Assume E_t are independent normal random variables with mean 0 and variance σ^2. Then (X_1, X_2) is bivariate normal. Write down the variance-covariance matrix for this distribution, and hence show that the maximum likelihood estimator of α is not the same as the least squares estimator. What condition will the ML method impose on $\hat{\alpha}$ that the LS method does not?

8. Suppose

$$Y_t = a + bt + E_t$$

Show that ∇Y_t is MA(1) with mean b, whereas $Y_t - (a + bt)$ is DWN.

9. Show that summing,

$$\sum_1^t,$$

is the inverse of differencing, ∇, if x_0 is taken to be 0 by demonstrating:

(i) $\nabla\left(\displaystyle\sum_{i=1}^t x_i\right) = x_t$

(ii) $\displaystyle\sum_{i=1}^t (\nabla x_i) = x_t$

10. Consider the model

$$Y_t = m_t + E_t$$

where E_t is zero mean DWN with variance σ_E^2, $E_0 = 0$, and m_t are monthly seasonal effects, i.e. m_1 represents the January effect, ..., m_{12} represents the December effect, and $m_{i+12} = m_i$. Then

$$\nabla_{12} Y_t = E_t - E_{t-12}$$

and

$$Y_{12} = Y_0 + E_{12}$$
$$Y_{24} = Y_{12} + E_{24} - E_{12} = Y_0 + E_{24}$$

and so on.

Now suppose that the model has been identified from a fairly short time series as

$$\nabla_{12} Y_t = E_t - 0.9E_{t-12}$$

Show that

$$Y_{36} = Y_0 + E_{36} + 0.1E_{24} + 0.1E_{12}$$

What is the variance of Y after n years?

11. Assume you have a time series $\{x_t\}$ for $t = 1, \ldots, n$ and wish to fit the model

$$X_t = (\alpha + \beta E_{t-1})X_{t-1} + E_t$$

Write e_t for the value taken by E_t.

(i)   Express e_t in terms of α, β, x_t, x_{t-1} and e_{t-1}. The function

$$\psi(\alpha, \beta, e_1) = \sum_{2}^{n} e_t^2$$

can then be minimized by a search procedure to give least squares estimates. The search would be reduced if e_1 is set equal to zero.

(ii) Repeat (i) for the model with βE_t in place of βE_{t-1}.

12. Take random numbers from $N(0, 1)$ and investigate realizations from the model

$$X_t = (0.6 + 0.3E_t)X_{t-1} + E_t$$

Repeat using random numbers from an $M(1)$ distribution with 1 subtracted.

13. Patching algorithm

We have mean annual flow data for three sites over 7 years.

	Site 1	Site 2	Site 3
Year1	29	54	115
Year 2	*	32	99
Year 3	24	52	75
Year 4	5	2	36
Year 5	32	*	*
Year 6	26	54	46
Year 7	43	50	121

(i)   Use the EM algorithm to estimate the missing values (*).

Step 0.   Standardize the time series for each site by subtracting the mean and dividing by the standard deviation. Replace missing values by 0.

Step 1.   Regress Site 1 on Site 2 and Site 3; Site 2 on Site 1 and Site 3; and Site 3 on Site 1 and Site 2. (This is the M-step.)

Step 2.   Use the regressions to re-estimate the missing values. (This is the E-step.)

Repeat Steps 1 and 2 together until convergence.

(ii) The regressions are fitted by least squares. What assumptions make the least squares estimates identical to maximum likelihood estimates?

8

Spectral Analysis and More on Dynamic Systems

8.1 Introduction

Although wind and wave forces affecting structures are random, they can be described in terms of an average frequency distribution. This is crucial information for designers, who need to consider the interaction between structures and the amplitudes and frequencies of disturbances. For example, if a structure can be realistically modelled by a lightly damped linear system designers must ensure that its natural frequencies are far away from the frequency range of the disturbances. Many civil engineering structures, such as suspension bridges, are best represented by non-linear models (Doole and Hogan, 1996), and their response to disturbances may have to be predicted from computer simulations and wind tunnel tests (Hay, 1992). Realistic modelling of wind forces, in terms of both amplitude and frequency distribution is vital. In a different context, we might wish to take advantage of the frequency composition of water waves, by tuning wave energy devices to the range containing most of the energy.

Uneven road or runway surfaces can also be described in terms of frequency, which is now measured in cycles per metre instead of cycles per second. This has implications for road speeds and suspension design. Again, we wish to avoid disturbance force frequency coinciding with natural frequencies of suspensions, despite the heavy damping. The disturbance frequency will be proportional to road speed.

8.2 The spectrum

8.2.1 The sample spectrum

We start with a time series of length n from a stationary stochastic process. It makes the algebra much simpler if we assume n is even, define $m = n/2$, and start measuring discrete time from $-m$. Then the sequence of observations is

$$\{x_t\} \quad \text{for} \quad t = -m, \ldots, 0, \ldots, m-1$$

The time between observations is known as the sampling interval. It must be short in comparison to the highest frequency components of the signal. We now fit a *finite Fourier[1] series* through these points. This is a continuous function of t, which is a sum of m harmonics at the following frequencies: 1 cycle per record length ($2\pi/n$ radians per sampling interval); 2 cycles per record length; and so on until m cycles per record length. A convenient parameterization is:

$$x(t) = R_0 + 2 \sum_{p=1}^{m-1} R_p \cos(p\omega_1 t + \phi_p) + R_m \cos m\omega_1 t$$

where $\omega_1 = 2\pi/n$, R_0 is a mean level, the other R_i are amplitudes of the harmonics and the ϕ_i are phase shifts.

The $2m$ parameters R_0, R_1, $\ldots, R_m$, ϕ_1, $\ldots, \phi_{m-1}$ are determined by requiring that $x(t)$ passes through all n data points. *Parseval's Theorem* for a finite Fourier series (see Appendix 13) can be stated as

$$\sum_{t=-m}^{m-1} x_t^2 / n = R_0^2 + 2 \sum_{p=1}^{m-1} R_p^2 + R_m^2$$

Since R_0 is $\bar{x}$, the result can be written as a partitioning of the variance over frequencies:

$$\sum_{t=-m}^{m-1} (x_t - \bar{x})^2 / n = 2 \sum_{p=1}^{m-1} R_p^2 + R_m^2$$

A plot of the contribution to the variance $(2R_p^2)$ against the frequency of the harmonic ($p2\pi/n$) is known as the *Fourier line spectrum*. It is much easier to find formulae for R_p in terms of $\{x_t\}$ if we use complex numbers. Proofs are given in Appendix 13, but you should remember the crucial result that

$$e^{i\theta} = \cos\theta + i\sin\theta$$

(see exercise 1(i)) is the complex representation of a harmonic wave. It will simplify the algebra if we assume $\{x_t\}$ has had the mean subtracted. Then[2]:

$$x_t = \sum_{p=-m}^{m-1} X_p \, e^{i2\pi pt/n}$$

where

$$X_p = (1/n) \sum x_t \, e^{-i2\pi pt/n}$$

---

[1] Named after Jean Baptiste Fourier (1768–1830), a professor at the Ecole Polytechnique from its foundation in 1795, who used trigonometric series to solve a heat conduction equation.
[2] Throughout this chapter lower case and upper case letters are used to denote time domain variables and transformed variables respectively.

so $X_{-p} = X_p^*$, and Parseval's Theorem can be written

$$\sum_{t=-m}^{m-1} x_t^2/n = \sum_{p=-m}^{m-1} |X_p|^2$$

We could plot $|X_p|^2$ against the frequency $2\pi p/n$ (radians/sampling interval), to show the contribution to the variance of the sequence (Figure 8.1). However, $\{x_t\}$ is just one realization of the random process and the discrete frequencies to which the variance is attributed are a direct consequence of the record length. Therefore, $|X_p|^2$ should be interpreted as an estimate of the contribution to the variance at frequencies between $2\pi(p-\frac{1}{2})/n$ and $2\pi(p+\frac{1}{2})/n$. If a rectangle of height $n|X_p|^2/2\pi$ and width $2\pi/n$, centred on $2\pi p/n$, is drawn, then the area equals $|X_p|^2$ and corresponds to the contribution to the variance between these frequencies. We define the *periodogram* as

$$\tilde{C}(\omega) = \frac{n}{2\pi} |X_p|^2 \qquad \text{for} \qquad 2\pi(p - \tfrac{1}{2})/n < \omega < 2\pi(p + \tfrac{1}{2})/n$$

and p from $-m$ to $m-1$.

If we now substitute for X_p we obtain a smooth function of ω

$$\tilde{C}(\omega) = \frac{n}{2\pi n} \left| \sum_{t=-m}^{m-1} x_t\, e^{-i\omega t} \right|^2 \qquad \text{for} \qquad -\pi < \omega < \pi \qquad (8.1)$$

where ω has units of radians per sampling interval, and varies continuously from $-\pi$ to π. Since negative frequency has the same physical interpretation as the

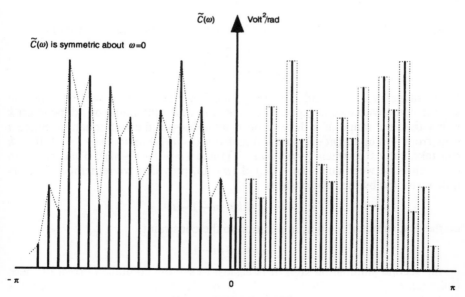

Figure 8.1   Fourier line spectrum and periodogram for a time series ($n = 52$)

corresponding positive frequency, $\tilde{C}(\omega)$ is symmetric about the vertical axis, it is usual to plot $2\tilde{C}(\omega)$ against ω for $0 < \omega < \pi$. Although $\tilde{C}(\omega)$ can be calculated for any ω in its domain, it is usual to calculate ordinates at $\omega = 2\pi p/n$ for integer p. The computational form of equation (8.1) is given in Exercise 1, but the fast Fourier transform algorithm is a more efficient means of calculation (see Firth, 1992 for example). The heights of these ordinates are asymptotically independent, whereas heights of ordinates at intermediate values of ω are not. The longer the time series the more independent ordinates we can calculate, but individual ordinates are no more precise. The resolution increases, but the precision does not. A compromise is to average q neighbouring ordinates, using a moving average, and this is known as a *Daniell window* of *bandwidth* $2\pi q/n$. This smoothed periodogram is known as the sample spectrum, which we will denote by $C(\omega)$. However, it does not lead directly to a satisfactory definition of the spectrum of the underlying random process. A definition emerges from the, perhaps rather surprising, relationship between the acvf and the spectrum.

8.2.2   The spectrum of a random process

For any complex number z, $|z|^2 = zz^*$, so we can write the periodogram as

$$\tilde{C}(\omega) = \frac{1}{2\pi n}\left\{\left(\sum_{t=-m}^{m-1} x_s\, e^{-i\omega s}\right)\left(\sum_{s=-m}^{m-1} x_t\, e^{i\omega t}\right)\right\}$$

and this can be rewritten as

$$= \frac{1}{2\pi n}\sum_{t=-m}^{m-1}\sum_{s=-m}^{m-1} x_s x_t\, e^{-i\omega s}\, e^{i\omega t}$$

If we now substitute $s = t + k$ we obtain a sum of lagged products. That is

$$= \frac{1}{2\pi n}\sum_{k=-m-t}^{m-1-t}\sum_{t=-m}^{m-1} x_{t+k}x_t\, e^{-i\omega k}$$

If the range of t is modified slightly, the inner summation will become the sample autocovariance at lag k. The trick is to make the range of t depend on k rather than the range of k depend on t, as it does in the last equation. Since t runs from $-m$ up to $(m-1)$ and k runs from $(-m-t)$ up to $(m-1-t)$, then k can take values between $(-m-(m-1))$ and $(m-1-(-m))$. That is, k can take values between and $-(n-1)$ and $(n-1)$, provided we remember that t is now restricted by the relationships

$$-m - t \leqslant k \leqslant m - 1 - t$$

on the first summation sign, in addition to the constraint

$$-m \leqslant t \leqslant m - 1$$

on the second summation sign. It is easiest to see the combined effect of these if the range of k is split into $-(n-1)$ up to -1, and 0 up to $(n-1)$. When k is in the former range, that is, when it is negative, the restriction

$$-m - t \leqslant k$$

bounds t below. When k is positive the restriction

$$k \leq m - 1 - t$$

bounds t above. Therefore

$$\tilde{C}(\omega) = \frac{1}{2\pi n} \left\{ \sum_{k=0}^{n-1} \sum_{t=-m}^{m-1-k} x_{t+k} x_t \, e^{-i\omega k} + \sum_{k=-(n-1)}^{-1} \sum_{t=-m}^{m-1} x_{t+k} x_t \, e^{-i\omega k} \right\}$$

Now remember that $\{x_t\}$ has a mean of zero, so

$$\tilde{C}(\omega) = \frac{1}{2\pi} \left\{ \sum_{k=0}^{n-1} c(k) \, e^{-i\omega k} + \sum_{k=-(n-1)}^{-1} c(k) \, e^{-i\omega k} \right\}$$

$$= \frac{1}{2\pi} \sum_{k=-(n-1)}^{n-1} c(k) \, e^{-i\omega k} \qquad \text{for} \qquad -\pi \leq \omega \leq \pi$$

The spectrum of the random process follows from letting n tend to infinity, but there is a subtlety. Although letting n tend to infinity will increase the number of independent ordinates, individual ordinates are no more precise and do not tend to a limiting value. We can evade this difficulty by taking the expected value over the ensemble. The spectrum of the process is defined by

$$\Gamma(\omega) = E\left[\lim_{n \to \infty} \tilde{C}(\omega) \right]$$

$$= \lim_{n \to \infty} \frac{1}{2\pi} \sum_{k=-(n-1)}^{n-1} E[c(k)] \, e^{-i\omega k}$$

$$= \lim_{n \to \infty} \frac{1}{2\pi} \sum_{k=-(n-1)}^{n-1} \gamma(k)(1 - |k|/n) \, e^{-i\omega k}$$

$$= \frac{1}{2\pi} \sum_{k=-\infty}^{\infty} \gamma(k) \, e^{-i\omega k} \qquad \text{for} \qquad -\pi < \omega < \pi \qquad (8.2a)$$

This is known as the *discrete Fourier transform* (see Appendix 13) of the autocovariance function. The discrete Fourier transform is defined if $\sum |\gamma(k)| < \infty$, which is a condition satisfied by stationary processes. The inverse transform is the relationship

$$\gamma(k) = \int_{-\pi}^{\pi} \Gamma(\omega) \, e^{i\omega k} \, d\omega \qquad (8.2b)$$

If we substitute $k = 0$ we obtain

$$\gamma(0) = \int_{-\pi}^{\pi} \Gamma(\omega) \, d\omega$$

which is a statement that the area under the spectrum equals the variance of the process. We defined the spectrum over a frequency range $(-\pi, \pi)$ to make the algebra easier, with the understanding that negative frequencies are physically equivalent to the corresponding positive frequency. In applications we usually use a one-sided spectrum

$$\Gamma(\omega) \quad \text{for} \quad 0 \leqslant \omega \leqslant \pi$$

It is defined as twice the height of the two-sided spectrum, so that the area underneath still equals the variance of the process. The spectra, acf and realizations of three simple autoregressive processes are shown in Figure 8.2. The spectrum of an AR(1) process is

$$\Gamma(\omega) = \sigma_E^2/[\pi(1 - 2\alpha \cos \omega + \alpha^2)] \quad \text{for} \quad 0 \leqslant \omega < \pi$$

where σ_E^2 is the variance of the DWN. The spectrum of an AR(2) process with parameters α_1 and α_2 is,

$$\Gamma(\omega) = \sigma_E^2/[\pi(1 + \alpha_1^2 + \alpha_2^2 - 2\alpha_1(1 - \alpha_2)\cos \omega - 2\alpha_2 \cos 2\omega)] \quad \text{for} \quad 0 \leqslant \omega < \pi$$

They are both obtained from equation (8.2a). The AR(2) case is more involved and Jenkins and Watts (1968) give details. The AR(2) process in Figure 8.2(c) has $\alpha_1 = 1$, $\alpha_2 = -0.5$, and hence an acf

$$\rho(k) = 2^{-k/2}(\cos(\pi k/4) + (1/3)\sin(\pi k/4))$$

First-order autoregressive processes with positive or negative parameters are intuitively 'low frequency' or 'high frequency' respectively, and their spectra confirm this. Second-order autoregressive processes can exhibit more variety in their spectra, and can model a single degree of freedom linear system with a noise input (see Section 8.3.6).

Many theoretical results are much more conveniently expressed in terms of a continuous time random process $\{x(t)\}$

$$\{x(t)\} \quad \text{for} \quad -\infty < t < \infty$$

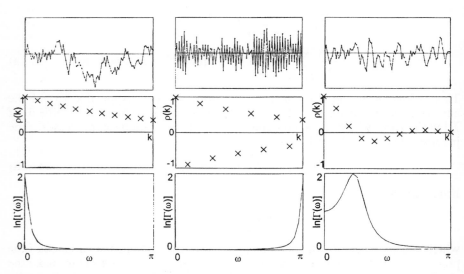

Figure 8.2 Spectra, acf, and realizations from (a) AR(1) with $\alpha = 0.9$; (b) AR(1) with $\alpha = -0.9$; (c) AR(2) with $\alpha = 1$ and $\alpha_2 = -0.5$

An analogous argument, with summation replaced by integration, leads to the definition of the spectrum of a continuous random process as the *Fourier transform* (Appendix 13) of the acvf. The acvf is

$$\gamma(\tau) = E[(x(t) - \mu)(x(t + \tau) - \mu)]$$

where the lag τ is on a continuous scale. The spectrum is

$$\Gamma(\omega) = \frac{1}{2\pi} \int_{-\infty}^{\infty} \gamma(\tau) \exp{(-i\omega\tau)}\, d\tau \qquad \text{for} \qquad -\infty < \omega < \infty \qquad (8.3a)$$

which has the inverse transformation

$$\gamma(\tau) = \int_{-\infty}^{\infty} \Gamma(\omega) \exp{(i\omega\tau)}\, d\omega \qquad (8.3b)$$

If a discrete random process is considered to be embedded in a continuous random process, the spectrum of the discrete process will equal that of the continuous process provided there are no frequencies higher than the reciprocal of twice the sampling interval (*Nyquist frequency*) in the continuous random process.

Example 8.1
The 396 data in Table A16.12 are from a wave tank. They are vertical distances (mm) from the still water level recorded at 0.1 second intervals by a probe at the centre of the tank. A spectrum, estimated by calculating moving averages of 15 neighbouring ordinates in the periodogram, is shown in Figure 8.3. The

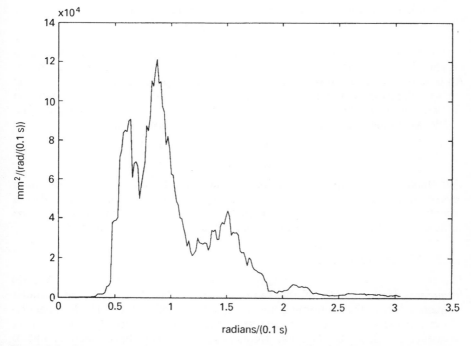

Figure 8.3   Spectrum for wave tank data, calculated with a Daniell window of bandwidth 0.24

periodogram was calculated from the formula in Exercise 1. An alternative method for calculating a sample spectrum, from the acvf, is given in Exercise 5.

Most engineers measure frequency in cycles per second (Hz). A frequency of 1 Hz is equivalent to 2π radians per second. You are asked to verify the scaling of the spectrum in Exercise 3. The next example is based on a consultancy, and the relevant British Standard uses Herz to estimate exposure to vibration.

Example 8.2
Personnel who operate earth moving equipment and pile drivers are exposed to mechanical vibration. The British Standard Guide BS 6841: 1987 gives methods for quantifying vibration and repeated shocks. The estimated vibration dose value ($eVDV$) is defined as:

$$eVDV = [(1.4 \times \tilde{a})^4 \times T]^{1/4}$$

where $\tilde{a}$ is the root mean square value of frequency weighted acceleration, and T is the duration in seconds. Let $\Gamma_{xx}(f)$, $\Gamma_{yy}(f)$ and $\Gamma_{zz}(f)$ be the spectra of linear acceleration measured in the forward (x), sideways (y) and vertical (z) directions measured at a point on the driving seat. Oscillatory translations in these directions are known as surge, sway and heave. The guide includes a weighting function $W(f)$, related to human responses, and the surge component is given by:

$$\tilde{a}_x^2 = \int \Gamma_{xx}(f)W^2(f)\,\mathrm{d}f$$

The other two components are defined in a similar way and

$$\tilde{a} = (\tilde{a}_x^2 + \tilde{a}_y^2 + \tilde{a}_z^2)^{1/2}$$

A typical application was the calculation of the $eVDV$ for the driver of a digger who was using the bucket to cut soft rock and coal in an open cast mine. An accelerometer was positioned on the seat. The spectrum of the signal in the vertical direction is shown in Figure 8.4, together with the weighting function $W^2(f)$, and $\tilde{a}_z^2 = 0.16$ m^2/s^4. The estimated value of $\tilde{a}$ was 0.42 m/s^2. The $eVDV$, assuming continuous exposure over an eight hour shift, was

$$eVDV = [(1.4 \times 0.42)^4 \times 60 \times 60 \times 8]^{1/4}$$
$$= 7.7 \text{ m s}^{-1.75}$$

The standard states that dose values in the region of 15 m s$^{-1.75}$ will cause severe discomfort. Oscillatory rotations (yaw, roll and pitch) are also a potential risk. The point about which the digger pitches was thought to be sufficiently removed from the cab for the linear motion to predominate.

8.2.3   Nyquist frequency

The frequency of π radian/(sampling interval) is known as the *Nyquist frequency*. If the length of the sampling interval is Δ seconds the Nyquist frequency is π radians/sampling interval or equivalently $1/(2\Delta)$ Hz. Frequencies which are higher than the Nyquist frequency will be indistinguishable from frequencies below the Nyquist frequency, once they have been sampled. Consider a harmonic wave with a frequency of 1 Hz, which is sampled at intervals of 0.2 s. This wave goes through 1/5 of a cycle between sampling

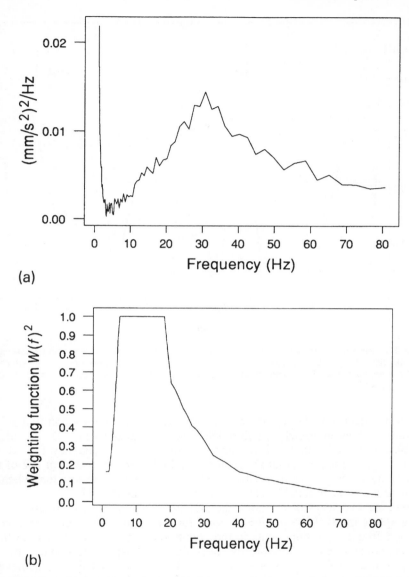

Figure 8.4   (a) Spectrum of vertical acceleration at driver's seat and (b) weighting function to calculate eVDV

points. It is therefore indistinguishable from waves that go through $\frac{1}{5}$th $\pm\, k$ cycles between sampling points, where k is any integer. The higher frequencies $1 \pm 5k$ Hz are known as alias frequencies because they are indistinguishable from 1 Hz. If we take $k = 1$ and a negative sign we have an alias frequency of -4 Hz, which is physically equivalent to 4 Hz. Figure 8.5 shows a wave of frequency 1 Hz and a wave of frequency 4 Hz coinciding at every sampling point. In this example the Nyquist frequency is 2.5 Hz. Aliasing could lead to

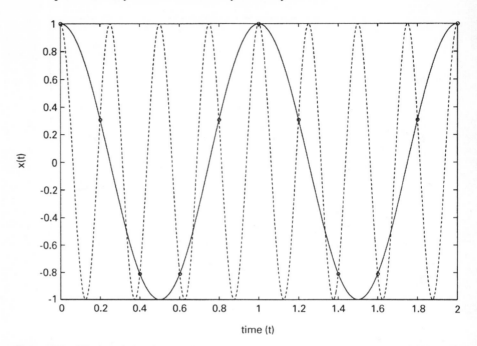

Figure 8.5 Aliasing of signals at 1 Hz and 4 Hz when sampling every 0.2 s. The frequency of 1 Hz has alias frequencies $1 \pm 5k$ for any k. If $k = -1$, the alias frequecy is -4, which is physically equivalent to 4 Hz (shown as a broken line). The Nyquist frequeny is $1/(2 \times 0.2)$ which equals 2.5 Hz

poor design decisions. For example, suppose a linear model for a lightly damped structure, such as an electricity pylon, has natural frequencies Ω_1, Ω_2, ..., Ω_n. It is subjected to a harmonic disturbance at some higher frequency Ω_H. If we sample the disturbance so that Ω_H is close to a high frequency alias of one of the lower frequencies, we would predict drastic fluctuations of the structure when, if the linear model is reasonable, it would be relatively unaffected.

There are two ways of avoiding the aliasing problem. One is to ensure that our sampling interval is sufficiently small for the Nyquist frequency to exceed any high frequency components in the continuous signal. The other is to filter out components whose frequency exceeds the Nyquist frequency by an analogue filter before the signal is sampled. It is not possible to achieve a perfect cut-off with physical filters. Typical spectral analysers use analogue pre-filters, which are designed to remove frequencies above 25 kHz. This cannot be achieved exactly so the sampling rate is set at 100 000 data per second, which makes an allowance for imperfections in the filter and the leakage which occurs with finite length signals. This is because frequencies near the Nyquist frequency will 'leak' into high frequencies, which will then act as alias frequencies (see Exercise 2). We can reduce the leakage effect by using a sampling interval that is somewhat less than that specified by the Nyquist frequency, e.g. between $1/(4f_H)$ and $1/(2.5f_H)$, where f_H (Hz) is the highest frequency that is expected to occur in the signal.

It is impossible to remove the effects of aliasing with digital filters. If the sampling frequency used to digitize the signal is too low, the between sampling points information has already been lost. However, modern electronic equipment can sample at rates which are faster than one million per second, and these are probably well in excess of anything needed for most civil engineering applications.

8.3 Linear dynamic systems

The simplest model for vibration of a structure is a one-dimensional (single degree of freedom) linear system. Applications include the modelling of wind loading on road sign gantries (Hay, 1992), wave power from the Salter duck, vibration of a footbridge (Jones *et al.*, 1981), and vortex shedding induced vibration of cables (Liu, 1991). However, it should be noted that the failure of the Tacoma Narrows bridge on 7 November 1940 cannot convincingly be attributed to resonance of a linear system (Billah and Scanlan, 1991).

8.3.1 Impulse response

Let $y(t)$ be the response of a linear system, which is initially at rest, to some force $u(t)$. Then

$$y(t) = \int_{-\infty}^{t} h(t - \tau)u(\tau)\,d\tau \qquad (8.4)$$

and $h(t)$ is the impulse response. To see why this is so, let $u(t)$ be a unit impulse at time 0. Then $u(t)$ can be approximated by a short pulse

$$u(t) = \begin{cases} 1/(2\delta) & \text{for } -\delta < t < \delta \\ 0 & \text{otherwise} \end{cases}$$

where δ is small. Then

$$y(t) = \int_{-\delta}^{\delta} h(t - \tau)(1/(2\delta))\,d\tau$$
$$= (2\delta)h(t)(1/(2\delta)) = h(t)$$

If we let $\delta \rightarrow 0$ we obtain the exact result that $h(t)$ is the response after time t to a unit impulse at time 0. The general result follows by considering $u(t)$ as a sequence of pulses, since the response of the linear system will be the sum of responses to individual pulses. The condition for a linear system to be stable is that $\int |h(\tau)|\,d\tau < \infty$. A useful alternative form of equation (8.4)

$$y(t) = \int_{0}^{\infty} h(\theta)u(t - \theta)\,d\theta$$

follows from the substitution $\theta = t - \tau$.

8.3.2 The convolution theorem

Equation (8.4) is known as a convolution integral.

If we now consider the Fourier transforms of $u(t)$ and $y(t)$ we can provide frequency dependent relationships between the input and output. In particular, we may show that

$$Y(\omega) = H(\omega)U(\omega) \tag{8.5}$$

where $H(\omega)$ is the Fourier transform of $h(t)$ (see Appendix 13). $H(\omega)$ is often referred to as the transfer function.

To prove this, we start from the definition of the Fourier transform of $y(t)$, i.e.

$$Y(\omega) = \int_{-\infty}^{\infty} \exp(-i\omega t) \left\{ \int_{-\infty}^{\infty} h(t-\tau)u(\tau) \, d\tau \right\} dt$$

Next, introduce the variable θ defined by

$$\theta = t - \tau$$

That is

$$Y(\omega) = \int_{-\infty}^{\infty} \left\{ \int_{-\infty}^{\infty} \exp(-i\omega\theta)\exp(-i\omega\tau)h(u)u(\tau) \, d\tau \right\} d\theta$$

since the Jacobian of the transformation is 1. Rearranging we have

$$Y(\omega) = \int_{-\infty}^{\infty} h(\theta)\exp(-i\omega\theta) \left\{ \int_{-\infty}^{\infty} u(\tau)\exp(-i\omega\tau) \, d\tau \right\} d\theta$$

and since the inner integral is a function of ω only

$$Y(\omega) = \int_{-\infty}^{\infty} h(\theta)\exp(-i\omega\theta) \, d\theta \int_{-\infty}^{\infty} u(\tau)\exp(-i\omega\tau) \, d\tau$$

that is

$$Y(\omega) = H(\omega)U(\omega)$$

8.3.3   Cross-covariance function and cross-spectrum

The cross-covariance function (ccvf) is defined by

$$\gamma_{uy}(\tau) = E[(u(t) - \mu_u)(y(t+\tau) - \mu_y)]$$

Notice that it is not symmetric about 0, since u determines future y whereas y does not affect future u.

The cross-spectrum is given by

$$\Gamma_{uy}(\omega) = \int \gamma_{uy}(\tau)\exp(-i\omega\tau)d\tau$$

which has a real and imaginary part. The sample ccvf is calculated as

$$c_{xy}(k) = \sum_{t=1}^{n-k} (x_t - \bar{x})(y_{t+k} - \bar{y})/n$$

An estimate of the cross-spectrum can be obtained by smoothing

$$\tilde{C}_{xy}(\omega) = \frac{n}{2\pi} X_m Y_m^*$$

or from the sample ccvf (see Exercise 5).

8.3.4   Relationships between spectra and the transfer function

It is convenient to assume the input $u(t)$ has zero mean for the derivation of the following results, although they are still applicable if $u(t)$ has a non-zero mean because all covariance functions are defined in terms of deviations from the mean. The first result shows the cross-spectrum to be the product of the spectrum of the input and the frequency response function.

$$\gamma_{uy}(\tau) = E[u(t)y(t+\tau)]$$

Substitution for $y(t+\tau)$ leads to

$$\gamma_{uy}(\tau) = E[u(t) \int h(\theta)x(t+\tau-\theta)\,d\theta]$$

$$= \int h(\theta)\, E[u(t)u(t+\tau-\theta)]\, d\theta$$

$$= \int h(\theta)\, \gamma_{uu}(t-\theta)\, d\theta$$

The result follows from the use of the convolution theorem, i.e.

$$\Gamma_{uy}(\omega) = H(\omega)\Gamma_{uu}(\omega) \tag{8.6}$$

The sample equivalent of this result is the usual way of estimating the transfer function, and is relatively unaffected by measurement noise on the response signal. The second result relates the input and output spectra. It is used to describe the response of a linear structure to a known input. The sample equivalent of the result can be used to estimate the modulus of the transfer function, but phase information is lost. It is more susceptible to noise on the response signal, but less prone to underestimate the height of spectral peaks.

The autocovariance function for the output signal y is

$$\gamma_{yy}(\tau) = E[y(t)y(t+\tau)]$$

If we substitute for $y(t)$ and $y(t+\tau)$ we can write

$$\gamma_{yy}(\tau) = E[\int h(\theta)u(t-\theta)\,d\theta \int h(\phi)u(t+\tau-\phi)\,d\phi]$$

$$= \iint h(\theta)h(\phi)E[u(t-\theta)u(t+\tau-\phi)]\, d\theta\, d\phi$$

That is

$$\gamma_{yy}(\tau) = \iint h(\theta)h(\phi)\gamma_{uu}(\tau+\theta-\phi)\, d\theta\, d\phi$$

Taking the Fourier transform of both sides gives

$$\Gamma_{uy}(\omega) = \iiint h(\theta)h(\phi)\gamma_{uu}(\tau+\theta-\phi)\exp(-i\omega\tau)\, d\theta\, d\phi\, d\tau$$

Use of the identity

$$\exp(-i\omega\tau) = \exp(-i\omega(\tau+\theta-\phi))\exp(+i\omega\theta)\exp(-i\omega\phi)$$

allows $\Gamma_{xy}(\omega)$ to be re-expressed as

$$\Gamma_{uy}(\omega) = \iiint \{h(\theta)\exp(i\omega\theta)\}\{h(\phi)\exp(-i\omega\phi)\}$$

$$\{\gamma_{uu}(\tau + \theta - \phi)\exp(-i\omega(\tau + \theta - \phi))\}d\theta\, d\phi\, d\tau$$

That is

$$\Gamma_{yy}(\omega) = H^*(\omega)H(\omega)\Gamma_{uu}(\omega) = |H(\omega)|^2\Gamma_{uu}(\omega) \qquad (8.7)$$

Example 8.3
Some wave energy devices, the Salter nodding duck for example, can reasonably be modelled as a single degree of freedom linear system.

$$M\ddot{y} + D\dot{y} + Ky = u$$

The easiest way to find $H(\omega)$ is to: substitute $u = e^{i\omega t}$ into the right-hand side of the differential equation; assume $y = A\exp[i(\omega t + \phi)]$ and substitute into the left-hand side; and then take modulus of both sides to find A as a function of frequency, i.e. $A(\omega)$. The relationship

$$H(\omega) = A(\omega)\quad e^{i\phi(\omega)}$$

follows from equation (8.5), and the result that the Fourier transform of $e^{i\theta t}$ is a pulse of area 2π located at the point $\omega = a$. This can easily be verified by substitution in the formula for the inverse transform. Hence

$$|H(\omega)|^2 = |A(\omega)|^2 = \frac{1}{[(K - M\omega^2)^2 + D^2\omega^2]}$$

Shaw (1982) shows that the spectrum of the power output

$$\Gamma_{output}(\omega) = k\omega^2|H(\omega)|^2\Gamma_{uu}(\omega)$$

where u is the wave elevation, and k is some constant of proportionality. A model of the Salter duck was moored in the wavetank of Example 1. The parameters K, M, D and k are assumed to be 1, 0.01, 0.08 and 1 respectively. The input spectrum, squared transfer function, displacement output spectrum and power output spectrum are shown in Figure 8.6.

Example 8.4
Hay (1992) applied an analysis of bridge response in a turbulent flow, by Scanlan and Gade (1977), to the Severn Bridge in the UK. The spectrum of displacement of a point on the deck in the vertical direction (z) was related to the spectra of components of wind velocity in the transverse (u) and vertical (w) directions.

$$\Gamma_{zz}(\omega) \propto (a\Gamma_{uu}(\omega) + \Gamma_{ww}(\omega))$$

The velocity spectra were assumed to be of a form (Exercise 4(a))

$$\Gamma(\omega) = b(1 + 3c\omega^2)/(1 + c\omega^2)^{11/6}$$

where b and c depended on the wind speed (v in the y direction) across the bridge. The variance of vertical displacement is given by

$$\sigma_z^2 = \int \Gamma_{zz}(\omega)\, d\omega$$

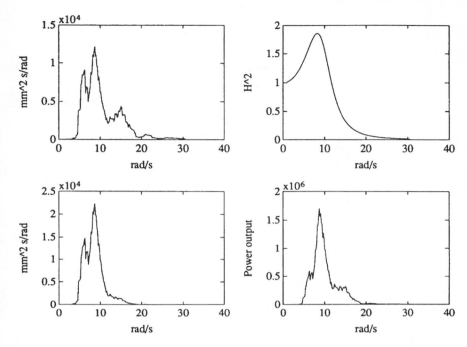

Figure 8.6 Input spectrum, squared transfer function, displacement output spectrum and power output spectrum for a model of the Salter duck in a wavetank

Using parameter values for the Severn Bridge, the values of σ_z were calculated to be 133 mm and 426 mm for wind speeds of 15 m s^{-1} and 30 m s^{-1} respectively.

Example 8.5
Paskalov *et al.* (1981) carried out forced vibration studies on an arch concrete dam as part of a Yugoslav research project on earthquake resistant designs. Two vibration generators were located 11 m either side of the middle of the crest. Accelerometers were positioned at various points along the crest and the downstream face, to determine natural frequencies and mode shapes (Thomson, 1993). A typical frequency response is shown in Figure 8.7.

8.3.5 Matrix solution of a linear dynamic system

Any linear dynamic system can be expressed in *state space* form:

$$\dot{x}(t) = Ax(t) + Bu(t)$$

$$y(t) = Cx(t)$$

where $u(t)$ is the vector of inputs, $x(t)$ is the *state vector*, and $y(t)$ is the observation vector. To solve the equation we need to know initial conditions $x(0) = x_0$

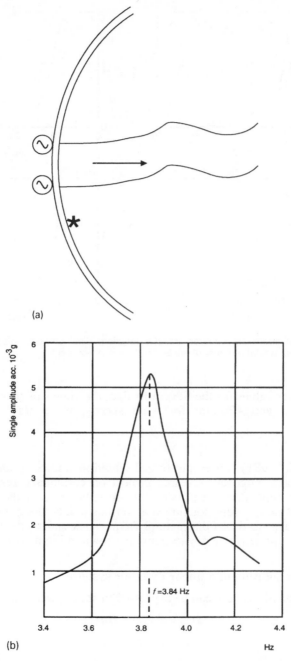

(a)

(b)

Single amplitude acc. 10^{-3} g

$f = 3.84$ Hz

Hz

Figure 8.7 Frequency response at a point location on a concrete arch dam to vibration generators near the middle of the crest (after Paskalov *et al.*, 1981)

Example 8.6
The standard form of a single degree of freedom linear system is

$$\ddot{y} + 2\xi\Omega\dot{y} + \Omega^2 y = u$$

where Ω is the natural frequency, ξ is the damping factor, and u is the forcing function. This can be expressed in state space form by setting

$$\boldsymbol{x}^{\mathrm{T}}(t) = (\,y,\,\dot{y}\,) \quad \boldsymbol{u}(t) = u$$

$$A = \begin{pmatrix} 0 & 1 \\ -\Omega^2 & -2\xi\Omega \end{pmatrix} \quad B = \begin{pmatrix} 0 \\ 1 \end{pmatrix}$$

If we measure the displacement, then $C = (1\ \ 0)$.

Matrix exponential   If A is a square matrix the matrix exponential of A is defined as

$$e^{At} = I + At + A^2 t^2/2! + A^3 t^3/3! + \dots$$

It follows that

$$\frac{\mathrm{d}}{\mathrm{d}t}(e^{At}) = A\, e^{At}$$

If A is diagonizable, i.e.

$$AM = M\Lambda$$

and hence

$$A = M\Lambda M^{-1}$$

where Λ is a diagonal matrix of eigenvalues, the matrix exponential can be calculated from

$$e^{At} = Me^{\Lambda t}M^{-1}$$

because

$$A^2 = (M\Lambda M^{-1})(M\Lambda M^{-1}) = M\Lambda^2 M^{-1}$$

etc.
   This is convenient because the exponential of a diagonal matrix is obtained by taking the exponential of the diagonal elements. To solve the system, rewrite as

$$\dot{x} - Ax = Bu$$

and multiply by the integrating factor e^{-At} to obtain

$$\frac{\mathrm{d}}{\mathrm{d}t}(e^{-At}x) = e^{-At}\,Bu$$

Now integrate both sides

$$e^{-At}x - e^{-A0}x(0) = \int_0^t e^{-A\tau}Bu\,\mathrm{d}\tau$$

and rearrange to obtain

$$x(t) = e^{At}x_0 + \int_0^t e^{A(t-\tau)}Bu(\tau)\,\mathrm{d}\tau$$

The matrix exponential is called the *state transition matrix* and is a generaliz-ation of the impulse response function. Although the continuous time representation of the system is more convenient for many theoretical develop-ments, we usually use sampled data. We now establish the link between discrete time and continuous time representations.

8.3.6   Discrete time model for a linear system

Assume the system is observed every T seconds, i.e. the measurements $y(0)$, $y(T)$, $y(2T)$, $y(3T)$... are available. Also assume the input u only changes in value at the sampling instants so that throughout the interval $kT < \tau < (k+1)T$, $u(\tau)$ has a constant value $u(kT)$. Then

$$x(kT) = e^{AkT}x_0 + \int_0^{kT} e^{A(kT-\tau)}Bu(\tau)\,d\tau \tag{8.8a}$$

$$x((k+1)T) = e^{A(k+1)T}x_0 + \int_0^{(k+1)T} e^{A((k+1)T-\tau)}Bu(\tau)\,d\tau \tag{8.8b}$$

Pre-multiply equation (8.8a) by e^{AT}, subtract from equation (8.8b), and rearrange to obtain

$$x((k+1)T) = e^{AT}x(kT) + \int_{kT}^{(k+1)T} e^{A((k+1)T-\tau)}Bu(\tau)\,d\tau$$

Now, $u(\tau)$ is assumed to have the constant value $u(kT)$ over the range of integration, so making the substitution $\theta = ((k+1)T - \tau)$ gives

$$x((k+1)T) = [e^{AT}]x(kT) - \left(\int_0^T e^{A\theta}B\,d\theta\right)u(kT)$$

Also

$$y(kT) = Cx(kT)$$

Example 8.7
The state space equations for Salter's duck in Example 8.3 are:

$$\begin{bmatrix} \dot{x}_1 \\ x_2 \end{bmatrix} = \begin{bmatrix} 0 & 1 \\ -1 & -0.4 \end{bmatrix}\begin{bmatrix} x_1 \\ x_2 \end{bmatrix} + \begin{bmatrix} 0 \\ 1 \end{bmatrix}u$$

$$y = \begin{bmatrix} 1 & 0 \end{bmatrix}\begin{bmatrix} x_1 \\ x_2 \end{bmatrix}$$

Suppose we observe the system every 0.1 s, so $T = 0.1$. An eigenvalue analysis, which is easy if you have access to Matlab, gives

$$A = M\Lambda M^{-1}$$

where

$$M = \begin{bmatrix} 0.6928 - 0.1414i & 0.6928 + 0.1414i \\ 0.7071i & -0.7071i \end{bmatrix}$$

$$\Lambda = \begin{bmatrix} -0.2 + 0.9798i & 0 \\ 0 & -0.2 - 0.9798i \end{bmatrix}$$

Then $e^{AT} = \begin{bmatrix} 0.9951 & 0.0979 \\ -0.0979 & 0.9559 \end{bmatrix}$

$$\int_0^T e^{A\theta} B \, d\theta = M \int_0^T e^{\Lambda\theta} \, d\theta M^{-1} B$$

$$= M(e^{0.1\Lambda} - I)M^{-1}B$$

$$= [0.0979 \quad -0.0441]^T$$

8.3.7   Kalman Filter

When we monitor a dynamic system we will not usually be able to measure, directly, all the variables we need for an analysis or for a control system. For example, we may measure displacements but also need velocities for a controller. The Kalman filter is a method of estimating all the states of a linear system from a reduced set of measurements. The question of whether or not a particular set of measurements will suffice is answered by the observability criterion. This is mentioned briefly below, and Kwakernaak and Sivan (1972), for example, give a detailed account. The principle is shown in Figure 8.8(a), and a derivation of a discrete time version of the filter is given below. The general dynamic linear model of Section 6.4 includes Kalman filtering, but we will repeat it here in the notation of this section.

Observability   Before attempting to set up a filter to estimate all the system states from a limited set of observations, we should check whether it is possible to do so. A test for *observability* is whether the matrix

$$\begin{pmatrix} C \\ CA \\ CA^2 \\ \vdots \\ CA^{\text{dimension of } x - 1} \end{pmatrix}$$

is of full rank.

Background to the Kalman filter   There are many variations of the Kalman filter (Kalman and Bucy, 1961) but we will set one up for the discrete linear system:

$$x_{t+1} = Ax_t + Bu_t + Gw_t$$

$$y_{t+1} = Cx_{t+1} + v_{t+1}$$

where x_t is the state vector, y_t is the observation, u_t is the control, w_t is a random disturbance acting on the system and v_{t+1} is measurement noise. The objective is to estimate the state vector x_{t+1} from the observation vector y_{t+1}. The notation $\hat{x}_{t+1\,|\,t}$ will be used to denote an estimate of x at time $t+1$ made at time t; but to simplify it somewhat, $\hat{x}_{t+1\,|\,t+1}$, for example, will be abbreviated to $\hat{x}_{t+1}$. The matrices are all assumed to be constant and of the appropriate

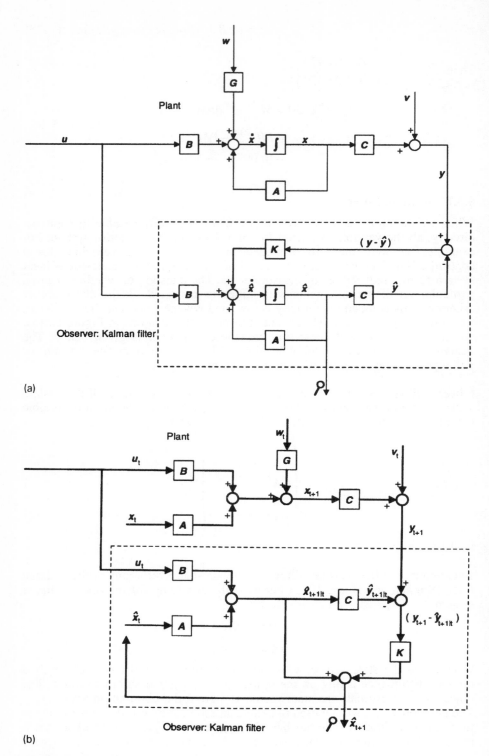

Figure 8.8 Principle of Kalman filter: (a) continuous time; (b) discrete time

dimensions. The random disturbances and measurement noise are assumed to have zero mean, and to be uncorrelated over time. However, the elements of w_t are generally correlated and similarly for v_t:

$$E[w_t w_t^T] = Q \qquad E[v_t v_t^T] = R$$

Finally, w_t and v_t are assumed independent.

We start by assuming that $\hat{x}_t$ is available. Then

$$\hat{x}_{t+1 \mid t} = A\hat{x}_t + Bu_t$$

$$\hat{y}_{t+1 \mid t} = C\hat{x}_{t+1 \mid t}$$

Now we make a measurement y_{t+1}. The principle is that we can improve $\hat{x}_{t+1 \mid t}$ by adding a proportion of the difference between the observation at time $t + 1$ and the predicted value of this observation made at time t. Thus

$$\hat{x}_{t+1} = \hat{x}_{t+1 \mid t} + K(y_{t+1} - \hat{y}_{t+1 \mid t})$$

and this is shown schematically in Figure 8.8(b). It remains to find an optimal choice for K.

Define the estimation error

$$e_{t+1} = x_{t+1} - \hat{x}_{t+1}$$

Substitution from the preceding equations gives

$$e_{t+1} = (I - KC)(Ae_t + Gw_t) - Kv_{t+1}$$

Write the covariance of the estimation error as

$$P_{t+1} = E[e_{t+1} e_{t+1}^T]$$

which, using the last expression for e_{t+1}, is

$$(I - KH)P_{t+1 \mid t}(I - KH)^T + KRK^T$$

where

$$P_{t+1 \mid t} = AP_t A^T + GQG^T$$

Now Q, P_t and $P_{t+1 \mid t}$ are all symmetric, and algebraic rearrangement leads to:

$$P_{t+1} = (K - V)(CP_{t+1 \mid t}C^T + R)(K - V)^T - V(CP_{t+1 \mid t}C^T + R)V^T + P_{t+1 \mid t}$$

where

$$V(CP_{t+1 \mid t}C^T + R) = P_{t+1 \mid t}C^T$$

This form implies that the optimal value of K is V.

Summary of Kalman filter equations

$$P_{t+1 \mid t} = AP_t A^T + GQG^T$$

$$K_{t+1} = P_{t+1 \mid t}C^T(CP_{t+1 \mid t}C^T + R)^{-1}$$

$$\hat{x}_{t+1 \mid t+1} = (I - K_{t+1}C)(A\hat{x}_t + Bu_t) + K_{t+1}y_{t+1}$$

$$P_{t+1} = (I - K_{t+1}C)P_{t+i \mid t}$$

We start the filter by making reasonable guesses for $\hat{x}_1$ and P_1. We also have to make some assessment of R and Q, which may have to be subjective. If the system is observable, K_t will converge to a constant value.

Extended Kalman filter   Kalman filters are available for systems with time varying matrices. The extended Kalman filter can be used with non-linear systems. This relies on Taylor expansions about the current best estimates. It is worth remembering that a linear system with unknown parameters can be treated as a non-linear system and an extended Kalman filter may succeed in estimating both states and unknown parameters of the original system. Robins (1982) gives a very clear introduction.

Example 8.8
In many situations it is desirable to set up a state feedback control for a linear system. If all the states cannot be measured directly, or if measurements are subject to error, a Kalman filter can be used to provide state estimates. These state estimates are then used in the control law. A hydrological application concerns biochemical oxygen demand (x_1) and oxygen deficit (x_2). The Streeter-Phelps equations lead to the linear model:

$$\begin{pmatrix} \dot{x}_1 \\ x_2 \end{pmatrix} = \begin{pmatrix} -k_1 & 0 \\ k_1 & -k_2 \end{pmatrix} \begin{pmatrix} x_1 \\ x_2 \end{pmatrix} + \begin{pmatrix} 1 & 0 \\ 0 & -1 \end{pmatrix} \begin{pmatrix} u_1 \\ u_2 \end{pmatrix}$$

where u_1 represents controlled biochemical oxygen demand additions/ reductions and u_2 represents re-aeration. The system is observable from x_2 alone but not from x_1 alone. The system can be controlled by u_1 and u_2 but not by re-aeration alone.

8.3.8   Non-linear systems

Dalzell (1974) expresses a wave excitation force $y(t)$ in terms of the wave elevation $x(t)$ by the first two terms of the Volterra series.

$$y(t) = \int g_L(\tau)x(t - \tau)\, d\tau + \iint g_Q(\tau_1, \tau_2)x(t - \tau_1)x(t - \tau_2)\, d\tau_1\, d\tau_2$$

The $g_L(\tau)$, which is referred to as the first order Volterra kernel, is the impulse response function for a linear approximation of the force. The second-order kernel $g_Q(\tau_1, \tau_2)$ improves the approximation by including quadratic terms which allow for the non-linear drag force. It is often more convenient to work in the frequency domain and the linear and quadratic transfer functions are defined by

$$G_L(\omega) = \int g_L(\tau)\, e^{-i\omega\tau}\, d\tau$$

and

$$G_Q(\omega_1, \omega_2) = \iint g_Q(\tau_1, \tau_2)\, e^{-i(\omega_1\tau_1 + \omega_2\tau_2)}\, d\tau_1\, d\tau_2$$

Hearn and Metcalfe (1995), for example, give more details and applications.

Wray and Green (1994) use an artificial neural network (ANN) to calculate the Volterra kernels of a discrete time representation of a non-linear system. That is, they work in the time domain and estimate the coefficients in a model of the form

$$y_t = a_0 x_t + a_1 x_{t-1} + a_2 x_{t-2} + \dots$$
$$+ a_{11} x_{t-1}^2 + a_{12} x_{t-1} x_{t-2} \dots$$

Their algorithm gives cubic and higher terms, but these are unlikely to make a significant contribution. ANN relate outputs to inputs in a very general structure. The nomenclature is quite colourful and Figure 8.9 is the architecture of a multilayer perceptron (MLP) that might be used to classify the production of oil wells, for example. The inputs could be: depth, geological variables, production at any neighbouring sites, and operating variables. The values of y_j^h at the hidden nodes are given by

$$y_j^h = f^h \left(\sum_{\text{inputs}} w_{ji} x_i \right)$$

and the outputs are given by

$$y_k = f \left(\sum_{\text{hidden}} w_{kj} y_j^h \right)$$

The functions f^h and f are usually chosen to limit the output to a finite range, often approximating a threshold function which is 0 if the sum is below a critical value and 1 otherwise. The tanh function and $f(x) = 1/(1 + e^{-x})$ are commonly used. The w_{ji} and w_{kj} are weights which have to be determined empirically from the *training set*, i.e. data for which x and y are known. The output variables can be on a continuous scale or be indicators of a classification into poor, moderate and good. There are many substantial advances from these basic principles, and there are usually many coefficients to be estimated, empirically, by a non-linear least squares algorithm. Large data sets are needed

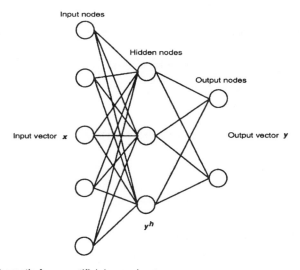

Figure 8.9   Schematic for an artificial neural net

to fit ANN, and drawbacks include the possibility of using an excessive number of parameters and the lack of a physical interpretation. Wray and Green (1995) have combined the optimization efficiency of an ANN algorithm with a physical interpretation. It is also an application in which a large number of parameters are likely to be needed, if no autoregressive terms (i.e. lagged y values) are included. ANN seem to be particularly effective for classification problems (Michie *et al.*, 1994).

8.4 Wave loading on offshore structures

Najafian *et al.* (1995) give a good review of the published work on this subject. In this section we will look at some of the statistical issues.

8.4.1 Spectral density of a derived process

The relationship between the spectra of a random process x and the derived process $\dot{x}$ (used in Example 8.3) is

$$\Gamma_{\dot{x}\dot{x}}(\omega) = \omega^2 \Gamma_{xx}(\omega)$$

This is intuitively reasonable if we think of the mean squared value of a sine wave at frequency ω and the mean squared value of its derivative. It is important because we can express the spectra of water particle velocities and accelerations as a product of attenuation factors and the spectrum of the water surface elevation. The proof is easier to follow if we assume the process $x(t)$ has a mean of 0. Then

$$\gamma(\tau) = E[x(t)x(t + \tau)]$$

$$\frac{\partial}{\partial \tau}\, \gamma(\tau) = E[x(t)\dot{x}(t + \tau)]$$

If we assume that the process is stationary we can write the last equation as

$$\frac{\partial}{\partial \tau}\, \gamma(\tau) = E[\, x(t - \tau)\dot{x}(t)]$$

and then

$$\frac{\partial^2}{\partial \tau^2}\, \gamma(\tau) = E[-\dot{x}(t - \tau)\dot{x}(t)]$$

so

$$\gamma_{\dot{x}\dot{x}}(\tau) = -\frac{\partial^2}{\partial \tau^2}\, \gamma_{xx}(\tau)$$

We now use equation (8.3b) to express

$$\gamma_{xx}(\tau) = \int \Gamma(\omega)\, e^{i\omega\tau}\, d\omega$$

and differentiate partially with respect to τ, twice, to obtain the result

$$\gamma_{\dot{x}\dot{x}}(\tau) = \int \omega^2 \Gamma(\omega)\, e^{i\omega\tau}\, d\omega$$

8.4.2 Morison's equation and the Pierson Holmes distribution

Morison *et al.* (1950) devised the following equation for the wave-induced horizontal force per unit length (Y) on a vertical submerged cylinder, with a diameter less than about one fifth of the wavelength

$$Y = k_d u\,|\,u\,| + k_i \dot{u}$$

where u and $\dot{u}$ are undisturbed horizontal components of water particle velocity and acceleration at the centre of the cylinder. Multiplication by the coefficients k_d and k_i gives the drag and inertial components of the force. The former is proportional to the cylinder diameter and the latter is proportional to the square of the diameter. Hence, as we might expect, the drag becomes more important as the diameter decreases. It is usual to assume u and $\dot{u}$ are independent Gaussian random variables with means of zero, and Y is of the form

$$Y = x_1 + x_2\,|\,x_2\,|$$

where x_1 and x_2 have independent normal distributions with mean 0. The transformation of variables method used in Section A1.8 of Appendix 1 gives the pdf of Y as

$$f(y) = \frac{1}{2\pi\sigma_{x_1}\sigma_{x_2}} \int_{-\infty}^{\infty} \exp\left[-\frac{1}{2}\left(\frac{(y - x_2|x_2|)^2}{\sigma_{x_1}^2} + \frac{x_2^2}{\sigma_{x_2}^2} \right) \right] dx_2$$

This was obtained by Pierson and Holmes (1965) and is known as the P-H distribution. Since x_1 and x_2 have means of zero, the mean, and all other odd moments of the distribution, are zero. The variance of Y is

$$\mathrm{var}(Y) = \sigma_{x_1}^2 + 3\sigma_{x_2}^4$$

and the kurtosis of Y is

$$k = 3 + 78(3 + \sigma_{x_1}^2/\sigma_{x_2}^4)^{-1}$$

The kurtosis has a considerable effect on the distribution of peak loadings, and has a range from 3 to nearly 12.

8.4.3 Crossing analysis and distribution of peaks

A stochastic process is narrow banded if it is predominantly at a single frequency. A harmonic wave with a randomly varying amplitude satisfies the requirement exactly. The essential feature of narrow bandedness, for crossing analysis, is that nearly all the peaks occur above the mean level and nearly all the troughs occur below it. Suppose $y(t)$ is a narrow band process with a mean of zero, and let a be some positive level. The process will make an upwards crossing of a in time $[t_p, t_p + \delta t]$ if, and only if:

$$y(t_p) < a \quad \text{and} \quad \dot{y}(t_p)\delta t > a - y(t_p)$$

Therefore

$$\text{Pr(upwards crossing of } a \text{ in time } \delta t) = \int_{\dot{y}=0}^{\infty} \left\{ \int_{y=a-\dot{y}(t_p)\delta t}^{a} f(y, \dot{y}) \, dy \right\} d\dot{y}$$

which for small δt reduces to

$$\int_0^{\infty} f(a, \dot{y}) \dot{y} \delta t \, d\dot{y}$$

Now let ν_a^+ be the average frequency of upwards crossings of level a in unit time. Since δt is small

$$\text{Pr(upwards crossing of } a \text{ in time } \delta t) = \nu_a^+ \delta t$$

and hence

$$\nu_a^+ = \int_0^{\infty} f(a, \dot{y}) \dot{y} \, d\dot{y}$$

If we assume the marginal distributions of y and $\dot{y}$ are independent Gaussian

$$\nu_a^+ = \frac{1}{2\pi} \frac{\sigma_{\dot{y}}}{\sigma_y} e^{-a^2/(2\sigma_y^2)}$$

The ratio $(\sigma_{\dot{y}}/\sigma_y)$ implies that ν_a^+ is proportional to an average frequency. The distribution of peaks of the process follows from the expression for ν_a^+ in two steps. To begin with

$$\text{Pr(peak exceeds } a) = \nu_a^+/\nu_0^+$$

since ν_a^+/ν_0^+ is the proportion of peaks, which are assumed all to be positive, which cross a level a. The second step is to note that the cdf is

$$\text{Pr(peak} < a) = F(a) = 1 - \nu_a^+/\nu_0^+$$

If y and $\dot{y}$ are assumed to be independent Gaussian we can see that the peaks have a Rayleigh distribution. Tickell (1977) obtained results for Morison loading.

8.5 Wavelets

Spectral analysis is apt when the time series to be analysed is from a stationary random process. It is also used to detect any deterministic harmonic components in noisy signals, and signature analysis of rotating machinery is an application. However, structures may be subject to earthquakes or other shocks which lead to transient responses. This prompted Goupillaud *et al.* (1984) to investigate how the frequency composition of earthquake signals changed with time.

The discrete wavelet transform is a means of analysing a signal over time and frequency. We start with a time series of length n. The length should be some power of 2 so we can write $n = 2^m$. There are $m + 1$ wavelet levels which represent increasing frequency. At each level, wavelets are displaced with respect to their neighbour, representing time. The pattern is summarized below.

Level	
−1	constant
0	2^0 wavelet
1	2^1 wavelets displaced by $n/2$ with respect to neighbour
2	2^2 wavelets displaced by $n/4$ with respect to neighbour
⋮	⋮
$m-1$	2^{m-1} wavelets displaced by $n/2^{m-1}$ with respect to neighbour

The discrete wavelet transform consists of the

$$1 + 1 + 2 + 2^2 + \cdots + 2^{m-1} = 2^m = n$$

coefficients of the wavelets which make them add to the original signal at the n sampling points. Let a_0 be the coefficient of the constant; a_1 be the coefficient of the level 0 wavelet; a_2 and a_3 be the coefficients of the level 1 wavelets; a_4, a_5, a_6 and a_7 be the coefficients of the level 2 wavelets; and so on. The variance of the signal is given by

$$a_1^2 + \tfrac{1}{2}(a_2^2 + a_3^2) + \tfrac{1}{4}(a_4^2 + a_5^2 + a_6^2 + a_7^2) + \tfrac{1}{8}(a_8^2 + \cdots + a_{15}^2) + \cdots$$

The mean square map shows the contribution to the variance, displacement on the horizontal scale and level on the vertical scale.

In this section we will use Haar (D2) wavelets which are simple square waves. They are a special case which do not share some of the more interesting features of higher order wavelets. The D4 wavelet is described in Appendix 14. Newland (1993) gives a more detailed account. The trace in Figure 8.10 is based on uplift of a boundary column in an earthquake simulation test (Wolfgram *et al.*, 1985). The 512 points of a sampled signal are also shown. The shapes of Haar wavelets are shown in Figure 8.11, and the mean square map is given in Figure 8.12.

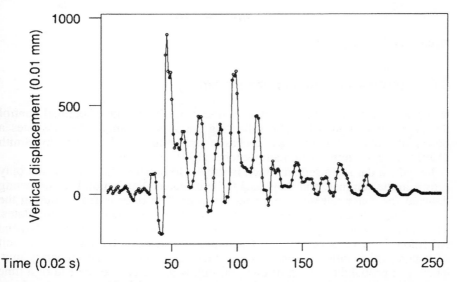

Figure 8.10   Trace and sampled trace of uplift of a column during an earthquake simulation

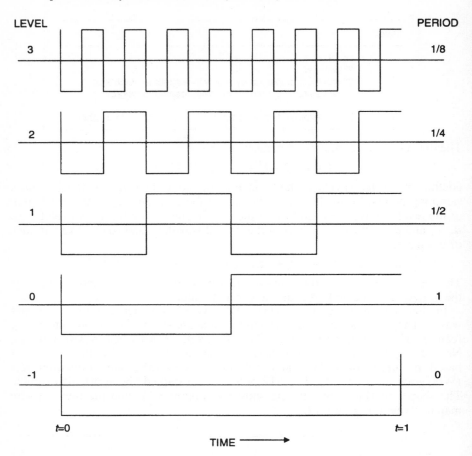

Figure 8.11   Haar wavelets

8.6 Stochastic dynamic programming

Many water resource systems are non-linear, so the results of optimal control theory are not applicable. The following simplified example demonstrates a powerful technique known as stochastic programming. The book by Smith (1991) is a good introduction and Butcher (1971) is a case study.

Imagine a dam in a region which is subject to harsh winters. Crops are only grown in the summer, and they need to be irrigated. Navigation is a competing demand for water releases during the summer. The river is frozen during the winter. The dam can contain 0 or 1 or 2 units of water, and these are the states. The value of releasing one unit of water, which would be used for essential irrigation, is two monetary units. The marginal value of releasing a second unit of water, which might be used to maintain navigation, is one monetary unit. There is a probability of 0.5 of one unit inflow to the dam at the end of a time step, i.e. it is available for use during the next period. The alternative is no

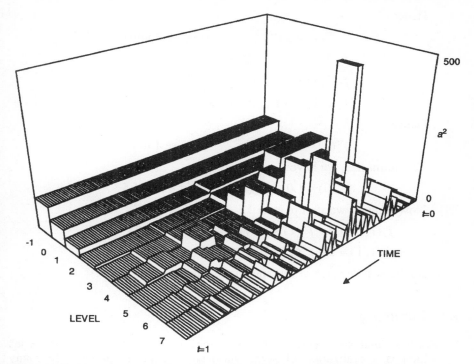

Figure 8.12   Mean square map for uplift of column during an earthquake simulation

input. We will model four time steps. Assume that time 4 corresponds to the start of the harsh winter during which no crops are grown and navigation ceases. Hence the optimum policy at time step 3 is to empty the dam.

The following notation will be used. $f_t^*(i)$ is the value of the optimum decision at time t if we are in state i. $f_t(i, j)$ is the expected value if we release j, if we are in state i at time t.

(i)   At time step 3 we will empty the dam. So optimum decisions are

$$f_3^*(0) = 0 \qquad \text{(nothing to release)}$$
$$f_3^*(1) = 2 \qquad \text{(release 1 unit)}$$
$$f_3^*(2) = 3 \qquad \text{(release 2 units)}$$

(ii)   Now move back to time step 2. Then

$$f_2(0, 0) = 0 + f_3^*(0) \times \tfrac{1}{2} + f_3^*(1) \times \tfrac{1}{2} = 1$$

because there is a 0.5 probability of no inflow and a 0.5 probability of an inflow of 1 unit. In a similar fashion

$$f_2(1, 0) = 0 + f_3^*(1) \times \tfrac{1}{2} + f_3^*(2) \times \tfrac{1}{2} = 2.5$$
$$f_2(1, 1) = 2 + f_3^*(0) \times \tfrac{1}{2} + f_3^*(1) \times \tfrac{1}{2} = 3.0$$

Hence

$$f_2^*(1) = 3.0 \qquad \text{(release 1 unit)}$$

$$f_2(2, 0) = 0 + f_3^*(2) = 3$$

$$f_2(2, 1) = 2 + f_3^*(1) \times \tfrac{1}{2} + f_3^*(2) \times \tfrac{1}{2} = 4.5$$

$$f_2(2, 2) = 3 + f_3^*(0) \times \tfrac{1}{2} + f_3^*(1) \times \tfrac{1}{2} = 4.0$$

Hence

$$f_2^*(2) = 4.5 \qquad \text{(release 1 unit)}$$

To summarize, we have:

$$f_2^*(0) = 1.0, \quad f_2^*(1) = 3.0, \quad f_2^*(2) = 4.5$$

(iii)   Now suppose we start in state 1

$$f_1(1, 0) = 0 + f_2^*(1) \times \tfrac{1}{2} + f_2^* \times \tfrac{1}{2} = 0 + 1.5 + 2.25 = 3.75$$

$$f_1(1, 1) = 2 + f_2^*(0) \times \tfrac{1}{2} + f_2^*(1) \times \tfrac{1}{2} = 2 + 0.5 + 1.5 = 4.00$$

So the best decision at this stage is to release 1 unit.

8.7 Summary

1. Spectral analysis describes the frequency composition of a time series which is a realization of some stationary stochastic process. In many applications this is the essential feature of the disturbance.

2. The spectrum is the Fourier transform of the acvf. The acvf can be obtained from the spectrum by the inverse Fourier transform. For a discrete process:

$$\Gamma(\omega) = \frac{1}{2\pi} \sum \gamma(k) e^{-i\omega k} \qquad \text{for} \qquad -\pi < \omega < \pi$$

$$\gamma(k) = \int_{-\pi}^{\pi} \Gamma(\omega) e^{i\omega k} \, d\omega$$

For a continuous process:

$$\Gamma(\omega) = \frac{1}{2\pi} \int \gamma(\tau) e^{-i\omega\tau} \, d\tau \qquad \text{for} \qquad -\infty < \omega < \infty$$

$$\gamma(\tau) = \int \Gamma(\omega) e^{i\omega\tau} \, d\tau$$

3. The transfer function of a linear system can be estimated from

$$\Gamma_{xy}(\omega) = H(\omega)\Gamma_{xx}(\omega)$$

The spectrum of the response of a linear system is related to the spectrum of the input by

$$\Gamma_{yy}(\omega) = |H(\omega)|^2 \Gamma_{xx}(\omega)$$

4. Wavelet analysis gives a time varying frequency representation of a transient signal.

Exercises

1   (i)   The important result

$$e^{i\theta} = \cos\theta + i\sin\theta$$

can be obtained by substituting into the Maclaurin series (Taylor expansion about 0), which remains valid for complex numbers. Provide an alternative justification by showing that $y = e^{i\theta}$ and $y = \cos\theta + i\sin\theta$ both satisfy $dy/d\theta = iy$ with $y(0) = 1$.

   (ii) Show that $\tilde{C}(\omega)$ can be calculated for any value of ω between $-\pi$ and π from

$$\tilde{C}(\omega) = \frac{1}{2\pi n} \left\{ \left(\sum_{t=-m}^{m-1} x_t \cos\omega t \right)^2 + \left(\sum_{t=-m}^{m-1} x_t \sin\omega t \right)^2 \right\}$$

A one-sided periodogram is given $2\tilde{C}(\omega)$ for $0 \leqslant \omega < \pi$.

2   Sample the three signals
   (i) $\sin(\pi t/2)$
   (ii) $\sin(3\pi t/4)$
   (iii) $\sin(5\pi t/8)$
at times $t = 0, 1, 2, \ldots, 7$. Compare their Fourier line spectra. Notice that those for (i) and (ii) are spikes of height 0.5 at the frequency of the sine wave, whereas that for (iii) has leaked to other frequencies.

3   Let $\Gamma_{\text{rad}}(\omega)$ be a one-sided spectrum, expressed as a function in rad/s, of a random process with a variance σ^2. Then

$$\int_0^\infty \Gamma_{\text{rad}}(\omega)\, d\omega = \sigma^2$$

Now suppose we wish to define a spectrum $\Gamma_{\text{Hz}}(f)$ as a function of frequency expressed in Hz. Show that to preserve

$$\int_0^\infty \Gamma_{\text{Hz}}(f)\, df = \sigma^2$$

$$\Gamma_{\text{Hz}}(f) = 2\pi\Gamma_{\text{rad}}(2\pi f)$$

4   (a)   Sketch the spectrum given by

$$\Gamma(\omega) = (1 + 3\omega^2)/(1 + \omega^2)^2$$

Show that the maximum of this function is 1.125 when $\omega = \pm 1/\sqrt{3}$.

   (b)   The Pierson–Moskowitz spectrum has the form

$$\Gamma(\omega) = a\omega^{-5}e^{-b\omega^{-4}}$$

Sketch this function when a and b equal 1.

5   In Section 8.2.2, it was shown that the periodogram

$$\tilde{C}(\omega) = \frac{1}{2\pi} \sum_{k=-(n-1)}^{n-1} c(k)e^{-i\omega k} \qquad \text{for} \qquad -\pi < \omega < \pi$$

(i)   Explain why a one-sided periodogram is given by

$$\tilde{C}(\omega) = \frac{1}{\pi} \left[c(0) + 2 \sum_{k=1}^{n-1} c(k)\cos \omega k \right] \qquad \text{for} \qquad 0 \leqslant \omega < \pi$$

(ii)   Explain why a smoothed estimate of the spectrum is given by

$$\tilde{C}(\omega) = \frac{1}{\pi} \left[c(0) + 2 \sum_{k=1}^{L} w(k)c(k)\cos \omega k \right] \qquad \text{for} \qquad 0 \leqslant \omega < \pi$$

where $w(k)$ is a *lag window*, such as the Tukey window defined by

$$w(k) = \tfrac{1}{2}(1 + \cos(\pi k/L)) \qquad \text{for} \qquad L < n$$

The bandwidth is $2.67\pi/L$.
[Hint: Compare this strategy with dividing the time series into sub series, calculating $\tilde{C}(\omega)$ for the sub series and then averaging.]

6   Verify that the general result for the spectrum of an $AR(p)$ process:

$$\frac{\sigma_E^2}{2\pi |1 - a_1 e^{-i\omega} \cdots - a_p e^{-i\omega p}|^2} \qquad \text{for} \qquad -\pi < \omega < \pi$$

reduces to the results quoted in Section 8.2.2.

7   Show that the discrete white noise (DWN) has a flat spectrum.

9

Design of Experiments

9.1 Introduction

There are a few general principles which should be applied when planning experiments. First, decide which variable you wish to predict and thereby maximize, minimize, or keep within a specification. This is the *response variable*.

Second, list the variables which may affect the response. Split them into those which you can set to given values, the *control* variables, and those over which you have no control but can measure in some way, the *concomitant* variables.

Third, allow for known sources of variation by choosing an appropriate experimental design. The relatively simple strategy of *blocking*, a generalization of the paired comparison procedure for comparing two treatments, is often used. You should try, during the course of your experiment, to cover the variation that will be met in normal operating conditions.

Fourth, randomize the experimental materials to runs and the order of runs, subject to the constraints imposed by the experimental design. The reason for randomizing is that it makes the independence assumptions about the errors in the regression model plausible, and hence provides the basis for assessing the accuracy of the results. You should not rely on randomizing to even out known sources of variation. It may not, and there is no point in continuing with an obviously biased experiment because you would obtain the right answer if you were to repeat the experiment, with independent randomizations, millions of times and average the results. Even when randomization does turn out much as you would have liked, the estimates, for a given amount of resources, will be less precise than if an appropriate experimental design had been used.

9.2 Completely randomized design (CRD) with one factor

In mining and quarrying operations charges are laid to explode in carefully planned sequences. The delay times in detonators should be as small, and as consistent, as possible. Anderson and McLean (1974) described an experiment to compare three different packing techniques, in which eight detonators had been prepared with each technique. This required 24 empty detonator cases, which were randomly sampled from stock. The cases were randomly allocated to each packing technique, subject to the restriction that each technique was used eight times. They were then fired in a random order. Randomization serves two purposes. First, it provides justification for the assumption that errors are independent. Secondly, when combined with replication it should assist in averaging out the effects of unknown extraneous factors that may be present. The design of this experiment is an example of a completely randomized design (CRD) with one factor, which is the packing technique.

The firing times (coded) are given in Table 9.1. Before starting any analysis we should plot the data, see Figure 9.1. The standard deviations are fairly similar for all three techniques. The ratio of the highest variance (technique C) to the lowest variance (technique B) is $17.15/9.54 = 1.78$. The upper 10% point of $F_{7,7}$ is 2.78 so we have no evidence against a hypothesis that the variances, and hence standard deviations, are equal. It is customary to compare the means by assuming the following model, which assumes equal variances. In this model Y_{ij} is the firing time for the jh replicate using packaging technique i

Table 9.1   *Firing times for detonators*

Packing technique

A	12.2	11.8	13.1	11.0	3.9	4.1	10.3	8.4
B	10.9	5.7	13.5	9.4	11.4	15.7	10.8	14.0
C	12.7	19.9	13.6	11.7	18.3	14.3	22.8	20.4

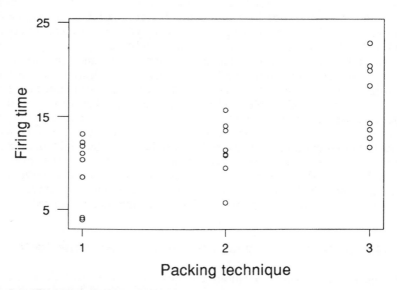

Figure 9.1   Firing times for quarrying detonators

(where 1, 2 and 3 represent techniques A, B and C)

$$Y_{ij} = \mu + \alpha_i + E_{ij} \qquad \text{for} \qquad i = 1, 2, 3 \text{ and} \\ j = 1, \ldots, 8$$

where $\sum_{i=1}^{3} \alpha_i = 0$ and $E_{ij} \sim N(0, \sigma^2)$ and are independent.

This model assumes that we have random samples of size 8 from three normal distributions A, B, C with means $\mu + \alpha_1$, $\mu + \alpha_2$ and $\mu + \alpha_3$ respectively.

Population A is the population of all such detonators packed using technique A. The quantity α_1 is referred to as the *effect* of packing technique A.

Similar remarks apply to populations B and C. A regression model with indicator variables is equivalent (Exercise 1), but it becomes less convenient as the number of populations to be compared increases.

The null hypothesis is that there is no difference between packing techniques

$$H_0: \alpha_1 = \alpha_2 = \alpha_3 = 0$$

and the alternative is

$$H_1: \text{at least one of the } \alpha_i \neq 0$$

and we will test the null hypothesis using an analysis of variance (ANOVA) procedure and one F-test. This avoids the imprecise overall significance level which arises if we perform the three possible two-sample t-tests. That is: if we reject H_0 if any of the three tests is significant at the ε level, then the overall significance level $< 3\varepsilon$, but the lack of independence of the tests makes it difficult to be more precise.

The model implies that the variance of the firing time for each packing technique is the same value σ^2. We can then make three independent estimates of σ^2, one from the data for each technique, and pool them to obtain a within samples estimate (s_w^2) of this common population variance σ^2

$$s_w^2 = \frac{s_A^2 + s_B^2 + s_C^2}{3} = \frac{12.85 + 9.54 + 17.15}{3} = 13.18$$

If H_0 is true, another unbiased estimate of σ^2 is the following between samples estimate (s_b^2). The null hypothesis is that the samples are all from the same population so the variance of $\overline{Y}$, based on a sample of size 8, is $\sigma_{\overline{Y}}^2 = \sigma^2/8$. This is estimated by

$$s_{\overline{Y}}^2 = \frac{(9.350 - 12.496)^2 + (11.425 - 12.496)^2 + (16.712 - 12.496)^2}{(3 - 1)}$$

$$= 14.41$$

Hence, the alternative estimate of σ^2, which is only valid if H_0 is true, is $s_b^2 = 8 \times 14.41 = 115.29$, with two degrees of freedom.

If H_0 is not true $E[S_b^2] > \sigma^2$.

If H_0 is true, we have two independent estimates of the same variance and $S_b^2/S_w^2 \sim F_{2,21}$.

Note that if H_0 is not true we would expect large values of the calculated F ratio. The critical region for testing H_0 against H_1 at the 1% level is values of F exceeding 5.78. In this case the calculated value of F is 8.74 so we have evidence to reject H_0 at the 1% level. We can obtain a precise P-value from Minitab:

$$\Pr(S_b^2/S_w^2 > 8.74 \mid H_0 \text{ true}) = 0.0017$$

The data provide evidence of a difference between firing times for the different packing techniques; in particular, that the time for switches packed with technique A is shorter than the time with technique C.

Estimators of the parameters in the model   Consider the more general case of samples of size n from k populations. Remember the constraint that $\sum a_i = 0$

$$\bar{Y}_{i.} = \sum_{j=1}^{n} Y_{ij}/n = \mu + a_i + \bar{E}_{i.}$$

$$\bar{Y}_{..} = \sum_{i=1}^{k}\sum_{j=1}^{n} Y_{ij}/nk = \sum_{i=1}^{k}\bar{Y}_{i.}/k = \sum_{i=1}^{k}(\mu + a_i + \bar{E}_{i.})/k = \mu + \bar{E}_{..}$$

$$E[\bar{Y}_{i.}] = \mu + a_i \qquad E[\bar{Y}_{..}] = \mu$$

The least squares estimators $\hat{\mu}$, $\hat{a}_i$ of μ and the a_i are obtained by minimizing

$$\psi = \sum_{i=1}^{k}\sum_{j=1}^{n}(Y_{ij} - \mu - a_i)^2$$

with respect to μ and $a_1, \ldots, a_k$. They are

$$\hat{\mu} = \bar{Y}_{..} \quad \text{and} \quad \hat{a}_i = \bar{Y}_{i.} - \bar{Y}_{..}$$

For any value of i

$$\sum_{j=1}^{n}(Y_{ij} - \bar{Y}_{i.})^2/(n-1)$$

is an unbiased estimator of σ^2. Hence

$$S_w^2 = \sum_{i=1}^{k}\sum_{j=1}^{n}(Y_{ij} - \bar{Y}_{i.})^2/k(n-1)$$

is also unbiased for σ^2 and will henceforth be written as S^2 for the estimator and s^2 for the estimate. In the previous example $\hat{\mu} = 12.496$, $\hat{a}_1 = -3.146$, $\hat{a}_2 = -1.071$, $\hat{a}_3 = 4.216$ and $s^2 = 13.18$.

Partitioning of the sum of squares   It is convenient to define S_{yy} as a shorthand for the sum of squared deviations of y_{ij} from the overall mean. This can be partitioned as follows

$$S_{yy} = \sum_{i=1}^{k}\sum_{j=1}^{n}(y_{ij} - \bar{y}_{..})^2$$

$$= \sum_{i=1}^{k}\sum_{j=1}^{n}(y_{ij} - \bar{y}_{i.} + \bar{y}_{i.} - \bar{y}_{..})^2$$

$$= \sum_{i=1}^{k}\sum_{j=1}^{n}(y_{ij} - \bar{y}_{i.})^2 + \sum_{i=1}^{k}\sum_{j=1}^{n}(\bar{y}_{i.} - \bar{y}_{..})^2 + 2\sum_{i=1}^{k}\sum_{j=1}^{n}(y_{ij} - \bar{y}_{i.})(\bar{y}_{i.} - \bar{y}_{..})$$

The last of these three terms is zero since

$$\sum_{i=1}^{k}\sum_{j=1}^{n}(y_{ij}-\bar{y}_{i.})(\bar{y}_{i.}-\bar{y}_{..}) = \sum_{i=1}^{k}(\bar{y}_{i.}-\bar{y}_{..})\sum_{j=1}^{n}(y_{ij}-\bar{y}_{i.}) = 0$$

The final result is

$$S_{yy} = \sum\sum(y_{ij}-\bar{y}_{i.})^2 + n\sum_{i=1}^{k}(\bar{y}_{i.}-\bar{y}_{..})^2$$

The first term on the right is referred to as the *within samples* (corrected) *sum of squares* and the second as the *between samples* (corrected) *sum of squares*.

Expected value of the between samples sum of squares   Consider

$$\sum_{i=1}^{k}(\bar{Y}_{i.}-\bar{Y}_{..})^2 = \sum_{i=1}^{k}(\alpha_i + \bar{E}_{i.}-\bar{E}_{..})^2 = \sum_{i=1}^{k}(\alpha_i^2 + (\bar{E}_{i.}-\bar{E}_{..})^2 + 2\alpha_i(\bar{E}_{i.}-\bar{E}_{..}))$$

Taking expectation gives

$$\mathrm{E}\left[\sum_{i=1}^{k}(\bar{Y}_{i.}-\bar{Y}_{..})^2\right] = \sum_{i=1}^{k}\alpha_i^2 + (k-1)\sigma^2 \Big/ n$$

The expected value of the between samples sum of squares is the product of this expression with n.

Calculations for ANOVA   The calculations described in the preceding two subjections can be summarized in an analysis of variance (ANOVA) table. There are k samples each of size n.

Source of variation	(Corrected) sum of squares	d.f.	Mean square	Expected value of the mean square
Between samples	$n\sum_{i=1}^{k}(\bar{y}_{i.}-\bar{y}_{..})^2$	$k-1$	$\dfrac{CSS}{k-1}=s_b^2$	$\sigma^2 + n\sum_i \alpha_i^2/(k-1)$
Within samples	$\sum_{i=1}^{k}\sum_{j=1}^{n}(y_{ij}-\bar{y}_{i.})$	$k(n-1)$	$\dfrac{CSS}{k(n-1)}=s_w^2\,(=s^2)$	σ^2
Total	$\sum_{i=1}^{k}\sum_{j=1}^{n}(y_{ij}-\bar{y}_{..})^2$	$nk-1$		

The model is a special case of the general multiple regression model, and the residuals are the differences between observations and their fitted values. In this context:

$$y_{ij} - (\hat{\mu}+\hat{\alpha}_i) = y_{ij} - \bar{y}_{i.}.$$

The within samples sum of squares is just the sum of squared residuals, and is often referred to as 'residual' or 'error'.

Multiple comparisons If H_0 is rejected it remains to find where the differences lie. The simplest way is to use *least significant differences* (LSD). This technique involves making all possible *t*-tests, so the probability of declaring a statistically significant difference if there is none will be higher than the nominal level. Such a procedure was avoided at the start of the analysis, but we do now have a significant *F*-test.

The standard deviation of the difference of any two group means, for the detonator example, is estimated by

$$\sqrt{[s_w^2(1/n + 1/n)]} = \sqrt{[13.18(1/8 + 1/8)]} = 1.815$$

The relevant percentage points of the *t*-distribution for nominal 5% and 1% significance levels are

$$t_{21,0.025} = 2.08, \quad \text{and} \quad t_{21,0.005} = 2.83 \text{ respectively}$$

Therefore the least difference in means to be significant at a nominal 5% level is $1.815 \times 2.08 = 3.78$. The least significant difference at the nominal 1% level is $1.815 \times 2.83 = 5.14$. These can be written concisely as:

$$\text{LSD}(5\%) = 3.78, \quad \text{and} \quad \text{LSD}(1\%) = 5.14$$

For the detonator data

$$\bar{y}_A = 9.35 \qquad \bar{y}_B = 11.43 \qquad \bar{y}_C = 16.71$$

There is evidence that technique C leads to a longer firing time than A or B. There is no evidence of any difference between techniques A and B.

9.3   Randomized block design

In many experimental problems, it is necessary to design the experiment so that variability arising from known extraneous sources can be controlled. A mining company uses a process known as bacterial leaching to extract copper from ore and wishes to compare the yield obtained from four strains of thiobaccilus A, B, C and D. The yield is known to depend on the quality of the ore. There are sufficient experimental resources to use each strain of thiobaccilus with three batches of ore. For a CRD, 12 batches of ore would be required. Three would be randomly selected and assigned to each strain and the 12 trials would be carried out in a random order. There is a potentially serious problem with this procedure. If the batches of ore differ in their quality, this will contribute to the variability observed in the yields. The company would like to make the experimental error as small as possible. A way of accomplishing this is to take large batches of ore from sites at which the local variability is relatively small. Each large batch is divided into four smaller batches which are randomly allocated to the four strains of thiobaccilus. This is an example of a *randomized block design* (RBD) where the *blocks* are the three large batches of ore. Within each block there is a random allocation of ore to the four strains. Since randomization is only carried out within blocks the blocks represent a restriction on randomization. The results of the experiment are given in Table 9.2 as kg/tonne yields, and the data are plotted in Figure 9.2. The sites were in different mines.

Table 9.2 *Yields of copper from ore (kg/tonne)*

Thiobaccilus strain	location of ore		
	Site 1 in mine 1	Site 2 in mine 2	Site 3 in mine 3
A	22	28	25
B	31	33	26
C	20	25	24
D	25	28	26

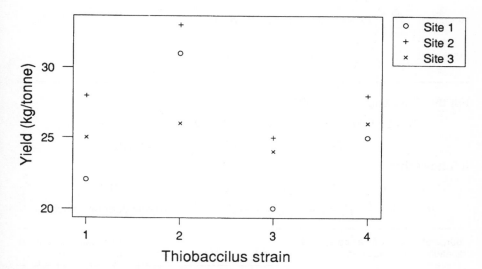

Figure 9.2 Yields of copper in bacterial leaching extraction

If there is just one observation on each of k treatments in each of b blocks, a *two way classification fixed effects model* can be used. This is:

$$Y_{ij} = \mu + \alpha_i + \beta_j + E_{ij}$$

where

$$\sum_{i=1}^{k} \alpha_i = 0, \qquad \sum_{j=1}^{b} \beta_j = 0$$

the $E_{ij} \sim N(0, \sigma^2)$ and are independent of the α_i, β_j and each other, μ is the overall mean, α_i, $i = 1, \ldots, k$ are the treatment effects, β_j, $j = 1, \ldots, b$ are the block effects.

The hypothesis of interest is

$$H_0: \alpha_1 = \cdots = \alpha_k = 0$$

The least squares estimates of the parameters in the model are

$$\hat{\mu} = \bar{y}_{..}, \qquad \hat{\alpha}_i = \bar{y}_{i.} - \bar{y}_{..}, \qquad \hat{\beta}_j = \bar{y}_{.j} - \bar{y}_{..}$$

The general form of the ANOVA table is

Source of variation	(Corrected) sum of squares	d.f.	Mean square	Expected value of the mean square
Treatments	$b \sum_{i=1}^{k} (\bar{y}_{i.} - \bar{y}_{..})^2$	$k-1$	CSS divided by the degrees of freedom	$\sigma^2 + b \sum_{i=1}^{k} \alpha_i^2/(k-1)$
Blocks	$k \sum_{j=1}^{b} (\bar{y}_{.j} - \bar{y}_{..})$	$b-1$		$\sigma^2 + k \sum_{j=1}^{b} \beta_j^2/(b-1)$
Residual	$\sum_{i=1}^{k} \sum_{j=1}^{b} (y_{ij} - \bar{y}_{i.} - \bar{y}_{.j} - \bar{y}_{..})^2$	$(k-1)(b-1)$		σ^2
Total	$\sum_{i=1}^{k} \sum_{j=1}^{b} (y_{ij} - \bar{y}_{..})^2$	$kb-1$		

For the above data, $\bar{y}_{..} = 26.083$

$$\bar{y}_{1.} = 25.000 \qquad y_{2.} = 30.000 \qquad \bar{y}_{3.} = 23.000 \qquad \bar{y}_{4.} = 26.333$$
$$\bar{y}_{.1} = 24.500 \qquad \bar{y}_{.2} = 28.500 \qquad \bar{y}_{.3} = 25.250$$

it follows that, $\hat{\mu} = 26.1$

$$\hat{\alpha}_1 = -1.1 \qquad \hat{\alpha}_2 = 3.9 \qquad \hat{\alpha}_3 = -3.1 \qquad \hat{\alpha}_4 = 0.2$$
$$\hat{\beta}_1 = -1.6 \qquad \hat{\beta}_2 = 2.4 \qquad \hat{\beta}_3 = -0.8 \quad \text{correct to one decimal place}$$

Source of variation	Corrected sum of squares	d.f.	Mean square	Expected value of the mean square
Strains	78.25	3	26.08	$\sigma^2 + 3 \sum \alpha_i^2/3$
Blocks	36.17	2	18.08	$\sigma^2 + 4 \sum \beta_j^2/2$
Residual	26.50	6	4.42	σ^2
Total	140.92	11		

If H_0 is true the ratio of the strains mean square to the residual mean square will have an F-distribution with 3, 6 degrees of freedom. If there are differences between strains the ratio will tend to be larger. In this case the calculated value of this statistic is $26.08/4.42 = 5.90$. Since $F_{3,6,0.05}$ equals 4.76, there is evidence to reject H_0 at the 5% level. If we use the LSD procedure with a 5% level, there is evidence that strain B gives systematically higher yields than the other strains. No other differences are significant.

9.4   Completely randomized design with two factors

Precast concrete paving slabs, known as flags, are the usual surface for pavements. They have been in common use since the 1890s, when they started to replace the high cost stone flags. They were popular with local authorities because they were cheaper to buy and to lay. The savings in construction costs were due mainly to the higher dimensional accuracy of the concrete flags. Unfortunately, pavements are sometimes over-ridden by heavy vehicles and this can damage the surface. The costs of pavement maintenance is considerable and there has been a need for research into methods of limiting the damage caused by misuse (Bull, 1988). Sahabandu (1988) investigated the effects of flag size and joint width on the load transfer of pavements.

We will start by analysing the experiment summarized in Table 9.3 by a two-way analysis of variance, and by multiple regression. The model for two-way analysis of variance looks the same as the model for the RBD but the interpretation is different (see Section 9.5). The assumptions underlying the analysis of variance technique and the standard regression analysis are equivalent, but we need to code indicator variables for the regression analysis. For example, with the coding at the start of the following Minitab session, the coefficient of B is an estimate of the difference between size A and size B flags. We also have an imprecise overall significance level if we look at all the estimates of coefficients. Edited Minitab (Release 7.2) calculations, with some annotation, follow.

MTB > print c1-c7

Row	loadtran	size	joint	B	C	joint6	joint10
1	0.200	1	1	0	0	0	0
2	0.120	1	2	0	0	1	0
3	0.121	1	3	0	0	0	1
4	0.530	2	1	1	0	0	0
5	0.280	2	2	1	0	1	0
6	0.180	2	3	1	0	0	1
7	0.327	3	1	0	1	0	0
8	0.142	3	2	0	1	1	0
9	0.158	3	3	0	1	0	1

MTB > twoway c1 c2 c3

ANALYSIS OF VARIANCE   loadtran

SOURCE	DF	SS	MS	F-ratio	approx P-value
size	2	0.05197	0.02599	5.14	0.09
joint	2	0.06997	0.03498	6.91	0.05
Error	4	0.02026	0.00506		
Total	8	0.14220			

$$F_{2,4,0.10} = 4.325 \quad F_{2,4,0.05} = 6.944$$

Table 9.3 *Maximum load transfer (tonnes) of flags in one replicate of a CRD with 2 factors*

Flag type	Joint 3 mm	Joint 6 mm	Joint 10 mm
A	0.200	0.120	0.121
B	0.530	0.280	0.180
C	0.327	0.142	0.158

MTB > regress c1 4 c4-c7

The regression equation is

loadtran = 0.271 + 0.183B + 0.0620C − 0.172 joint6 − 0.199 joint10

Predictor	Coef	Stdev	t-ratio	p
Constant	0.27067	0.05305	5.10	0.007
B	0.18300	0.05811	3.15	0.035
C	0.06200	0.05811	1.07	0.346
joint6	−0.17167	0.05811	−2.95	0.042
joint10	−0.19933	0.05811	−3.43	0.027

$s = 0.07117$ $R - sq = 85.8\%$ $R - sq(adj) = 71.5\%$

Analysis of Variance

SOURCE	DF	SS	MS	F	p
Regression	4	0.121943	0.030486	6.02	0.055
Error	4	0.020259	0.005065		
Total	8	0.142202			

There are only nine data, and the P-values are too high for definitive results, but there is an indication that increasing the joint widths decreases the load transfer, and that flag type B gives a higher load transfer than flag type A. The design is balanced, since each flag type is tested in combination with each joint width the same number of times. A consequence of the balance of the design is that the indicator variables for flag type are uncorrelated with the indicator variables for joint width. So, the point estimates of the effect of flag type would not change if the joint width variables were removed from the regression.

MTB > corr c4-c7

	B	C	joint6
C	−0.500		
joint6	0.000	0.000	
joint10	0.000	0.000	−0.500

Table 9.4   *Maximum load transfer (tonnes) of flags*

Flag type	Joint 3 mm	Joint 6 mm	Joint 10 mm
600 by test 1	0.232	0.208	0.217
600 by test 2	0.200	0.120	0.121
63 test 3	0.154	0.089	0.087
450 by test 1	0.542	0.324	0.250
450 by test 2	0.530	0.280	0.180
100 test 3	0.530	0.255	*
400 by test 1	0.330	0.210	0.256
400 by test 2	0.327	0.142	0.158
65			

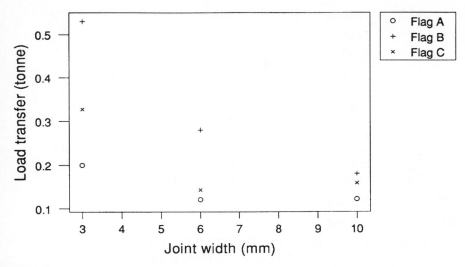

Figure 9.3 Load transfer of flagstones

There is a straightforward explanation for the correlation between the indicator variables for joint types B and C. If both indicator variables are in the regression we compare joint type B with type A, and joint type C with type A. If only the indicator variable for joint type B is in the regression we would be comparing type B with the average of types A and C.

The design is described as a completely randomized design with two factors because the order of testing the factor combinations is randomized.

The actual experiment involved two or three tests for each factor combination. The results are given in Table 9.4, and the data are plotted in Figure 9.3. The tests were carried out under slightly different conditions, so we will treat the test number as a concomitant variable. The design is no longer exactly balanced, because some factor combinations were only tested twice. It is convenient to analyse the data by multiple regression.

ROW	loadtran	B	C	joint6	joint10	test1	test3
1	0.232	0	0	0	0	1	0
2	0.200	0	0	0	0	0	0
3	0.154	0	0	0	0	0	1
4	0.208	0	0	1	0	1	0
5	0.120	0	0	1	0	0	0
6	0.089	0	0	1	0	0	1
7	0.217	0	0	0	1	1	0
8	0.121	0	0	0	1	0	0
9	0.087	0	0	0	1	0	1
10	0.542	1	0	0	0	1	0
11	0.530	1	0	0	0	0	0
12	0.530	1	0	0	0	0	1

continued

ROW	loadtran	B	C	joint6	joint10	test1	test3
13	0.324	1	0	1	0	1	0
14	0.280	1	0	1	0	0	0
15	0.255	1	0	1	0	0	1
16	0.250	1	0	0	1	1	0
17	0.180	1	0	0	1	0	0
18	0.330	0	1	0	0	1	0
19	0.327	0	1	0	0	0	0
20	0.210	0	1	1	0	1	0
21	0.142	0	1	1	0	0	0
22	0.256	0	1	0	1	1	0
23	0.158	0	1	0	1	0	0

MTB > regress c1 6 c2-c7

The regression equation is

$$\text{loadtran} = 0.248 + 0.192B + 0.0664C - 0.152 \text{ joint6} - 0.163 \text{ joint10}$$
$$+ 0.0568 \text{ test1} - 0.0080 \text{ test3}$$

Predictor	Coef	Stdev	t-ratio	p
Constant	0.24760	0.03510	7.05	0.000
B	0.19239	0.03186	6.04	0.000
C	0.06636	0.03620	1.83	0.085
joint6	−0.15212	0.03251	−4.68	0.000
joint10	−0.16343	0.03394	−4.81	0.000
test1	0.05678	0.03065	1.85	0.083
test3	−0.00802	0.03822	−0.21	0.836

$$s = 0.06503 \qquad R-\text{sq} = 82.6\% \qquad R-\text{sq(adj)} = 76\%$$

Analysis of Variance

SOURCE	DF	SS	MS	F	p
Regression	6	0.320725	0.053454	12.64	0.000
Error	16	0.067654	0.004228		
Total	22	0.388379			

MTB > corr c2-c7

	B	C	joint6	joint10	test1
C	−0.434				
joint6	0.042	−0.018			
joint10	−0.086	0.037	−0.483		
test1	−0.024	0.132	−0.024	0.051	
test3	0.058	−0.313	0.058	−0.120	−0.423

There is strong evidence that flag type B gives a higher load transfer, and that increasing the width between flags reduces the load transfer. Most of the correlations between indicator variables, between categories, are small. The highest, in terms of absolute value, is between test 3 and flag type C which was not included in the third round of tests. Inclusion of the test variables compensates for type C flags being disadvantaged if results in test 3 had tended to be higher.

If we have replicate tests for each factor combination, we can investigate possible interactions. For example, flag type interacts with the joint width if the effect of the flag type depends on the joint width. Interactions can be incorporated into the regression model by calculating product terms. There will be four of these for the flag type interactions with joint width.

MTB > let c24 = c2*c4
MTB > let c25 = c2*c5
MTB > let c34 = c3*c4
MTB > let c35 = c3*c5
MTB > regress c1 10 c2-c7 c24 c25 c34 c35

The regression equation is

$$\text{loadtran} = 0.185 + 0.339B + 0.115C - 0.0563 \text{ joint6} - 0.0537 \text{ joint10}$$
$$+ 0.0568 \text{ test1} - 0.0260 \text{ test3} - 0.191 \text{ intB*j6} - 0.283 \text{ intB*j10}$$
$$- 0.0962 \text{ intC*j6} - 0.0678 \text{ intC*j10}$$

Predictor	Coef	Stdev	t-ratio	p
Constant	0.18508	0.01627	11.38	0.000
B	0.33867	0.02044	16.57	0.000
C	0.11503	0.02331	4.94	0.000
joint6	−0.05633	0.02044	−2.76	0.017
joint10	−0.05367	0.02044	−2.63	0.022
test1	0.05678	0.01180	4.81	0.000
test3	−0.02601	0.01493	−1.74	0.107
intB*j6	−0.19133	0.02891	−6.62	0.000
intB*j10	−0.28347	0.03100	−9.14	0.000
intC*j6	−0.09617	0.03232	−2.98	0.012
intC*j10	−0.06783	0.03232	−2.10	0.058

$$s = 0.02504 \qquad R-\text{sq} = 98.1\% \qquad R-\text{sq(adj)} = 96.4\%$$

There are some very significant interactions between the joint width and flag type. We will only get the full advantage of flags type B, if we lay them so that the joint width is 3 mm. Ignoring interactions can lead to misleading recommendations.

9.5   Difference between a RBD design and a CRD design with two factors

The model looks the same in both cases but there is an important difference in interpretation. In Section 9.3 we can estimate the block effects, but we should not attribute them to the corresponding mines. Batches of ore were taken from one site in each mine. We have no information about the variability within mines. Differences between blocks may be due to a systematic difference between mines, but they may just have arisen because of variability between sites. If we are only interested in comparing the leaching treatments, this does not matter.

If we wanted to compare leaching treatments and mines in a CRD, we would have to take ore from four random sites within each mine. We would then randomize the allocation of ore from these sites to the four treatments. The order of testing the 12 factor combinations should be randomized. In the RBD we just randomize the allocation of bacterial leaching treatments, and order of testing, within each batch. The advantage of the CRD is that we obtain information on differences between mines. The RBD has the advantage that estimators of treatment effects will be more precise, if there is variation between sites in the same mine, because we take all the ore from one site.

9.6 2^k factorial designs

A company makes cement in a rotary kiln. An engineer wished to investigate the effect of four control variables: fuel rate (x_1), rotation speed (x_2), feed rate (x_3), and fan speed (x_4), on the free lime in a cement product. She started by deciding how much each variable could safely be changed from the normal operating conditions. She referred to half these maximum changes as $+1$ or -1 unit and ran one replicate of a 2^4 factorial experiment. Every possible factor combination was tried once, in a random order. The composition of the meal, fed into the kiln, was kept as consistent as possible, but some natural variation was inevitable and two concomitant variables were monitored: burnability (x_5) and water content (x_6). The results are given in the upper part of Table 9.5. You are asked to analyse these, using multiple regression, in Exercise 2.

Composite design   The 2^k design enables us to estimate linear effects and interactions, but not quadratic effects. We can augment our original design with a star design. In this example we would need nine additional points arranged thus:

x_1	x_2	x_3	x_4
$-\alpha$	0	0	0
$+\alpha$	0	0	0
0	$-\alpha$	0	0
0	$+\alpha$	0	0
0	0	$-\alpha$	0
0	0	$+\alpha$	0
0	0	0	$-\alpha$
0	0	0	$+\alpha$
0	0	0	0

The distance of the original points from the centre can be found by Pythagoras' theorem; it is

$$\sqrt{(1^2 + 1^2 + 1^2 + 1^2)} = 2$$

So with four control variables, an α of 2 will give equal precision of estimation in all directions. This is the preferred value. It is also common to replicate the observation at the centre, and the design can even be made orthogonal if it is replicated enough times: 12 times for four control variables

Table 9.5 *Percentage free lime (percentage above 1% × 100) in a cement product (other units code)*

Fuel	Rot	Feed	Fan	Burn	Water	Lime
2^4 factorial design results						
−1	−1	−1	−1	4	7	37
1	−1	−1	−1	5	4	7
−1	1	−1	−1	7	7	131
1	1	−1	−1	3	6	14
−1	−1	1	−1	11	−6	131
1	−1	1	−1	1	4	116
−1	1	1	−1	2	−2	99
1	1	1	−1	2	−24	73
−1	−1	−1	1	3	16	102
1	−1	−1	1	3	−4	−12
−1	1	−1	1	1	−18	−2
1	1	−1	1	0	12	−87
−1	−1	1	1	10	5	175
1	−1	1	1	−2	10	−28
−1	1	1	1	−4	−15	13
1	1	1	1	−3	−3	−46
Star design results						
−2	0	0	0	4	−13	137
2	0	0	0	−4	−2	−16
0	−2	0	0	9	12	78
0	2	0	0	5	−15	15
0	0	−2	0	3	−3	49
0	0	2	0	5	1	148
0	0	0	−2	5	−1	136
0	0	0	2	2	−3	−5
0	0	0	0	−3	0	−23

(Montgomery, 1991). The results are given in the lower part of Table 9.5 (see Exercise 3).

Fractional factorial design If k is large, a full 2^k design would involve many runs. If we are prepared to do without estimates of high order interactions we can reduce the size of the experiment. You are asked to work out a half replicate of the 2^4 design in Exercise 4. The same principle can be applied to get 2^{k-m} designs and Montgomery (1991) gives more details. The DEX software (Greenfield and Savas, 1992) will generate these designs, and analyse the results using multiple regression. It is particularly easy to use.

9.7 *D*-optimum designs

This section is based on part of a description of a research proposal, made by Tony Greenfield, to investigate the effects of geometry, weld, environment and corrosion protection techniques on the lifetimes of welded plate joints. The response variable was the logarithm of the number of cycles to failure. The control variables, and their codings were:

Pre-weld heat treatment (x_1)	no (-1), yes $(+1)$
Weld improvement (x_2)	none (-1), toe ground $(+1)$
Applied stress (x_3)	coded $-1.2, -1.0, 0.0, 1.0, 1.2$
Cathodic protection (x_4)	0.0 V (-1), -0.85 V $(+1)$
Temperature (x_5)	$5°$ C (-1), $20°$C $(+1)$
Variation in amplitude of oscillatory load (x_6)	narrow S1 (-1), wide S4 $(+1)$

The research team thought that the applied stress might interact with any of the other variables, and they also wished to investigate any quadratic or cubic effects of applied stress. Therefore, they needed to fit a regression model which included:

$$x_7 = x_1 \times x_3$$

$$x_8 = x_2 \times x_3$$

$$x_9 = x_4 \times x_3$$

$$x_{10} = x_5 \times x_3$$

$$x_{11} = x_6 \times x_3$$

$$x_{12} = x_3^2$$

$$x_{13} = x_3^3$$

That is

$$Y_i = \beta_0 + \beta_1 x_{1i} + \cdots + \beta_{13} x_{13i} + E_i$$

In matrix terms

$$Y = XB + E$$

where X is called the design matrix. The strategy is to choose a number of runs and then, conditional on this choice, find the corresponding values of $x_1, \ldots, x_6$ so that $\det(X^TX)$ is a maximum. This is known as a *D*-optimum design. The covariance matrix of $\hat{B}$ is

$$\sigma^2 (X^TX)^{-1}$$

and the criterion corresponds to minimizing the hypervolume of the hyperellipsoidal simultaneous confidence interval for the parameters. This is rather less daunting in the special case when the x are orthogonal. Then X^TX is diagonal and the determinant of X^TX is the reciprocal of the product of the elements along the diagonal of $(X^TX)^{-1}$, which are proportional to the variances of the estimators.

It had been decided that 20 joints would be tested for this experiment. There are $2 \times 2 \times 5 \times 2 \times 2 \times 2 = 160$ different values for the array of control variables $x_1, \ldots, x_6$, and the values of x_7 up to x_{13} will be determined by this. It remains to select the 20, from these 160 possibilities, that maximize the determinant. There are 1.4×10^{25} choices, so checking them all is not going to be possible! Some sort of algorithm, even if it cannot guarantee finding the overall maximum, is essential. We will refer to a particular array

$$x_i = [1 \ \ x_{1i} \ \ x_{2i} \ \ \ldots \ \ x_{13i}]$$

as a design point.

Greenfield used the following algorithm.

Step 1. Start with any 14 design points for which the rows of the 14×14 design matrix X are linearly independent. This can be achieved by using the Gram–Schmidt orthogonalization procedure, which is described in most texts on linear algebra (for example, Press *et al.*, 1992).

Step 2. Try adding each of the 146 remaining design points in turn, and choose the one for which the determinant criterion is a maximum. This is greatly facilitated by the following well-known and extremely useful result. Define

$$\Delta_m = \det(X_m^T X_m)$$

where m is the number of rows in the $m \times 14$ design matrix X. Now if

$$X_{m+1} = \begin{pmatrix} X_m \\ x \end{pmatrix}$$

is an $(m + 1) \times 14$ matrix, we have

$$\Delta_{m+1} = \Delta_m \det(1 + x(X_m^T X_m)^{-1} x^T)$$

Step 3. Now try leaving out each of the 15 design points, and drop the one which gives the least decrease in the criterion.

Step 4. Iterate Steps 2 and 3 until one of the design points which has been added in Step 2 is dropped in Step 3.

Table 9.6 *Experimental design for welded plate joints in sea water with cathodic protection and variable loading*

	PWHT	Weld improvement	Applied stress	Cathodic protection (V)	Temp (°C)	Variable load
1	N	N	−1.0	0.0	5	S1
2	Y	Y	1.2	0.85	5	S4
3	Y	N	−1.0	0.85	20	S1
4	Y	N	1.0	0.0	5	S4
5	N	Y	1.2	0.85	20	S1
6	N	Y	−1.2	0.85	20	S1
7	Y	N	1.2	0.85	5	S1
8	Y	Y	1.0	0.0	20	S1
9	N	Y	−1.2	0.0	5	S4
10	N	N	1.0	0.85	20	S4
11	N	Y	−1.0	0.85	5	S1
12	N	Y	1.0	0.85	5	S1
13	N	Y	−1.2	0.85	20	S4
14	N	N	1.2	0.0	5	S1
15	Y	N	1.2	0.0	20	S4
16	Y	N	−1.0	0.0	5	S4
17	N	Y	1.0	0.0	20	S4
18	N	N	0.0	0.85	20	S4
19	Y	N	0.0	0.0	20	S4
20	Y	Y	0.0	0.85	5	S4

Step 5. Add design points, one at a time so that the criterion is maximized, until the required 20 points have been selected.

This procedure will not generally lead to the absolute maximum of the criterion, but it should give a reasonable design. The design obtained for the experiment to investigate the lifetimes of welded joints is given in Table 9.6. There is no reason why some particular design points should not be fixed, and the algorithm implemented subject to these points remaining in the design. Atkinson and Donev (1992) cover the subject of optimum design in detail. Their procedure starts from a single design point, and avoids the problem of a singular matrix by adding $\varepsilon^2 I$, where ε is a small number and I is the identity matrix.

9.8 Summary

Factorial designs   A 2^n factorial design allows you to investigate the linear effects and interactions of n control variables on some response variable. Although the control variables are usually continuous they are restricted to two levels, 'high' and 'low'.

Fractional factorial design   If n is large, the number of runs for a full factorial design soon becomes prohibitive. It is possible to generate fractions, involving 2^{n-m} runs, that can provide information on the linear effects and the low-order interactions which could be of practical importance.

Star design and composite design   A star design can be used as a follow-up to a factorial or fractional factorial design. It allows quadratic effects to be investigated. The overall design – factorial plus star – is known as a composite design. It is only practical for continuous control variables.

D-optimum design   This is a general technique for finding a design that can estimate chosen effects given a set of possible values for the control variables. There is no restriction on the number of levels.

Exercises

1. Use two indicator variables to set up a regression model for analysing the detonator data (Table 9.1).

2. Analyse the data in the upper part of Table 9.5. Start by fitting the regression model

$$Y_i = \beta_0 + \beta_1 x_1 + \beta_2 x_2 + \beta_3 x_3 + \beta_4 x_4 + \beta_5 x_5 + \beta_6 x_6$$
$$+ \beta_7 x_1 \times x_2 + \cdots + \beta_{12} x_3 \times x_4 + E_i$$

3. Analyse all the data in Table 9.5 by including quadratic terms in $x_1, \ldots, x_4$ (i.e. $x_1^2, x_2^2, x_3^2, x_4^2$) in your regression model.

4. Write down the sixteen $+1$, -1 combinations for $x_1, \ldots, x_4$ in a 2^4 design.

In a fifth column write the sign of the product $x_1 \times x_2 \times x_3 \times x_4$.

(i) Take the eight combinations which correspond to the product $x_1 \times x_2 \times x_3 \times x_4$ equalling one. This is a half replicate of the 2^4 design. There a $2^{4-1} = 8$ runs.

(ii) Show that you can fit a regression including: x_1, x_2, x_3, x_4, $x_1 \times x_2$, $x_1 \times x_3$, $x_1 \times x_4$.

(iii) Show that you cannot include, for example, both $x_1 \times x_2$ and $x_3 \times x_4$. These are said to be *aliased*.

(iv) Refer to Table 9.5 and analyse the two half replicates you could have obtained.

Appendix 1   Introduction to Statistical Methods

A1.1   Probability

Addition rule

$$\Pr(A \text{ or } B) = \Pr(A) + \Pr(B) - \Pr(A \text{ and } B) \qquad \text{(A1.1)}$$

In mathematics 'or' conventionally includes the possibility of both. If $\Pr(A \text{ and } B) = 0$, we say A and B are *mutually exclusive*.

Complement

$$\Pr(\text{not } A) = 1 - \Pr(A)$$

Conditional probability

$$\Pr(A \mid B) = \Pr(A \text{ and } B)/\Pr(B) \qquad \text{(A1.2)}$$

Multiplicative rule

$$\Pr(A \text{ and } B) = \Pr(A \mid B)\Pr(B) \qquad \text{(A1.3a)}$$

and also

$$= \Pr(B \mid A)\Pr(A) \qquad \text{(A1.3b)}$$

A and B are *independent* if, and only if

$$\Pr(A \text{ and } B) = \Pr(A)\Pr(B)$$

Decision trees

The *expected monetary value* (*EMV*) of some strategy is the sum of the products of possible monetary outcomes with their probabilities of occurrence.

This idea is the basis of *decision trees*. Imagine you have to decide whether to bid for a new overseas airport construction. The cost of preparing a tender would be £20 000. The profit would depend on ground conditions. You assess the probabilities of good (*G*), fair (*F*), poor (*P*) and bad (*B*) conditions as 0.1, 0.5, 0.2 and 0.2 respectively. You would price the tender so that the profit, in thousand pounds, would be 1000, 500, 200 and −500 (a loss) respectively. You think the chance of winning the contract, at this price, is 10%. A survey to determine ground conditions at the beginning of the tender would cost £30 000. The profit estimates have not allowed for the cost of preparing the tender or of a survey. The decision tree for this scenario is shown in Figure A1.1. Squares represent decisions and circles represent contingencies. We work backwards from the right-hand side; e.g. the EMV of the tender, if we have good ground conditions, is $0.1 \times (1000 - 20 - 30) + 0.9 \times (-20 - 30) = 50$. EMVs are shown in square brackets. The EMV of carrying out a survey, and only preparing a tender if the ground conditions are good or fair, is [−7], whereas the EMV of submitting a tender without a survey is [9]. The latter is preferable. The EMV criterion is only defensible if we can sustain the worst possible loss of −520.

Bayes' theorem

The following result, first published by Thomas Bayes in 1763, is the foundation of a system of inference with many engineering applications. Suppose we have *k* mutually exclusive and exhaustive events $\{A_1, \ldots, A_k\}$, and some observed event *B*. Then

$$\Pr(A_i|B) = \frac{\Pr(B|A_i)\Pr(A_i)}{\sum_{j=1}^{k} \Pr(B|A_j)\Pr(A_j)} \tag{A1.4}$$

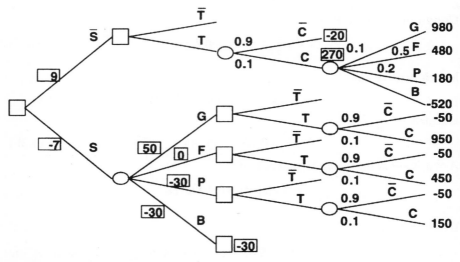

Figure A1.1   Decision tree for airport tender

Arrangements and choices (permutations and combinations)

There are

$$_nP_r = n(n-1) \cdots (n-r+1)$$

ways of arranging r objects from n (n choices for the first, $n-1$ choices for the second, and so on). There are

$$_nC_r = {_nP_r}/r!$$

choices of r objects from n, because each choice can be arranged in $r!$ different ways.

Gamma function

The *gamma function* is defined by

$$\Gamma(\alpha) = \int_0^\infty x^{\alpha-1}\,e^{-x}\,dx \qquad \text{usually for} \qquad 0 < \alpha \qquad (A1.5)$$

Integration by parts shows that it is a generalization of the factorial function, with

$$\Gamma(\alpha) = (\alpha - 1)!$$

A1.2   Sampling

Simple random samples

Many statistical procedures are based on an assumption that we have a *simple random sample* (*SRS*) from a well-defined population. An SRS of size n can be obtained by numbering all the members of a population and using random numbers to select n of them. If the population is a consignment of 100 000 pavers, we would not wish to number each paver explicitly. We would set up a rule for locating those which we selected, e.g. pallet 1 contains pavers 1 up to 1000; and so on. If the population is imaginary and infinite, such as all possible jars of water from the domestic supply if present conditions continue, we have to use an approximation to this procedure. When we analyse environmental data we are usually reduced to assuming a random sample in time.

Stratified sampling and weighted mean

A water company which operates in two divisions A and B, has decided to replace all lead communication pipes (between a property and the main) in its area with PVC pipes. It requires an estimate of the cost. The authority suspects there may be a smaller proportion of lead communication pipes in division B. The company will divide the population into two sub-populations, formally called strata: the N_A properties in A, and the N_B properties in B. A random sample of properties will be taken from A, and an independent random sample of properties will be taken from B. Estimates of the proportions of lead communication pipes in A and B can be made, $\hat{p}_A$ and $\hat{p}_B$. These two estimates can be combined, with the appropriate weighting, to give an estimate of the proportion for the whole area ($\hat{p}$)

$$\hat{p} = \hat{p}_A(N_A/(N_A + N_B)) + \hat{p}_B(N_B/(N_A + N_B))$$

The probability that a property in Division A is in the sample, n_A/N_A, does not have to equal the probability that a property in Division B is in the sample, n_B/N_B.

Size weighted means

Suppose, for example, that $\{x_i\}$ are periods between buses. The average wait per person is not $\frac{1}{2} \sum x_i/n$, but $\frac{1}{2} \sum x_i^2 / \sum x_i$. This is because the number of people arriving at the bus stop is proportional to the length of the periods between buses, and these should therefore be weighted according to their lengths. Hutchinson (1996) gives more examples.

A1.3   Descriptive statistics

Suppose we have an SRS $\{x_i\}$ of size n from a finite population of size N.

	Sample estimate of population parameter	Population parameter
Mean	$\bar{x} = \Sigma x_i/n$	$\mu = \Sigma x_i/N$
Variance	$s^2 = \Sigma(x_i - \bar{x})^2/(n-1)$	$\sigma^2 = \Sigma(x_i - \mu)^2/N$
Standard deviation	$s = \sqrt{s^2}$	$\sigma = \sqrt{\sigma^2}$
Skewness	$\hat{\gamma} = [\Sigma(x_i - \bar{x})^3/(n-1)]/s^3$	$\gamma = [\Sigma(x_i - \mu)^3/N]/\sigma^3$
Kurtosis	$\hat{\kappa} = [\Sigma(x_i - \bar{x})^4/(n-1)]/s^4$	$\kappa = [\Sigma(x_i - \mu)^4/N]/\sigma^4$

The sum of deviations from the mean is zero. Formally

$$\sum (x_i - \bar{x}) = \sum x_i - n\bar{x} = \sum x_i - n(\sum x_i/n) = 0$$

If a unit point mass is assigned to each datum, at its position on the number line, the mean is the centre of gravity. This follows from taking moments about the origin. The *skewness* is a measure of asymmetry of a distribution. Although $\sum (x_i - \bar{x}) = 0$, when we cube the distances, $\sum (x_i - \bar{x})^3$ will be positive if the distribution has a long tail to the right. *Kurtosis* can be thought of as a measure of the weight in the tails, and it is near 3 for data that have an approximate bell-shaped histogram. The coefficient of variation (CV) is the ratio of the mean to the standard deviation. Although the population parameters are properly defined, they cannot be calculated unless we know the values of the x_j for the entire population.

We model infinite populations by assuming the histogram tends to a smooth curve known as the *probability density function* (*pdf*) $f(x)$. The finite population averages tend to integrals (Figure A1.2), known as expected values.

$$\mu = E[X] = \int xf(x)\, dx$$

$$\sigma^2 = E[(X - \mu)^2] = \int (x - \mu)^2 f(x)\, dx$$

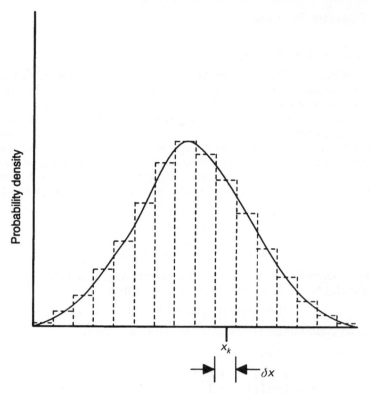

Figure A1.2 Expected value for a continuous distribution. Assume f_k of n data is interval centred on x_k. Then $\bar{x} \approx \sum x_k (f_k/n) \approx \sum x_k f(x_k) \, \delta x \rightarrow \int x f(x) \, dx$.

and

$$E[\phi(x)] = \int \phi(x) f(x) \, dx$$

The next result is often used

$$\sigma^2 = E[(X - \mu)^2] = E[X^2 - 2\mu X + \mu^2]$$
$$= E[X^2] - 2\mu E[X] + \mu^2$$

If this is rearranged we have

$$E[X^2] = \mu^2 + \sigma^2$$

A1.4   Binomial Distribution ($Bin(n, p)$)

There are n trials, the outcome of each trial is a success or a failure, and the probability of a success is p. If X is the number of successes in the n trials:

$$\Pr(X = x) = P(x) = {}_nC_x p^x (1 - p)^{n-x} \qquad \text{for} \quad x = 0, 1, \dots, n \qquad (A1.6)$$

The mean and variance of X are np and $np(1 - p)$ respectively.

A1.5   Poisson Process

Poisson distribution (Poisson (λt))

Assume that occurrences of events are random and independent, and that they occur at a constant average rate of λ per unit time. Let X be the number of occurrences in some chosen length of time t. The Poisson distribution gives the probabilities that X equals x, for integer values of x from 0 upwards. The result was published by Siméon Denis Poisson in 1837. The simplest derivation of the formula is to divide the time into a large number (n) of small increments of equal length (δt). The increments are so small that the probability of more than one event occurring within an interval is negligible. The exact result is obtained as the number of intervals tends to infinity, and their length tends to zero. Formally

$$t = n\delta t$$

and,

$$\Pr(1 \text{ event in time } \delta t) = \lambda \delta t$$

since the probability of more than one event is proportional to $(\delta t)^2$ and becomes negligible as δt tends to zero. It follows that

$$\Pr(0 \text{ events in times } \delta t) = 1 - \lambda \delta t$$

and X has an approximate $\text{Bin}(n, \lambda \delta t)$ distribution. That is

$$P(x) = {}_nC_x(\lambda\delta t)^x(1 - \lambda\delta t)^{n-x}$$

$$= {}_nC_x\left(\frac{\lambda t}{n}\right)^x\left(1 - \frac{\lambda t}{n}\right)^{n-x}$$

$$= \frac{n(n-1)\cdots(n-x+1)}{x!} \frac{(\lambda t)^x}{n^x} \left(1 - \frac{\lambda t}{n}\right)^n\left(1 - \frac{\lambda t}{n}\right)^{-x}$$

Now let n tend to infinity, remembering the result

$$\lim\left(1 - \frac{\lambda t}{n}\right)^n = e^{-\lambda t}$$

which follows easily from Taylor expansions, to obtain the formula

$$P(x) = \frac{(\lambda t)^x \, e^{-\lambda t}}{x!} \qquad \text{for} \qquad x = 0, 1, \ldots \tag{A1.7}$$

Notice that

$$\sum P(x) = e^{-\lambda t}e^{\lambda t} = 1$$

The mean and variance of the Poisson distribution are both λt. The proof of the result for the mean follows

$$E[X] = \sum_{x=0}^{\infty} x\, e^{-\lambda t}(\lambda t)^x/x! = \sum_{x=1}^{\infty} x\, e^{-\lambda t}(\lambda t)^x/x! = (\lambda t)\sum_{y=0}^{\infty} e^{-\lambda t}(\lambda t)^y/y! = \lambda t$$

The continuum can be space, i.e. an area or volume, rather than time. If the continuum is time the continuous distribution of times between occurrences is known as the *exponential distribution*.

Exponential distribution ($M(\lambda)$)

Let T be the time between occurrences in a Poisson process. Then

$$\Pr(T > t) = \Pr(\text{no occurrence in time } t) = e^{-\lambda t}$$

The cdf $F(t)$ is given by

$$F(t) = \Pr(T < t) = 1 - \Pr(T > t) = 1 - e^{-\lambda t} \qquad \text{for} \quad 0 \leqslant t$$

The pdf follows from differentiation

$$f(t) = \lambda e^{-\lambda t} \qquad \text{for} \quad 0 \leqslant t \tag{A1.8}$$

The mean is $1/\lambda$ and the variance is $(1/\lambda)^2$.

Since occurrences in a Poisson process are independent, the time from 'now' until the next occurrence is independent of the history of the process, and has the same distribution as the time between occurrences. This is known as the *Markov property*.

A1.6 Uniform Distribution ($U[a, b]$)

X is distributed uniformly over $[a, b]$ if

$$f(x) = 1/(b - a) \qquad \text{for} \quad a \leqslant x \leqslant b \tag{A1.9}$$

The mean and variance of X are $(a + b)/2$ and $(b - a)^2/12$ respectively.

A1.7 Normal Distribution ($N(\mu, \sigma^2)$)

The pdf is

$$f(x) = \frac{1}{\sigma\sqrt{2\pi}} e^{-((x-\mu)/\sigma)^2/2} \qquad \text{for} \qquad -\infty < x < \infty \tag{A1.10}$$

The parameters μ and σ are the mean and standard deviation of the distribution.

The cdf has to be evaluated numerically. However, μ and σ simply determine the location and scale, so it is only necessary to produce tables for one choice of these parameters, $\mu = 0$ and $\sigma = 1$ being the most convenient. The *standard normal distribution* has the pdf

$$\phi(z) = \frac{1}{\sqrt{2\pi}} e^{-z^2/2}$$

and its cdf is

$$\Phi(z) = \frac{1}{\sqrt{2\pi}} \int_{-\infty}^{z} e^{-\theta^2/2} \, d\theta$$

The *inverse* cdf (Φ^{-1}) is defined by:

$$z = \Phi^{-1}(p) \qquad \text{if and only if} \qquad \Phi(z) = p$$

The upper $\alpha \times 100\%$ percentage point is the value above which a proportion α of the distribution lies. That is

$$\Phi(z_\alpha) = 1 - \alpha$$

Derivation of the normal distribution

We will obtain the normal distribution in three steps. Here we prove that

$$(2\pi)^{-1/2} e^{-(1/2)z^2}$$

is a pdf. In the next section we find an alternative representation of the distribution known as a moment generating function. This facilitates a succinct proof of the Central Limit Theorem. This theorem justifies the claim that errors, which happen to be the sum of a large number of small discrepancies, have a normal distribution. Something of the form

$$e^{-1/2z^2}$$

has the potential to be a useful pdf. It is symmetric and decays faster than the exponential distribution. However, it needs scaling so that it has an area of 1. To find the area, A, we argue as follows

$$A = \int_{-\infty}^{\infty} e^{-1/2z^2}\, dz$$

$$A^2 = \int_{-\infty}^{\infty} e^{-1/2z^2}\, dz \int_{-\infty}^{\infty} e^{-1/2y^2}\, dy = \int_{-\infty}^{\infty} \int_{-\infty}^{\infty} e^{-1/2(z^2 + y^2)}\, dz\, dy$$

In polar coordinates

$$x = r \cos \theta \qquad y = r \sin \theta$$

$$x^2 + y^2 = r^2$$

and an element of area is $r\delta\theta\delta r$. So

$$A^2 = \int_{0}^{2\pi} \int_{0}^{\infty} e^{-1/2r^2} r\, dr\, d\theta = \int_{0}^{2\pi} [-e^{-1/2r^2}]_{0}^{\infty}\, d\theta = 2\pi$$

A1.8   Moment Generating Function

A continuous probability distribution can be characterized by either its pdf or its cdf, and a discrete distribution can similarly be defined by its probability function or the accumulated probabilities. An alternative representation, which is often useful for proving theoretical results, is the *moment generating function* (mgf). The mgf of a variable X is defined by

$$M_X(\theta) = E[e^{\theta x}] \tag{A1.11}$$

which holds for both discrete and continuous distributions. Provided it exists – which it does for the distributions we need – it determines the distribution uniquely. We will use it for two purposes. First, it can be a useful method for finding moments of a distribution

$$E[e^{\theta X}] = \int_{-\infty}^{\infty} e^{\theta x} f(x) \, dx \quad \text{(or summation for a discrete distribution)}$$

$$= \int_{-\infty}^{\infty} \left(1 + \theta x + \frac{(\theta x)^2}{2!} + \cdots \right) f(x) \, dx$$

$$= 1 + \theta E[X] + \frac{\theta^2}{2!} E[X^2] + \cdots$$

Hence

$$E[X^k] = \left(\frac{d^k}{d\theta^k} M_X(\theta) \right)_{\theta = 0}$$

The moments about the mean

$$E[(X - \mu)^k]$$

follow by expanding $(X - \mu)^k$ and using the expressions for $E[X^j]$ for j up to k.

The mgf of a normal distribution is

$$M_X(\theta) = e^{\mu\theta + (1/2)\sigma^2 \theta^2}$$

Proof

$$M_X(\theta) = \frac{1}{\sigma\sqrt{2\pi}} \int_{-\infty}^{\infty} \exp(\theta x) \exp\{-\tfrac{1}{2}[(x - \mu)/\sigma^2]\} \, dx$$

$$= \frac{1}{\sigma\sqrt{2\pi}} \int_{-\infty}^{\infty} \exp\{-\tfrac{1}{2}[(x^2 - 2\mu x + \mu^2 - 2\theta\sigma^2 x)/\sigma^2]\} \, dx$$

$$= \frac{1}{\sigma\sqrt{2\pi}} \int_{-\infty}^{\infty} \exp\left\{-\frac{1}{2}\left[\frac{(x - (\mu + \sigma^2\theta))^2 - 2\mu\sigma^2\theta - \sigma^2\theta^2}{\sigma^2}\right]\right\} \, dx$$

$$= e^{\mu\theta + 1/2\sigma^2\theta^2} \int_{-\infty}^{\infty} \frac{1}{\sigma\sqrt{2\pi}} \exp\{-\tfrac{1}{2}[(x - (\mu + \sigma^2\theta))^2/\sigma]^2\} \, dx$$

The integrand in the last expression is a normal pdf, with mean $\mu + \sigma^2\theta$ and variance σ^2, the integration is over the interval $(-\infty, \infty)$ and therefore the integral equals 1.

Moment generating function of a sum of independent variables

Let $X_1, \ldots, X_n$ be independent variables with mgf $M_{X_1}(\theta), \ldots, M_{X_n}(\theta)$ respectively. Define the sum T by

$$T = X_1 + \cdots + X_n$$

Then the mgf of T is the product of the mgfs of $X_1, \ldots, X_n$.

Proof

$$M_T(\theta) = E[e^{\theta T}] = E[e^{\theta(X_1 + \dots + X_n)}]$$

$$= E[e^{\theta X_1} e^{\theta X_2} \dots e^{\theta X_n}]$$

$$= \int \dots \int e^{\theta x_1} e^{\theta x_2} \dots e^{\theta x_n} f(x_1, \dots, x_n) \, dx_1 \dots dx_n$$

$$= \int e^{\theta x_1} f(x_1) \, dx_1 \dots \int e^{\theta x_n} f(x_n) \, dx_n \quad \text{(since they are independent)}$$

$$= M_{X_1}(\theta) \dots M_{X_n}(\theta)$$

We will use the next result as a step in the proof of the Central Limit Theorem.

Lemma

If a and b are constants, and b is positive, then

$$M_{(X+a)/b}(\theta) = e^{a\theta/b} M_X(\theta/b)$$

Proof   Define

$$W = (X + a)/b$$

By definition

$$\Pr(X < x) = F(x)$$

and it follows that

$$\Pr((X + a)/b < (x + a)/b) = F(x)$$

If we now use the substitution $w = (x + a)/b$

$$\Pr(W < w) = F(bw - a)$$

Now differentiate with respect to w to obtain

$$fW(w) = bf_X(bw - a)$$

where the subscripts emphasize which pdf corresponds to which variable. Next, from the definition

$$M_W(\theta) = \int e^{w\theta} f_W(w) \, dw$$

$$= \int e^{w\theta} b f_X(bw - a) \, dw$$

we now substitute $x = bw - a$ to obtain

$$\int e^{[(x+a)/b]\theta} fX(x) \, dx$$

The result follows.

A1.9   The Central Limit Theorem

Let $X_1, \dots, X_n$ be independent and identically distributed random variables with mean μ and finite variance σ^2. Define

$$U_n = \frac{\bar{X} - \mu}{\sigma/\sqrt{n}}$$

The distribution of U_n converges to the standard normal as $n \to \infty$.

Proof Define

$$Y_i = \frac{X_i - \mu}{\sigma}$$

Then Y_i has zero mean and unit variance and its mgf can be written as

$$M_Y(\theta) = 1 + \frac{\theta^2}{2} + \frac{\theta^3}{3!} E[Y_i^3] + \cdots$$

Also

$$U_n = \sum Y_i/\sqrt{n}$$

$$M_{U_n}(\theta) = \left(M_Y\left(\frac{\theta}{\sqrt{n}}\right) \right)^n = \left(1 + \frac{\theta^2}{2n} + \frac{\theta^3}{3!n^{3/2}} \xi + \cdots \right)^n$$

where

$$\xi = E[Y_i^3]$$

Now consider

$$\ln M_{U_n}(\theta) = n \ln\left(1 + \left(\frac{\theta^2}{2n} + \frac{\theta^3 \xi}{6n^{3/2}} + \cdots \right) \right)$$

and remember the Taylor series

$$\ln(1 + x) = x - x^2/2 + x^3/3 - \cdots$$

provided $|x| < 1$. Using this result

$$\ln M_{U_n}(\theta) = n\left(\left(\frac{\theta^2}{2n} + \frac{\theta^3 \xi}{6n^{3/2}} + \cdots \right) - \frac{1}{2}\left(\frac{\theta^2}{2n} + \frac{\theta^3 \xi}{6n^{3/2}} + \cdots \right)^2 + \cdots \right)$$

$$\lim_{n \to \infty} \ln M_{U_n}(\theta) = \frac{\theta^2}{2}$$

$$\lim_{n \to \infty} M_{U_n}(\theta) = e^{\theta^2/2}$$

which is the mgf for the standard normal random variable.

The theorem can be proved under considerably more general conditions, but the assumption that all the random variables have finite mean and finite variance is essential. In practical terms, the theorem justifies the following approximation.

If we have a large random sample from any distribution with finite mean and variance the distribution of $\overline{X}$ will be near-normal. If we imagine errors as made up of a large number of small, additive, independent components then the errors will have a normal distribution. This provides some justification for assuming normality in many applications.

A1.10   Multivariate Probability Distributions

If we have two variables we can draw a bivariate histogram with a total volume of one. As the sample size increases we imagine this tending towards a smooth surface given by the formula

$$z = f(x, y)$$

such that $f(x, y) \geq 0$ and

$$\iint f(x, y) \, dx \, dy = 1$$

This is known as a bivariate pdf (Figure A1.3). The bivariate cdf is $F(x, y)$ where

$$F(x, y) = \Pr(X < x \text{ and } Y < y)$$
$$= \int_{-\infty}^{x} \int_{-\infty}^{y} f(\zeta, \eta) \, d\zeta \, d\eta$$

It follows that

$$\frac{\partial^2 F}{\partial x \partial y} = f(x, y)$$

The variables X and Y are *independent* if and only if

$$f(x, y) = f_X(x) f_Y(y)$$

Marginal distributions

The *marginal distribution* of X is just the distribution of X, i.e. the information about Y is ignored

$$f_X(x) = \int f(x, y) \, dy$$

Expected value

The expected value of some function ϕ of X and Y is given by

$$E[\phi(X, Y)] = \iint \phi(x, y) f(x, y) \, dx \, dy$$

In particular, *covariance*, which is a measure of linear association, is defined by

$$\text{cov}(X, Y) = E[(X - \mu_X)(Y - \mu_Y)] = \iint (x - \mu_X)(y - \mu_Y) f(x, y) \, dx \, dy$$

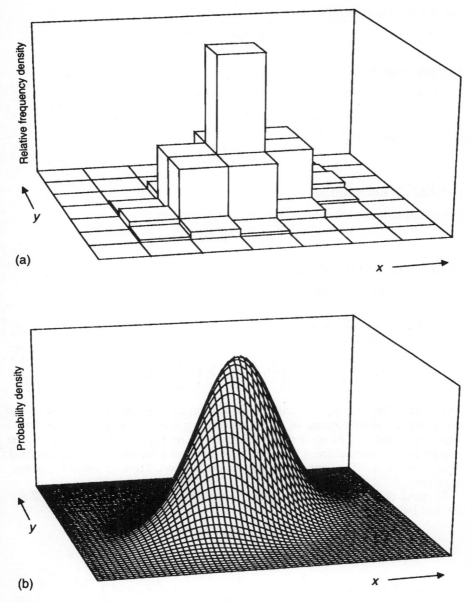

Figure A1.3   Bivariate histogram and pdf

The *correlation* is a dimensionless measure defined by $\rho = \operatorname{cov}(X, Y)/(\sigma_X \sigma_Y)$. If X and Y are independent $\operatorname{cov}(X, Y)$ equals zero. The converse is not necessarily true, e.g. there may be a quadratic relationship or the values that can be taken by X may depend on the value of Y. The following useful result follows from the definitions

$$E[XY] = E[X]E[Y] + \rho\sigma_X\sigma_Y$$

Sample covariance and correlation

If we have n data pairs (x_i, y_i), the sample covariance is

$$\hat{cov} = \Sigma\ (x_i - \bar{x})(\ y_i - \bar{y})/(n - 1)$$

and the sample correlation is

$$r = \hat{cov}/(s_x s_y)$$

Conditional distributions

The conditional distribution of Y given x is defined by

$$f(\ y\ |\ x) = f(x,\ y)/f(x)$$

Bivariate normal distribution and regression

The bivariate normal distribution has the rather formidable looking pdf

$$f(x,\ y) = \frac{1}{2\pi\sigma_X\sigma_Y\sqrt{(1-\rho^2)}}$$

$$\times \exp\left\{-\frac{1}{2(1-\rho^2)}\left[\left(\frac{x-\mu_X}{\sigma_X}\right)^2 - 2\rho\left(\frac{x-\mu_X}{\sigma_X}\right)\left(\frac{y-\mu_Y}{\sigma_Y}\right) + \left(\frac{y-\mu_Y}{\sigma_Y}\right)^2\right]\right\}$$

The parameters μ_X, μ_Y, σ_X and σ_Y are the means and standard deviations of the marginal distributions of X and Y. The parameter ρ is the correlation between X and Y. The easiest way to find the conditional distribution of y on x is to use the standardized distribution which has a mean of 0 and a standard deviation of 1. The general result then follows from a simple scaling argument.

Suppose that (X, Y) has a bivariate normal distribution. Then if W and Z are defined by

$$W = (X - \mu_X)/\sigma_X \qquad \text{and} \qquad Z = (Y - \mu_Y)/\sigma_Y$$

(W, Z) has a standardized bivariate normal distribution and the pdf is

$$f(w,\ z) = \frac{1}{2\pi\sqrt{(1-\rho^2)}}\exp\left\{\frac{-1}{2(1-\rho^2)}(w^2 - 2\rho wz + z^2)\right\}$$

The marginal distributions are standard normal. For example, write $(w^2 - 2\rho wz + z^2)$ as $[(z - \rho w)^2 + (1 - \rho^2)w]$ then substitute θ for $(z - \rho w)$ and integrate to obtain

$$f_W(w) = \frac{1}{\sqrt{2\pi}}\exp(-\tfrac{1}{2}w^2)$$

Some straightforward algebra leads to the conditional distribution

$$f(z|w) = \frac{1}{\sqrt{2\pi}\sqrt{(1-\rho^2)}} \exp\left\{\frac{-1}{2(1-\rho^2)}(z-\rho w)^2\right\}$$

This is a normal distribution with mean, $E[Z|w]$, equal to ρw and a variance of $(1-\rho^2)$. The regression line of z on w is

$$z = \rho w$$

This result can be rescaled so it is explicitly in terms of Y and x. First, use the relationship between Z and Y to write

$$E\left[\frac{Y-\mu_Y}{\sigma_Y}\middle| W = w_p\right] = \rho w_p$$

If W is w_p then X equals $\mu_X + \sigma_X w_p$, and we define x_p by $\mu_X + \sigma_X w_p$. Then

$$E\left[\frac{Y-\mu_Y}{\sigma_Y}\middle| X = x_p\right] = \rho \frac{x_p - \mu_X}{\sigma_X}$$

and finally

$$E[Y|X = x_p] = \mu_Y + \rho \frac{\sigma_Y}{\sigma_X}(x_p - \mu_X)$$

Furthermore, if the conditional distribution of Z given w has a variance of $(1-\rho^2)$, the conditional distribution of Y given x has a variance $(1-\rho^2)\sigma_Y^2$. Notice this variance does not depend on the value of x. The regression line of Y on x is

$$y = \mu_Y + \rho \frac{\sigma_Y}{\sigma_X}(x - \mu_X)$$

An identical argument leads to the regression line of X on y which is

$$x = \mu_X + \rho \frac{\sigma_X}{\sigma_Y}(y - \mu_Y)$$

The two regression lines are not the same. Since $|\rho| < 1$, the regression line of y on x is less steep than the major axis of the elliptical contours of the bivariate distribution shown in Fig A1.4, whereas the regression line of x on y is steeper.

Multivariate normal distribution

Let $x = (x_1, \dots, x_k)^T$

$$f(x) = (2\pi)^{-k/2}|V|^{-1/2} \exp(-\tfrac{1}{2}(x-\mu)^T V^{-1}(x-\mu)) \qquad \text{(A1.12)}$$

where μ is the mean and V is the variance–covariance matrix, often just called

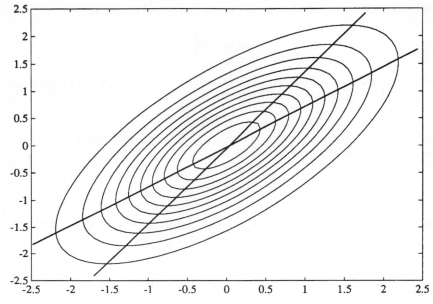

Figure A1.4 Bivariate normal distribution and regression lines

the covariance matrix. If the elements of V are denoted by σ_{ij}, then

$$\sigma_{ii} = \text{var}(x_i)$$

$$\sigma_{ij} = \text{cov}(x_i x_j) \qquad \text{for} \quad i \neq j$$

A1.11   Functions of variables

Linear function

Let X and Y be variables with means μ_X, μ_Y and variances σ_X^2, σ_Y^2 respectively. Let

$$W = aX + bY$$

where a and b are constants. Then

$$\mu_W = a\mu_X + b\mu_Y$$

and

$$\sigma_W^2 = a^2\sigma_X^2 + b^2\sigma_Y^2 + 2ab \, \text{cov}(X, Y)$$

Non-linear function

Let ϕ be an arbitrary function of X and Y and expand ϕ about their mean values. Then

$$\phi(X, Y) = \phi(\mu_X, \mu_Y) + \frac{\partial \phi}{\partial x}(X - \mu_X) + \frac{\partial \phi}{\partial y}(Y - \mu_Y)$$

$$+ \frac{1}{2!}\left(\frac{\partial^2 \phi}{\partial x^2}(X - \mu_X)^2 + 2\frac{\partial^2 \phi}{\partial x\,\partial y}(X - \mu_X)(Y - \mu_Y) + \frac{\partial^2 \phi}{\partial y^2}(Y - \mu_Y)^2\right) + \cdots$$

where all the partial derivatives are evaluated at (μ_X, μ_Y). We start by considering the question of bias. Take expectation of both sides to obtain the approximate result that

$$E[\phi(X, Y)] = \phi(\mu_X, \mu_Y) + \frac{1}{2}\frac{\partial^2 \phi}{\partial x^2}\sigma_X^2 + \frac{1}{2}\frac{\partial^2 \phi}{\partial y^2}\sigma_Y^2 + \frac{\partial^2 \phi}{\partial x\,\partial y}\text{cov}(X, Y)$$

Remember that the expected value of a constant is that constant, and the expected value of a variable minus its mean, for example $(X - \mu_X)$ is 0.

The expected value of a nonlinear function of variables is not that function of the expected values of the variables – for example, the mean of $\ln X$ is not $\ln(\mu_X)$ as you may remember from the lognormal distribution.

If the variables are independent the covariance term is zero and can be dropped. An approximation to the variance of $\phi(X, Y)$ is usually based on the linear terms only. If X and Y are independent

$$\text{var}(\phi(X, Y)) \simeq \left(\frac{\partial \phi}{\partial x}\right)^2 \sigma_X^2 + \left(\frac{\partial \phi}{\partial y}\right)^2 \sigma_Y^2$$

A useful application of this result gives the coefficient of variation of a ratio of two independent random variables. If X and Y are independent, and $W = X/Y$, then

$$[CV(W)]^2 \simeq [CV(X)]^2 + [CV(Y)]^2$$

Unbiased estimators and bias

An estimator X for a parameter θ is unbiased if

$$E[X] = \theta$$

Bias is defined as the difference between $E[X]$, μ say, and θ. Mean squared error (MSE) is

$$E[(X - \theta)^2] = E[((X - \mu) + (\mu - \theta))^2]$$
$$= E[(X - \mu)^2] + (\mu - \theta)^2$$
$$= \text{var}(X) + (\text{bias})^2$$

A biased estimator is acceptable if the bias is small compared with its variance. For example, S is a slightly biased estimator of σ. An estimator is said to be *consistent* if the bias and variance tend to zero as the sample size tends to infinity.

A1.12   Distribution of the Sample Mean

Let $\{X_i\}$ be a random sample from any distribution with a mean μ and variance σ^2. Define the sample total

$$T = X_1 + \ldots + X_n$$

Using the results for linear functions, $\mu_T = n\mu$ and $\sigma_T^2 = n\sigma^2$, since randomization makes an assumption of independence valid. Now $\overline{X} = T/n$ and it follows that: $\mu_{\overline{X}} = n\mu/n = \mu$; $\sigma_{\overline{X}}^2 = n\sigma^2/n^2 = \sigma^2/n$; and $\sigma_{\overline{X}} = \sigma/\sqrt{n}$.

A consequence of the Central Limit Theorem is that $\overline{X}$ is approximately normally distributed if the sample is reasonably large. This is an example of a *sampling distribution*, so called because it is the hypothetical distribution that would be obtained if we took sample means from an infinite number of samples of size n. The approximation is usually adequate if n exceeds 30 and, if the original population is itself near normal, any sample size will suffice. If we are sampling from a finite population of size N, the variance of $\overline{X}$ is reduced by a factor of $(1 - n/N)$, known as the *finite population correction*. It is usually ignored unless at least 10% of the population is sampled. We can use the result

$$\overline{X} \sim N\left(\mu, \left(\frac{\sigma}{\sqrt{n}}\right)^2\right)$$

to construct a confidence interval for μ.

$$\Pr\left(-z_{\alpha/2} < \frac{\overline{X} - \mu}{\sigma/\sqrt{n}} < z_{\alpha/2}\right) = 1 - \alpha$$

$$\Pr(\overline{X} - z_{\alpha/2}\sigma/\sqrt{n} < \mu < \overline{X} + z_{\alpha/2}\sigma/\sqrt{n}) = 1 - \alpha$$

Hence we are $(1 - \alpha) \times 100\%$ confident that the interval

$$\bar{x} \pm z_{\alpha/2}\sigma/\sqrt{n}$$

includes the population mean μ. The limitation of this result is that it assumes the population standard deviation is known. In practice we usually rely on a sample estimate s and the confidence interval becomes

$$\bar{x} \pm t_{n-1,\alpha/2}s/\sqrt{n}$$

where t is the upper $\alpha/2 \times 100\%$ point of Student's t-distribution with $n - 1$ degrees of freedom (see Section A1.17 for details).

A1.13   The Chi-square Distribution

The chi-square distribution was derived by Friedrich Robert Helmert (1843–1917) in 1876. The special cases of the gamma distribution with $c = \nu/2$ and $\lambda = 1/2$ are known as chi-square distributions with ν degrees of freedom (χ_ν^2). The mgf of a gamma distribution is given by

$$M(\theta) = \int_0^\infty e^{\theta x} \lambda^c x^{c-1} (\Gamma(c))^{-1} e^{-\lambda x} \, dx$$

and substituting $y = x(\lambda - \theta)$ gives

$$M(\theta) = (1 - \theta/\lambda)^{-c} \qquad \text{for} \quad \theta < \lambda$$

A chi-square distribution with ν degrees of freedom has mgf

$$M(\theta) = (1 - 2\theta)^{-\nu/2}$$

The mean and variance are ν and 2ν. It follows from the form of $M(\theta)$ that the sum of independent chi-square random variables with ν_1 and ν_2 of freedom have a chi-square distribution with $(\nu_1 + \nu_2)$ degrees of freedom.

A1.14 Sampling Distribution of Sample Variance

We start by showing that if $Z \sim N(0, 1)$ and $Y = Z^2$ then $Y \sim \chi_1^2$

$$\Pr(-z < Z < z) = 2 \int_0^z \frac{1}{\sqrt{2\pi}} e^{-u^2/2} \, du$$

and this is equivalent to $\Pr(Y < z^2)$. Now replace z^2 by y to obtain

$$F(y) = \Pr(Y < y) = 2 \int_0^{\sqrt{y}} \frac{1}{\sqrt{2\pi}} e^{-u^2/2} \, du \qquad \text{for} \quad 0 < y$$

Differentiate with respect to y to obtain

$$f(y) = \frac{1}{\sqrt{2\pi}} y^{-1/2} e^{(-1/2)y} \qquad \text{for} \quad 0 < y$$

We show next, that if $\bar{X}$ and S^2 are the mean and variance of a random sample of n from $N(\mu, \sigma^2)$ then

$$\frac{(n-1)S^2}{\sigma^2} \sim \chi_{n-1}^2$$

Proof

$$\sum (X_i - \mu)^2 = \sum ((X_i - \bar{X}) + (\bar{X} - \mu))^2$$
$$= \sum (X_i - \bar{X})^2 + 2(\bar{X} - \mu) \sum (X_i - \bar{X}) + n(\bar{X} - \mu)^2$$

and since $\sum (X_i - \bar{X}) = 0$ we have

$$\sum (X_i - \mu)^2 = \sum (X_i - \bar{X})^2 + n(\bar{X} - \mu)^2 \qquad \text{(A1.13)}$$

If we divide all through by σ^2 we can write

$$\sum \left(\frac{X_i - \mu}{\sigma} \right)^2 = \frac{(n-1)S^2}{\sigma^2} + \left(\frac{\bar{X} - \mu}{\sigma/\sqrt{n}} \right)^2$$

Since the $\{X_i\}$ are a random sample the $((X_i - \mu)/\sigma)^2$ are independent χ_1^2 and their sum is χ_n^2. Since

$$\left(\frac{\bar{X} - \mu}{\sigma/\sqrt{n}} \right)^2 \sim \chi_1^2$$

and $\bar{X}$ and S^2 are independent (an intuitively reasonable result proved in, for example, Hogg and Craig, 1978),

$$\frac{(n-1)S^2}{\sigma^2} \sim \chi^2_{n-1}$$

The result that $E[S^2] = \sigma^2$ follows either from the fact that the mean of a χ^2 distribution equals its degrees of freedom or, more directly, by taking expected value of both sides of equation (A1.13).

We can use the chi-square distribution to construct a $(1 - \alpha) \times 100\%$ confidence interval for σ^2, and hence σ

$$\Pr\left(\chi^2_{n-1,1-\alpha/2} < \frac{(n-1)S^2}{\sigma^2} < \chi^2_{n-1,\alpha/2}\right) = 1 - \alpha$$

Rearrangement gives

$$\Pr\left(\frac{(n-1)S^2}{\chi^2_{n-1,\alpha/2}} < \sigma^2 < \frac{(n-1)S^2}{\chi^2_{n-1,1-\alpha/2}}\right) = 1 - \alpha$$

and hence a $(1 - \alpha) \times 100\%$ confidence interval for σ^2 of

$$\left[\frac{(n-1)s^2}{\chi^2_{n-1,\alpha/2}}, \frac{(n-1)s^2}{\chi^2_{n-1,1-\alpha/2}}\right]$$

A corresponding confidence interval for σ is obtained by taking the square root of the interval for σ^2. Percentage points of the chi-square distribution are given in Table A15.3.

Example A1.1
In an experiment to study the effects of acceleration on passengers in Metro trains, the acceleration at which loss of balance occurred was measured for a random sample of 12 adults from the Metro staff. The sample mean, $\bar{x}$, and standard deviation, s, were 1.62 m/s^2 and 0.36 m/s^2 respectively. If a normal distribution of such accelerations is assumed, a 90% confidence interval for the variance in the corresponding population, σ^2, is

$$\left[\frac{(11)(0.36)^2}{19.68}, \frac{(11)(0.36)^2}{4.575}\right] = [0.0724, 0.3116]$$

The 90% confidence interval for σ is $[0.27, 0.56]$.

The corresponding population is all adult Metro staff. Any inferences made about the population of Metro travellers would be subjective.

A large sample approximation   Since the mean and variance of a chi-squared distribution with $(n-1)$ degrees of freedom are $(n-1)$ and $2(n-1)$

$$E[S^2] = \sigma^2 \qquad \text{and} \qquad \text{var}[S^2] = 2\sigma^4/(n-1)$$

It follows from the standard result for the mean and variance of a function of a random variable that, ignoring the bias term

$$E[S] \approx \sigma$$

and

$$\text{var}[S] \approx \left(\frac{1}{2\sigma}\right)^2 2\sigma^4/(n-1) \approx \sigma^2/(2n)$$

A rather rough approximation is

$$S \sim N(\sigma, \sigma^2/(2n))$$

Example A1.2
Water companies in the UK may be expected to show, with reasonable confidence, that 95% of PAH levels, in jars of water from properties thought to be at risk of high levels, are less than 180 ng/l. Probability plots of data suggest that an assumption of a lognormal distribution is reasonable. Suppose we have a sample of n PAH results $\{x_i\}$ for i from 1 up to n. We define $y_i = \ln(x_i)$ and assume that Y is normally distributed with mean μ and variance σ^2

$$Y \sim N(\mu, \sigma^2)$$

Then, since the upper 5% point of a standard normal distribution is 1.645, the upper 5% quantile of Y is:

$$\mu + 1.645\sigma$$

We do not know μ or σ so we replace them by $\bar{y}$ and s_y. The estimated upper 5% quantile of Y is:

$$\bar{y} + 1.645 s_y$$

Because we use $(\sigma^2/2n)$ for the variance of s_y, the sampling variance of this quantity is, rather approximately

$$\frac{\sigma^2}{n} + (1.645)^2 \frac{\sigma^2}{2n}$$

We now assume that the estimator of the upper 5% quantile of Y is normally distributed. This is another approximation, but both will improve as n increases. The upper 10% point of a standard normal distribution is 1.282. If we replace σ^2 in the sampling variance by s_y^2, we claim 90% confidence that the upper 5% quantile of Y is less than

$$\bar{y} + 1.645 s_y + 1.282 s_y \sqrt{\frac{1}{n} + \frac{(1.645)^2}{2n}}$$

In one case a company took 16 samples after a zone had been rehabilitated, and calculated $\bar{y}$ and s_y to be 4.2 and 0.4. The one-sided 90% confidence interval for the upper 5% quantile is

$$4.2 + 1.645 \times 0.4 + 1.282 \times 0.4 \sqrt{\frac{1}{14} + \frac{(1.645)^2}{28}} = 5.07$$

We are 90% confident that 95% of the distribution of $\ln X$ is less than 5.07, which corresponds to X being less than $\exp(5.07) = 159$. The criterion is satisfied.

A1.15   Student's *t*-distribution

If $Z \sim N(0, 1)$ and $W \sim \chi^2_\nu$ and Z and W are independent

$$X = \frac{Z}{\sqrt{W/\nu}}$$

has a pdf

$$f(x) = \frac{\Gamma((\nu + 1)/2)}{\sqrt{(\nu\pi)}\Gamma(\nu/2)} (1 + x^2/\nu)^{-(\nu+1)/2} \qquad \text{for} \quad -\infty < x < \infty$$

which is called Student's *t*-distribution with ν degrees of freedom (Student was William S. Gosset, see, for example the 1907 *Biometrika*). We need the following result for the transformation of variables to derive the pdf.

Transformation of variables

Let X_1, X_2 be random variables with joint pdf $f(x_1, x_2)$. Now suppose Y_1 and Y_2 are given by one-to-one functions of X_1 and X_2. That is

$$Y_1 = u_1(X_1, X_2) \qquad \text{and} \qquad Y_2 = u_2(X_1, X_2)$$

and since u_1 and u_2 are one-to-one

$$X_1 = v_1(Y_1, Y_2) \qquad \text{and} \qquad X_2 = v_2(Y_1, Y_2)$$

We will find the joint pdf of Y_1, Y_2, $g(y_1, y_2)$ say. Suppose (X_1, X_2) is in a set A if and only if (Y_1, Y_2) is in a set B

$$\text{Pr}((X_1, X_2) \text{ in } A) = \iint_A f(x_1, x_2) \, dx_1 \, dx_2$$

$$= \iint_B f(v_1(y_1, y_2), v_2(y_1, y_2)) |J| \, dy_1 \, dy_2$$

where J is the Jacobian (e.g. Spiegel, 1963)

$$J = \begin{vmatrix} \dfrac{\partial x_1}{\partial y_1} & \dfrac{\partial x_1}{\partial y_2} \\[2mm] \dfrac{\partial x_2}{\partial y_1} & \dfrac{\partial x_2}{\partial y_2} \end{vmatrix}$$

It follows that

$$g(y_1, y_2) = f(w_1(y_1, y_2), w_2(y_1, y_2)) |J|$$

The marginal pdf for Y_1 can be found by integrating over y_2.

Derivation of the *t*-distribution

Since Z and W are independent their joint pdf is given by

$$f(z, w) = \frac{1}{\sqrt{2\pi}} e^{(-1/2)z^2} \frac{1}{\Gamma(\nu/2)2^{\nu/2}} w^{(\nu/2)-1} e^{-w/2}$$

for $-\infty < z < \infty$ and $0 < w$. Now set $x = z/\sqrt{w/\nu}$ and $y = w$

$$
J = \begin{vmatrix} \dfrac{\partial z}{\partial x} & \dfrac{\partial z}{\partial y} \\[2ex] \dfrac{\partial w}{\partial x} & \dfrac{\partial w}{\partial y} \end{vmatrix} = \sqrt{y/\nu}
$$

The pdf is found by using the general result to write down $g(x, y)$ and then integrating over y, after substituting $\theta = (y/2)(1 + x^2/\nu)$.

Special case

Let $X_1, \ldots, X_n$ be a random sample from $N(\mu, \sigma^2)$. Then

$$
\frac{\bar{X} - \mu}{\sigma/\sqrt{n}} \sim N(0, 1)
$$

and

$$
\frac{(n-1)S^2}{\sigma^2} \sim X_{n-1}^2
$$

Therefore

$$
\frac{\bar{X} - \mu}{S/\sqrt{n}} \sim t_{n-1}
$$

This result justifies the confidence for μ given in Section A1.12.

A1.16 The *F*-distribution

If U and V are independent random variables having chi-square distributions with ν_1 and ν_2 degrees of freedom, then the distribution of

$$
W = \frac{U/\nu_1}{V/\nu_2}
$$

has pdf

$$
f(w) = \frac{\Gamma((\nu_1 + \nu_2)/2)}{\Gamma(\nu_1/2)\Gamma(\nu_2/2)} (\nu_1/\nu_2)^{\nu_1/2} w^{(\nu_1/2)-1} (1 + \nu_1 w/\nu_2)^{-(\nu_1+\nu_2)/2} \qquad 0 < w
$$

which is known as the *F*-distribution with ν_1 and ν_2 degrees of freedom. The *F*-distribution was obtained by G. W. Snedecor in 1934, who named the ratio '*F*' in honour of R. A. Fisher (1890–1962) who had already derived the distribution of $\frac{1}{2} \ln F$. The distribution can be derived by applying the transformation of variables technique. Since U and V are independent random variables with known pdf we can write down $f(u, v)$. If we now use the transformation of variable technique we used to derive the pdf of the *t*-distribution, with $w = (u/v_1)/(v/v_2)$ and $s = v$, we can write down the pdf of w and s. Integrating over s leads to the result. Notice that if W has an F-distribution with v_1 and v_2 degrees of freedom its reciprocal has an F-distribution with v_2 and v_1 degrees of freedom

$$\text{if} \qquad W \sim F_{v_1, v_2} \qquad \text{then} \qquad W^{-1} \sim F_{v_2, v_1}$$

Therefore only the upper tail is tabled (Table A15.4 for example).

Example A1.3
This example deals with laboratory tests of manufacturers' claims about the replicability of pressure measurements made with their transducers. Samples of 18 pressure transducers from manufacturer A, and 12 pressure transducers from manufacturer B were bought from a range of stockists in order to approximate random samples.

To construct a 90% confidence interval for the ratio of the standard deviation of the population of transducers from B to the standard deviation of transducers from $A(\sigma_B/\sigma_A)$ we adopt the following argument.

By definition

$$\Pr\left(F_{17,11,0.95} < \frac{S_A^2/\sigma_A^2}{S_B^2/\sigma_B^2} < F_{17,11,0.05}\right) = 0.90$$

and therefore

$$\Pr\left(\frac{S_B^2}{S_A^2} F_{17,11,0.95} < \frac{\sigma_B^2}{\sigma_A^2} < \frac{S_B^2}{S_A^2} F_{17,11,0.05}\right) = 0.90$$

Now s_A^2 and s_B^2 equal $(0.1931)^2$ and $(0.2657)^2$ respectively, and $F_{17,11,0.05}$ equals 2.71 (interpolating tables of the F-distribution). To find the lower 5% point we apply the reciprocal relationship noted earlier, namely if

$$\Pr\left(\frac{S_B^2/\sigma_B^2}{S_A^2/\sigma_A^2} < F_{11,17,0.05}\right) = 0.95$$

then

$$\Pr\left(\frac{S_A^2/\sigma_A^2}{S_B^2/\sigma_B^2} > 1/F_{11,17,0.05}\right) = 0.95$$

Therefore

$$F_{17,11,0.95} = 1/F_{11,17,0.05} = \frac{1}{2.415} = 0.414$$

It follows that the 90% confidence interval for σ_B^2/σ_A^2 is

$$[0.7838, 5.131]$$

Hence a 90% confidence interval for σ_B/σ_A is

$$[0.88, 2.27]$$

Relationship between the *t*-distribution and the *F*-distribution

The square of a variable which has a t_{n-1} distribution has an $F_{1,n-1}$ distribution

$$\frac{\bar{X}-\mu}{S/\sqrt{n}} = \frac{\bar{X}-\mu}{\sigma/\sqrt{n}} \bigg/ \frac{S}{\sigma} \sim t_{n-1}$$

$$\left(\frac{\bar{X}-\mu}{\sigma/\sqrt{n}}\right)^2 \bigg/ \frac{S^2}{\sigma^2} \sim F_{1,n-1}$$

Improved estimate of an upper percentile of a normal distribution

The upper $\alpha \times 100\%$ percentile of a normal distribution is

$$\mu + z_\alpha \sigma$$

The natural estimator of this quantity is

$$\bar{X} + z_\alpha S$$

which implicitly assumes that

$$\frac{X-\bar{X}}{S} \sim N(0, 1)$$

We will show below that, if X is normally distributed

$$\frac{X-\bar{X}}{S\sqrt{((n+1)/n)}} \sim t_{n-1}$$

and improved estimates would be given by

$$\hat{x}_\alpha = \bar{x} + t_{\alpha,n-1} s\sqrt{((n+1)/n)}$$

We now justify this claim.

Assume X is normally distributed with mean μ and variance σ^2. $\bar{X}$ is an independent estimate of μ with mean μ and variance σ^2/n. Therefore

$$X - \bar{X} \sim N(0, \sigma^2 + \sigma^2/n)$$

and

$$\frac{X-\bar{X}}{\sigma^2(1+1/n)} \sim \chi_1^2$$

S^2 is independent of $\bar{X}$ and

$$\frac{(n-1)S^2}{\sigma^2} \sim \chi_{n-1}^2$$

Hence

$$\frac{(X-\bar{X})^2}{S^2(1+1/n)} \sim F_{1,n-1}$$

and the result follows.

A1.17   Differences in Means

Independent samples

Populations A and B have means μ_A and μ_B, and standard deviations σ_A and σ_B respectively. These population means and standard deviations are unknown. Now suppose we have two independent random samples of sizes n_A and n_B from these populations. We rely on the central limit theorem, which is a good approximation if the sample is large or the population near-normal, to state that

$$\bar{X}_A \sim N(\mu_A, \sigma_A^2/n_A) \qquad \text{and} \qquad \bar{X}_B \sim N(\mu_B, \sigma_B^2/n_B)$$

Therefore, since the samples are independent

$$\bar{X}_A - \bar{X}_B \sim N(\mu_A - \mu_B, \sigma_A^2/n_A + \sigma_B^2/n_B)$$

An approximate $(1 - \alpha) \times 100\%$ confidence interval for $\mu_A - \mu_B$ is

$$\bar{x}_A - \bar{x}_B \pm t_{\nu, \alpha/2}(s_A^2/n_A + s_B^2/n_B)^{1/2}$$

where

$$\nu = \left(\frac{s_A^2}{n_A} + \frac{s_B^2}{n_B}\right)^2 \bigg/ \left(\frac{(s_A^2/n_A)^2}{n_A - 1} + \frac{(s_B^2/n_B)^2}{n_B - 1}\right)$$

Paired comparisons

Suppose we wish to compare two laboratories, A and B, for any systematic difference in cement content analyses. The cement contents of individual pavers can vary considerably, and this variation would probably swamp any difference between the laboratories. A good way to conduct the test, which avoids this problem, would be to crush each paver, mix the powder thoroughly, and send half, selected at random, to each laboratory. Let the cement contents for pavers from each laboratory be x_{Ai} and x_{Bi}. Then calculate the differences

$$d_i = (x_{Ai} - x_{Bi})$$

for each paver. If we have n pavers a $(1 - \alpha) \times 100\%$ confidence interval for the systematic difference in cement content measurements is

$$\bar{d} \pm t_{n-1, \alpha/2} s_d / \sqrt{n}$$

The cement contents (kg/m^3) from a trial using six pavers are

Paver Number	1	2	3	4	5	6
x_{Ai}	360.6	342.0	347.4	352.8	361.2	348.0
x_{Bi}	383.4	361.8	346.8	367.8	358.2	366.0
d_i	−22.8	−19.8	0.6	−15.0	3.0	−18.0
$\bar{d} = -12.0$				$s_d = 11.01$		

the 90% confidence interval is

$$-12 \pm 9.1$$

Despite the small sample we have some evidence that the measurements from Laboratory A are systematically lower.

A1.18   Proportions

Confidence interval for a proportion

Let p represent the proportion of defectives in the population. Take a random sample of size n from this population. Let X be the number of defectives in the sample

$$X \sim \text{Bin}(n, p)$$

Provided both np and $n(1 - p)$ exceed about five, X is approximately normally distributed. That is, to a good approximation

$$X \sim N(np, np(1 - p)) \qquad \text{and hence} \qquad \frac{X}{n} \sim N(p, p(1 - p)/n))$$

An approximate $(1 - \alpha) \times 100\%$ confidence interval for p is

$$\frac{x}{n} \pm z_{\alpha/2} \sqrt{\left. \frac{x}{n}\left(1 - \frac{x}{n}\right) \right/ n}$$

provided n exceeds about 30.

Example A1.4
A water company found that 38 out of a random sample of 200 properties had lead communication pipes to the main. A 95% confidence interval for the proportion in the corresponding population is

$$38/200 \pm 2.0 \times \sqrt{(38/200) \times (162/200)/200}$$

which gives

$$0.19 \pm 0.06$$

The company has decided that a more accurate estimate is needed and wishes to halve the width of the confidence interval. This means increasing the sample size by a factor of four, and a further 600 properties will be investigated.

Confidence interval for difference in proportions   Consider two large (or ∞) populations A, B and let the proportion of defectives be p_A, p_B respectively. Draw random samples of sizes n_A, n_B from the populations and let X and Y represent the number of defectives in each sample.

Provided $n_A p_A$, $n_A(1 - p_A)$, $n_B p_B$, $n_B(1 - p_B)$ all exceed 5, and n_A, n_B exceed at least 30, an approximate $(1 - \alpha) \times 100\%$ confidence interval for $p_A - p_B$ is given by

$$(x/n_A - y/n_B) \pm z_{\alpha/2}\sqrt{(x/n_A)(1 - x/n_A)/n_A + (y/n_B)(1 - y/n_B)/n_B}$$

A1.19 Sample Size

The widths of all the confidence intervals covered in this appendix are inversely proportional to the square root of the sample size. An appropriate sample size can be set by choosing a target width for the confidence interval and assuming some plausible values for standard deviations.

A1.20 Hypothesis Testing

A water company adds fluoride to the domestic supply, with the aim of reducing dental decay. The target content is 1.00 ppm and the standard deviation should not exceed 0.06 ppm. A public health inspector has taken a random sample of 16 jars over the past week. The inspector will issue a warning if there is evidence of non-compliance. This can be formalized as a *hypothesis test*. For example, the inspector will test a *null hypothesis* (H_0) that the mean of the population of all jars is $1.\dot{0}$ ppm, against a *two-sided alternative hypothesis* (H_1) that the mean is not $1.\dot{0}$ ppm where the dot signifies recurring. If there is no evidence against the hypothesis the inspector will accept it and act as though it were true, i.e. not issue any warning. We should not interpret acceptance of H_0 as a claim that it is true. In practice, the mean will not be exactly $1.\dot{0}$ ppm. Nevertheless, the argument starts with an assumption that H_0 is true. Then

$$\frac{\bar{X} - 1.000}{S/\sqrt{16}} \sim t_{15}$$

The inspector obtained a mean $\bar{x}$ of 1.035 ppm and a standard deviation of 0.104. The calculated value of the t-statistic is $(1.035 - 1.000)/(0.104/4) = 1.346$. The *P-value* is the probability of a value as extreme, or more extreme, than that obtained in the experiment, if H_0 is true. From Table A15.2, $t_{15,.10} = 1.341$ and the *P*-value is only slightly less than $2 \times 0.10 = 0.20$. We claim evidence to reject H_0 at the $\alpha \times 100\%$ level, where α is usually something less than 0.1, if the *P*-value is less than α. In this case we have no evidence to reject H_0 at the 10% level with a two-sided alternative. An equivalent procedure is to construct a 90% confidence interval for μ, and to accept H_0 if the interval includes 1.00.

$$1.035 \pm 1.753 \times 0.104/\sqrt{16}$$

gives

$$[0.99, 1.08]$$

The inspector would also test a null hypothesis about the variance

$$H_0: \sigma = 0.06$$

against a *one-sided alternative hypothesis*

$$H_1: \sigma > 0.06$$

If H_0 is true

$$\frac{15S^2}{(0.06)^2} \sim \chi_{15}^2$$

The calculated value of this statistic is $15 \times (0.104)^2/(0.06)^2 = 45.1$. Since the alternative is one-sided we only need the probability of exceedance of 45.1. The upper 0.05% point of χ^2_{15} is 39.72, so the P-value is less than 0.0005. There is strong evidence that the variance is too high.

A1.21 Goodness-of-fit Tests

The chi-square goodness-of-fit test

This test compares observed (O_k) and expected (E_k) frequencies over c class intervals. If a simple random sample was taken, the observed frequency is a realization of a Poisson variable with mean E_k. The variance of a Poisson variable equals its mean so, provided the expected value exceeds about 5, we have, approximately

$$\frac{O_k - E_k}{\sqrt{E_k}} \sim N(0, 1)$$

and

$$\frac{(O_k - E_k)^2}{E_k} \sim \chi^2_1$$

Their sum

$$W = \sum_{k=1}^{c} \frac{(O_k - E_k)^2}{E_k} \sim \chi^2_{c-1-p}$$

We lose one degree of freedom for each parameter of the distribution which we estimate from the data, and one more for the constraint that the sum of the observed frequencies equals the sum of the expected frequencies. A proof of this result, when the parameters are estimated by minimizing the value of W (*minimum chi-square estimators*), is given in advanced texts on mathematical statistics.

EDF tests

The empirical distribution function (edf) was defined, for a sample of size n, in Section 3.4.5 by

$$\hat{F}(x) = (\text{number of data} \leq x)/n$$

It is a step function which jumps by $1/n$ at the order statistics $x_{1:n}, x_{2:n} \ldots x_{n:n}$. Many statistics have been proposed for comparing the edf with a hypothetical cdf $(F_0(x))$. Two of the more commonly used are the Kolmogorov–Smirnov statistics and the Anderson–Darling statistic. Large values of the statistic are evidence against the hypothesised distribution. The Kolmogorov–Smirnov statistics are

$$D_n^+ = \sup_x [\hat{F}(x) - F_0(x)]; \qquad D_n^- = \sup_x [F_0(x) - \hat{F}(x)]$$

and

$$D_n = \max(D_n^+, D_n^-)$$

where $\sup_x$ means the greatest value over all possible x. A more convenient form for programming is

$$D_n^+ = \max_{1 \leqslant i \leqslant n}\left(\frac{i}{n} - F_0(x_{i:n})\right)$$

$$D_n^- = \max_{1 \leqslant i \leqslant n}\left(F_0(x_{i:n}) - \frac{i-1}{n}\right)$$

The Anderson–Darling statistic is

$$A^2 = n \int \{[\hat{F}(x) - F_0(x)]^2/[F_0(x)(1 - F_0(x))]\}f_0(x)\,\mathrm{d}x$$

and the computational form is

$$A^2 = -\sum_{i=1}^{n} \frac{2i-1}{n}\{\ln(F_0(x_{i:n})) + \ln(1 - F_0(x_{n+1-i:n}))\} - n$$

The Anderson–Darling statistic is designed to be sensitive to discrepancies in the tail of the distribution. The upper percentage points of these statistics depend on the form of F_0 and the number of parameters estimated from the data as well as the sample size n. Percentage points can be found by simulation, and some tables are also available. However, percentage points for the special case when the hypothetical distribution is completely specified are quite misleading when parameters are estimated from the data.

The following excerpt from Pearson and Hartley (1976) (D is D_n) gives upper percentage points for modified forms, $T(\ )$, when testing for exponentiality with an unspecified λ.

	10%	5%	1%
$T(D) = (D - 0.2/n)(\sqrt{n} + 0.26 + 0.5/\sqrt{n})$	0.990	1.094	1.308
$T(A^2) = A^2(1 + 1.5/n - 5/n^2)$	1.078	1.341	1.957

We need to remember the effect of sample size on goodness-of-fit tests. If the sample size is very large we are likely to have evidence against hypothetical distributions which are a good enough approximation for practical purposes. If the sample size is small we are unlikely to have evidence against a variety of distribution types.

Appendix 2   Random Number Generation

A2.1   An efficient and portable pseudo-random number generator

This is a brief description of Wichmann's and Hill's (1982) algorithm. It relies on the fact that the fractional part of the sum of n independent uniform random numbers remains uniform, despite the fact that their sum tends to normality. To see why, suppose $R_1 \sim U[0, 1]$ and add any number (a) between 0 and 1, to R_1. Then

$$R_1 + a \sim U[a, 1 + a]$$

Now, the interval $[a, 1 + a]$ can be split up as $[a, 1) + [1, 1 + a)$. If we take the fractional part of $(R_1 + a)$ the corresponding interval will be

$$[a, 1) + [0, a)$$

which is $[0, 1]$. Therefore, the fractional part of $R_1 + a$ has a $U[0, 1]$ distribution. In particular, a could be the value taken by an independent variable

$$R_2 \sim U[0, 1]$$

The general result follows by induction. (Note the the result would still apply if a has any value, and R_2 could have any distribution provided R_2 is independent of R_1.) The reason for combining numbers from several generators is that none is perfect and the combination is an improvement with a longer cycle length. The algorithm is: I_0, J_0 and K_0 can be arbitrarily set at any integer values within the range $1 - 30\,000$

$$I_{m+1} = 171 I_m \qquad (\text{mod } 30\,269)$$
$$J_{m+1} = 172 J_m \qquad (\text{mod } 30\,307)$$
$$K_{m+1} = 170 K_m \qquad (\text{mod } 30\,323)$$

which must be written in integer arithmetic

$$S_{m+1} = I_{m+1}/30\,269 + J_{m+1}/30\,307 + K_{m+1}/30\,323$$
$$R_{m+1} = S_{m+1} - \text{integer part of } S_{m+1}$$

The cycle length is the product of 30 269 with 30 307 and 30 323, which exceeds 2.7×10^{13}. The $\{R_m\}$ appear to be independently and uniformly distributed over the interval $[0, 1]$, insomuch as the usual tests, applied to quite long sequences, do not provide evidence against the hypothesis. However, Johnson (1987) claims that a generalized feedback shift register generator provides better n-space uniformity. This is presumably due to slight correlations, in sequences from congruential generators, which become more apparent in multi-dimensional work.

An adjustment to Wichmann's and Hill's algorithm to avoid the possibility of obtaining exactly zero, within the precision of the machine, is given by McLeod (1985).

A2.2   Acceptance–rejection algorithm

We can obtain a random number from a distribution with pdf $f(x)$ by first generating a random number from some other pdf $g(x)$, provided we can find a constant k such that:

$$f(x) \leqslant kg(x) \quad \text{for all } x$$

We choose $g(x)$ to be some distribution which it is relatively easy to sample from, and to be efficient k should be the smallest number which satisfies the inequality. The algorithm is justified by considering the probability of being in a small interval as proportional to the height of the pdf. The algorithm is as follows.

(i)   Let v be a random number from $g(x)$.
(ii)   Let u be a random number from $U[0, 1]$.
(iii)   Calculate $c = kug(v)$.
(iv)   If $c > f(v)$ reject v and return to (i).
        If $c \leqslant f(v)$ accept v and take

$$X = v$$

as a random number from f return to (i).

Example
Generate random samples from a standard folded normal distribution, i.e.

$$f(x) = (2/\pi)^{1/2} \exp(-x^2/2)$$

A suitable $g(x)$ is the exponential with $\lambda = 1$, i.e.

$$g(x) = \exp(-x)$$

The smallest value for k is $\sqrt{(2e/\pi)}$, and the pdf touch when $x = 1$.

Appendix 3   Maximum Likelihood

Let $X_1, \ldots, X_n$ be a random sample from a distribution with pdf $f(x)$ and parameters θ. We can emphasize the dependence on the parameters by writing $f(x \mid \theta)$. The random sampling justifies an assumption that the X_i are independent, and the joint pdf is the product

$$\Pi f(x_i \mid \theta)$$

This probability that the X_i are within δx of x_i is proportional to the joint pdf. If the joint pdf is regarded as a function of θ it is called the likelihood function $\mathscr{L}(\theta)$. The maximum likelihood estimator (MLE) of θ is the value of θ which minimizes $\mathscr{L}(\theta)$, or equivalently $\ln(\mathscr{L}(\theta))$, and is denoted $\hat{\theta}$ (see Section 1.2.1 and Section 1.3.2 for examples).

Rao–Cramér Inequality

The following proof is based on that given by Hogg and Craig (1978). Let $X_1, \ldots, X_n$ denote a random sample from a distribution with pdf $f(x \mid \theta)$. Let $Y = u(X_1, \ldots, X_n)$ be an unbiased estimator of θ. We will show that the variance of Y exceeds

$$\left\{ n\mathrm{E}\left[\left(\frac{\partial \ln f(x \mid \theta)}{\partial \theta} \right)^2 \right] \right\}^{-1}$$

This is known as the Rao–Cramér lower bound (RCLB).

Proof   Since $f(x \mid \theta)$ is a pdf

$$1 = \int f(x_i \mid \theta)\, \mathrm{d}x_i \qquad \text{for} \quad i = 1, \ldots, n \tag{A3.1}$$

Since y is an unbiased estimator of θ

$$\theta = \int y f_Y(y \mid \theta)\, \mathrm{d}y$$

$$= \int \ldots \int u(x_1, \ldots, x_n) f(x_1 \mid \theta) \ldots f(x_n \mid \theta)\, \mathrm{d}x_1 \ldots \mathrm{d}x_n \tag{A3.2}$$

We now differentiate both sides of these equations with respect to θ, and use the fact that for any function, h say, of θ: $dh/d\theta = [d(\ln h)/d\theta] h$.

We assume that the limits of integration do not depend on θ. Differentiating equation (A3.1) gives

$$0 = \int \frac{\partial f(x_i|\theta)}{\partial \theta} \, dx_i = \int \frac{\partial \ln f(x_i|\theta)}{\partial \theta} f(x_i|\theta) \, dx_i \qquad \text{(A3.3)}$$

Differentiating equation (A3.2) gives

$$1 = \int \cdots \int u(x_1, ..., x_n) \left(\sum \frac{1}{f(x_i|\theta)} \frac{\partial f(x_i|\theta)}{\partial \theta} \right) f(x_1|\theta) ... f(x_n|\theta) \, dx_1 ... dx_n$$

$$= \int \cdots \int u(x_1, ..., x_n) \left(\sum \frac{\partial \ln f(x_i|\theta)}{\partial \theta} \right) f(x_1|\theta) ... f(x_n|\theta) \, dx_1 ... dx_n \qquad \text{(A3.4)}$$

Now define the random variable

$$W = \sum \frac{\partial \ln f(x_i|\theta)}{\partial \theta}$$

Equation (A3.3) implies that $E[\partial \ln f(x_i|\theta)/\partial \theta] = 0$, and hence $E[W] = 0$. Also, since the $\partial \ln f(x_i|\theta)/\partial \theta$ are independent

$$\text{var}(W) = nE\left[\left(\frac{\partial \ln f(x_i|\theta)}{\partial \theta} \right)^2 \right]$$

Equation (A3.4) implies that $E[YW] = 1$. For any random variables, from the definition of covariance,

$$E[YW] = E[Y]E[W] + \rho \sigma_Y \sigma_W$$

Since $\rho^2 \leqslant 1$, and $E[W] = 0$

$$\frac{1}{\sigma_Y^2 \sigma_Z^2} \leqslant 1$$

which is the required result.

Asymptotic efficiency of MLE

We solve the equation

$$W(\hat{\theta}) = 0$$

to obtain $\hat{\theta}$. We have already proved that

$$E[W(\hat{\theta})] = 0$$

For large n

$$E[W(\hat{\theta})] \simeq W(E[\hat{\theta}])$$

and hence $E[\hat{\theta}] \approx \theta$. It is plausible that ρ^2 tends to 1, and the variance of the estimator approaches the RCLB. The efficiency of an estimator is defined as the ratio of the RCLB to the actual variance of the estimator.

It is sometimes easier to find

$$-E\left[\frac{\partial^2 \ln f(x\,|\,\theta)}{\partial\theta^2}\right] \qquad \text{than} \qquad E\left[\left(\frac{\partial \ln f(x\,|\,\theta)}{\partial\theta}\right)^2\right]$$

They can be shown to be equivalent by differentiating equation (A3.3) with respect to θ. We estimate the variance of $\hat{\theta}$ by replacing θ in the RCLB with $\hat{\theta}$.

Example A3.1
We will find the RCLB for the MLE estimator of the rate parameter λ of an exponential distribution.

$$f(x\,|\,\lambda) = \lambda e^{-\lambda x}$$
$$\ln f(x\,|\,\lambda) = \ln \lambda - \lambda x$$
$$\frac{\partial^2 \ln f(x\,|\,\lambda)}{\partial\lambda^2} = \frac{-1}{\lambda^2}$$

Hence the estimated RCLB is $\hat{\lambda}^2/n$.

If we have more than one parameter, the RCLB for the covariance matrix is

$$\left\{-nE\left[\frac{\partial^2 \ln f(x\,|\,\theta)}{\partial\theta_k\,\partial\theta_l}\right]\right\}^{-1}$$

It must be remembered that the asymptotic efficiency of MLE depends on an assumption that the domain of x does not depend on any of the parameters which are being estimated. For example, the result does not apply to estimation of a lower, or upper, bound of a distribution.

Example A3.2
The RCLB does not apply to the MLE of the bounds for a uniform distribution. The MLE are just the smallest and largest sample values and are not very plausible. They are certainly unsatisfactory for small samples. Formally, for a $U[a, b]$ distribution

$$\hat{a} = x_{1:n} \quad \text{and} \quad \hat{b} = x_{n:n}$$

Unbiased estimators are

$$\hat{a} = m - (n+1)h/(n-1) \quad \text{and} \quad \hat{b} = m + (n+1)h/(n-1)$$

where m is the mid range, $(x_{1:n} + x_{n:n})/2$, and h is the semi-range, $(x_{n:n} - x_{1:n})/2$.

Likelihood ratio test

Let $X_1, \ldots, X_n$ be a random sample from a distribution with pdf $f(x)$ and parameters $\theta_1, \ldots, \theta_k$. Let $\mathcal{L}(\omega)$ be the likelihood if $\theta_{k-r+1}, \ldots, \theta_k$ are restricted to values specified by a null hypothesis H_0. The specified values will often be zeros. Let $\mathcal{L}(\Omega)$ be the likelihood when $\theta_1, \ldots, \theta_k$ are not restricted. The likelihood ratio is

$$\phi = \mathcal{L}(\hat{\omega})/\mathcal{L}(\hat{\Omega})$$

where, for example, $\hat{\omega}$ means that $\theta_1, \ldots, \theta_{k-r}$ have been replaced by their MLE under the hypothesis H_0. Since Ω includes ω, ϕ must be less than one. It can be shown that if H_0 is true, then as the sample size tends to infinity

$$-2 \ln \phi \sim \chi_r^2$$

Large values of $-2 \ln \phi$ are taken as evidence against H_0.

Appendix 4   Bootstrap Methods for Confidence Intervals

The basic principle will be demonstrated with the following example. The ratios (r_i) of out-turn cost to predicted cost for eight water distribution schemes are:

$$1.16 \quad 0.90 \quad 1.09 \quad 0.95 \quad 0.75 \quad 1.50 \quad 1.48 \quad 1.67$$

The mean ratio $(\bar{r})$ is 1.29 and the standard deviation s_r is 0.34. The individual ratios are unlikely to be well approximated by a normal distribution, but the construction for a confidence interval for μ_r using a t-distribution is not very sensitive to this assumption. In particular, a 95% confidence interval for μ_r is given by

$$1.19 \pm 2.365 \times 0.33/\sqrt{8}$$

which gives [0.91, 1.46]. We will compare this interval with one obtained using a bootstrap procedure. The simplest method is as follows.

Step 1
   Use random numbers, to re-sample, with replacement, 8 data from the set of 8 ratios.
   For example, I took x_i from $U[0, 1]$ and let

$$m_i = \text{round}(x_i \times 8 + 0.501)$$

   for $i = 1, \ldots, 8$. This gave me

$$3 \quad 2 \quad 3 \quad 2 \quad 5 \quad 8 \quad 4 \quad 6$$

   and the first *bootstrap sample* was

$$1.09 \quad 0.90 \quad 1.09 \quad 0.90 \quad 0.75 \quad 1.67 \quad 0.95 \quad 1.50$$

   Calculate the mean ratio for the bootstrap sample, 1.11 in this case.

Steps 2...1000
   Repeat step 1, a large number of times. A thousand will suffice.

Final step

Sort the 1000 bootstrap sample means into ascending order. For a 95% confidence interval take the 25th and 975th value. I obtained

$$[0.97, 1.40]$$

This is narrower than the interval based on a t-distribution, as tends to be the case with small samples. It is closer to the interval we would have obtained if we had used $z_{0.025}$ instead of $t_{7, 0.025}$. An improvement is to use what are called *pivotal* methods (see, for example, Efron and Gong, 1983).

An important modification, which is often useful in more complicated situations, is the *parametric bootstrap*. The device is to estimate the parameters of an assumed distribution, from the original sample, and then to generate bootstrap samples from this distribution rather than by re-sampling the original sample. This idea can be extended to complex stochastic models.

Bootstrap methods are a means of assessing the precision of estimates, and require programming ability, rather than statistical expertise, for their implementation. Intuitively, it seems that they will tend to give confidence intervals that are somewhat narrower than they should be for a given level, but good enough for most engineering purposes. The methods can be justified by mathematical arguments. More details about practical applications, including Minitab code, can be found in Taffe and Garnham (1996).

Appendix 5 Optimization Algorithms

The objective is to minimize some function f with respect to m parameters. Nothing is lost by referring only to minimization because minimizing $-f$ is equivalent to maximizing f. The following optimization methods can be used to search for a local minimum. They have two limitations. First, when there are constraints imposed on parameter values the least value taken by a function need not be at a minimum. Even if there are no explicit constraints the parameter values will be constrained by the search region. Secondly, a local minimum is not necessarily the global minimum. A plot of $f(t) = e^{2t} \sin t$ for t from 0 to 5.49π illustrates both of these remarks. If m is only one or two there may not be any practical problem, but difficulties often arise with larger values of m. *Simulated annealing* and *genetic algorithms* aim to reduce the problem by allowing some probability of changes in parameter values that lead to a local increase. Another strategy is to start deterministic algorithms from different guesses of parameter values. The descriptions below are abridged from Everitt (1987).

Simplex method

A *simplex* is a geometric figure formed by $m + 1$ points in an m-dimensional space, e.g. a triangle when $m = 2$, and a tetrahedron when $m = 3$. We begin with $(m + 1)$ parameter points $\theta_1, \ldots, \theta_{m+1}$ (each of which has m elements corresponding to the m parameters) and evaluate $f(\theta_i)$ for each point. Find the greatest of these function values, and suppose it is at θ_p. We move to a point θ_r which is the reflection of θ_p along a line joining θ_p to the centroid of the other points in the simplex (θ_0). That is

$$\theta_r = \theta_0 + (\theta_0 - \theta_p)$$

where

$$\theta_0 = \sum_{\substack{i=1 \\ i \neq p}}^{m+1} \theta_i / m$$

The routine is repeated for the simplex obtained by replacing θ_p with θ_r Figure A5.1. When the procedure leads to a point θ_r for which $f(\theta_r)$ exceeds all the $f(\theta_i)$, excepting $i = p$, the size of the simplex is reduced. The usual convergence criterion is to stop when

$$\sum_{i=1}^{m+1} (f(\theta_i) - f(\theta_0))^2 < \varepsilon$$

for some suitably small ε.

Gradient methods using first derivatives

The *gradient* of a function, $f(\theta)$, will be denoted by

$$g^{\mathrm{T}} = \left(\frac{\partial f}{\partial \theta_1}, ..., \frac{\partial f}{\partial \theta_m} \right)$$

In the *method of steepest descent* we move from a point θ_i to a point θ_{i+1} which corresponds to the minimum of f in the direction of steepest descent. This is found by a linear search. So, writing g_i for $g(\theta_i)$

$$\theta_{i+1} = \theta_i - \lambda_i g_i$$

where λ_i is found by a linear search, e.g. the simplex method with $m = 1$. The method is often slow, and needs substantial modification to be useful. An important improvement is the *conjugate gradients method*. The initial step is in the direction of steepest descent and subsequent directions are found

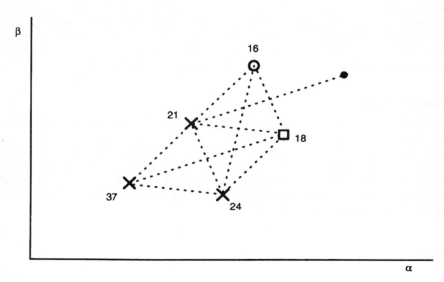

Figure A5.1 Simplex algorithm. A function f has two parameters α and β. The X at which the function equals 37 is reflected to give a function value of 18 at the square. The cross (24) reflects to circle (16), the cross (21) then reflects to the dot

from

$$d_{i+1} = -g_{i+1} + (|g_{i+1}|^2/|g_i|^2)d_i$$

These directions are conjugate (Press *et al.*, 1992). The concept is related to orthogonality, and a formal definition will be given after the *Hessian* matrix has been defined in the next sub-section. Surprisingly, the Hessian does not need to be evaluated (Press *et al.*, 1992), and this is one of the attractions of the conjugate gradients method.

Gradient methods using second derivatives

Another improvement on the method of steepest descent is to use the information on curvature contained in second derivatives. A quadratic function can be expressed in the form

$$q(\theta) = a + \theta^T b + \tfrac{1}{2}\theta^T B\theta$$

If this has a minimum, equating the $\partial f/\partial \theta_i$ to zero shows that it is at the point

$$\theta^* = -B^{-1}b$$

We approximate a function $f(\theta)$ by using a Taylor expansion as far as quadratic terms.

$$q(\theta) = f(\theta_0) + (\theta - \theta_0)^T g(\theta_0) + \tfrac{1}{2}(\theta - \theta_0)^T H(\theta_0)(\theta - \theta_0)$$

where $g(\theta_0)$ is the gradient vector and H is the *Hessian* matrix, evaluated at θ_0, with elements $h_{ij} = \partial^2 f/\partial \theta_i\, \partial \theta_j$. The *Newton–Raphson method* uses θ^* as the next approximation, i.e.

$$\theta_{i+1} = \theta_i - H^{-1}(\theta_i)g(\theta_i)$$

The Newton–Raphson method can diverge if the initial guess is a long way from the minimum. Marquardt (1963) proposed a method of alternating between the steepest descent method and the Newton–Raphson method, and Marquardt's algorithm is often used in non-linear least squares routines.

We can now define conjugate gradients. The directions d_{i+1} and d_i are *conjugate* with respect to the symmetric positive definite matrix H if

$$d_{i+1}^T H d_i = 0$$

Appendix 6 Theoretical Background to CUSUM Charts

Derivation of the Sequential Probability Ratio Test (SPRT)

Suppose a random variable has a distribution $f(x, \theta)$ and we wish to decide between

$$H_0: \theta = \theta_0 \quad \text{and} \quad H_1: \theta = \theta_1$$

We also require

$$\Pr(\text{accept } H_0 \mid H_0 \text{ true}) = 1 - \alpha$$

$$\Pr(\text{accept } H_1 \mid H_1 \text{ true}) = 1 - \beta$$

We start with a single observation from the distribution and carry out a test. On the basis of this test we will either accept H_0 or H_1, or take another random observation. In the latter case we carry out a similar test on the sample of size 2 and so on. Let n be the current sample size. Wald's SPRT (Wald, 1943) operates as follows. Define ℓ_n by

$$\ell_n = \prod_{i=1}^{n} \frac{f(x_i, \theta_1)}{f(x_i, \theta_0)}$$

$$A = (1 - \beta)/\alpha, \quad B = \beta/(1 - \alpha)$$

If $B < \ell_n$ continue sampling.
If $\ell_n < B$ reject θ_1 and accept θ_0.
If $A < \ell_n$ reject θ_0 and accept θ_1.

Proof At time n there are four possible outcomes to the test.

Outcome	Required probability
(i) θ_0 is true and θ_0 accepted	$1 - \alpha$
(ii) θ_0 is true and θ_1 accepted	α
(iii) θ_1 is true and θ_1 accepted	$1 - \beta$
(iv) θ_1 is true and θ_0 accepted	β

Denote the probability of an observed set of results $x_1, \ldots, x_n$ when θ_i is true as

$$p_i(x_1, \ldots, x_n)$$

Then for any set of results $x_1, \ldots, x_n$ leading to a decision in favour of θ_0

$$p_1(x_1, \ldots, x_n) < B p_0(x_1, \ldots, x_n) \qquad (A6.1)$$

by the stopping rule. Denote by D_i the set of all possible sets of results $(x_i, \ldots, x_n)$ for all n which lead to a decision in favour of θ_i. Then by definition

$$\sum_{D_0} p_0(x_1, \ldots, x_n) = 1 - \alpha$$

$$\sum_{D_0} p_1(x_1, \ldots, x_n) = \beta$$

Using equation (A6.1) we have

$$\sum_{D_0} p_1(x_1, \ldots, x_n) < B \sum_{D_0} p_0(x_1, \ldots, x_n)$$

and hence $\beta/(1 - \alpha) < B$. Since sampling is stopped as soon as $\ell_n < B$ the inequality in equation (A6.1) is nearly an equality and

$$B \simeq \frac{\beta}{1 - \alpha}$$

Similarly

$$A \simeq \frac{1 - \beta}{\alpha}$$

Example A6.1
Assume we are sampling from a normal distribution with variance σ^2 and we wish to decide between

$$H_0: \mu = 0 \quad \text{and} \quad H_1: \mu = \delta\sigma$$

$$\ell_n = \prod_{i=1}^{n} \frac{e^{-1/2((x_1 - \delta\sigma)/\sigma)^2}}{e^{-1/2(x_i/\sigma)^2}}$$

$$\ln \ell_n = \frac{1}{2} \sum \left(\frac{x_i}{\sigma} \right)^2 - \frac{1}{2} \sum \left(\frac{x_i - \delta\sigma}{\sigma} \right)^2$$

Now consider the equation

$$\ell_n = A \Longrightarrow \ln \ell_n = \ln A \Longrightarrow \sum x_i = \frac{\sigma}{\delta} \ln A + \tfrac{1}{2} \delta\sigma n$$

It follows that if $\sum x_i$ is plotted against n the accept H_1 boundary is a straight line. Similarly for the accept H_0 boundary.

Sequential test of three hypotheses

Assume we have three hypotheses, that observations are independently and normally distributed with distributions as follows

$$H_{-1}:N(-\delta\sigma, \sigma^2); \quad H_0:N(0, \sigma^2); \quad H_1:N(\delta\sigma, \sigma^2)$$

Suppose that we want a probability $(1 - 2\alpha_0)$ of accepting H_0 if it is true, and a probability $(1 - \alpha_1)$ of accepting H_1 or H_{-1} if they are true. The sequential test of these three hypotheses can be thought of as the superposition of two SPRTs of two hypotheses with the 'ambiguous' intersection region included in the 'undecided' region. This is shown in Figure A6.1.

The outer boundaries are

$$\sum_{i=1}^{n} x_i = \left\{ \frac{1}{\delta} \ln[(1 - \alpha_1)/\alpha_0] + \tfrac{1}{2}\delta n \right\}\sigma$$

$$\sum_{i=1}^{n} x_i = -\left\{ \frac{1}{\delta} \ln[(1 - \alpha_1)/\alpha_0] + \tfrac{1}{2}\delta n \right\}\sigma$$

Wetherill (1975) gives more details.

Johnson's approximate approach for CUSUM chart

First reverse the CUSUM chart and look at it as if it were proceeding backwards. Now regard the outer arms of the V-mask as boundaries of a test of

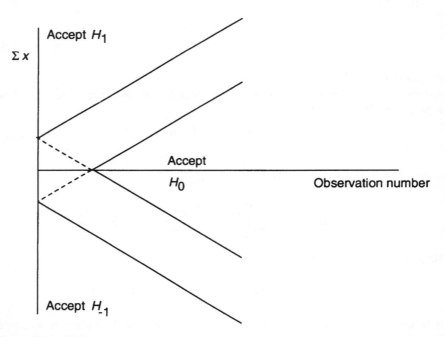

Figure A6.1 SPR test

three hypotheses using the SPRT (Figure A6.2). Now look at Figure A6.3, in which 1 unit on the horizontal scale is represented by the same length as 2σ units on the vertical scale. The outer boundaries of the V-mask are

$$\sum_{i=1}^{n} x_i = 2\sigma \tan \theta (n + d)$$

$$\sum_{i=1}^{n} x_i = -2\sigma \tan \theta (n + d)$$

If we identify these with the outer boundaries in the previous section we shall have a V-mask in which, approximately, the probability of a path crossing an outer boundary is $2\alpha_0$, when the process is in control. By identifying these pairs of equations we obtain

$$\tan \theta = \delta/4$$
$$d = 2 \ln[(1 - \alpha_1)/\alpha_0]/\delta^2$$

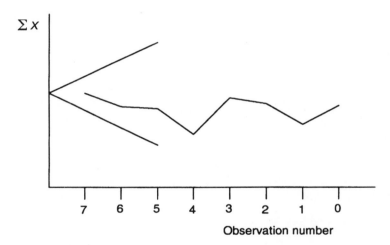

Figure A6.2   CUSUM V-mask

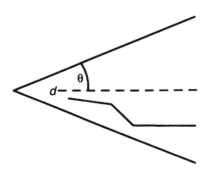

Figure A6.3   Geometry of CUSUM mask

There is no accept H_0 boundary when using the CUSUM chart but α_1 is usually small so

$$d \simeq -2(\ln \alpha_0)/\delta^2$$

Johnson (1961) points out that since CUSUM charts are like a three-hypothesis SPRT without a middle boundary, and there is no decision to 'accept H_0', a path which would have been terminated on a SPRT could go on and cross one of the decision boundaries. He adds that paths which cross the decision boundaries a long way from the vertex should be regarded with suspicion.

Appendix 7   Derivation of EV1 Distribution

Let $X_1, \ldots, X_n$ be independent random variables with cdf $F(x)$. Define the maximum in the sample

$$X_{(n)} = \max(X_1, \ldots, X_n)$$

and denote the cdf of this variable by $G(x)$. Then

$$
\begin{aligned}
G(x) &= \Pr(X_{(n)} < x) \\
&= \Pr(\max(X_1, \ldots, X_n) < x) \\
&= \Pr(X_1 < x, \ldots, X_n < x) \\
&= [F(x)]^n
\end{aligned}
$$

We now suppose that $G(x)$ will tend towards some standard form as n tends to infinity. If this is so, the distribution of the maximum of a random sample from G will have the same form of distribution, although the location and spread will change. That is

$$[G(x)]^n = G(a_n x - b_n)$$

where the subscript on the a_n and b_n emphasizes that they depend on n.

The extreme value type 1 distribution (EV1) is obtained by taking a_n as 1. It can be shown that the distribution of the maximum of samples of size n from distributions whose upper tail decays at least as fast as a negative exponential function does tend to a G such that

$$[G(x)]^n = G(x - b_n) \tag{A7.1}$$

Leadbetter *et al.* (1982) give a modern proof of this result.

It is relatively easy to show that if equation (A7.1) holds then $G(x)$ has the form given in Section 4.2. The trick is to notice that the maximum of m maxima of n variables is the same as the maximum of all mn variables. Therefore

$$[G(x)]^{mn} = \{[G(x)^n]\}^m = [G(x - b_n)]^m = G(x - b_n - b_m)$$

But

$$[G(x)]^{mn} = G(x - b_{mn})$$

so

$$b_{mn} = b_m + b_n$$

and b_i is essentially a logarithm. That is

$$b_n = \theta \ln(n) \tag{A7.2}$$

for some constant θ. The next step is to take logarithms of equation (A7.1) twice, remembering to multiply by -1 to obtain positive quantities between log operations:

$$\ln(n) + \ln(-\ln G(x)) = \ln(-\ln(G(x - b_n))) \tag{A7.3}$$

It is now convenient to define,

$$c(x) = \ln(-\ln G(x)) \tag{A7.4}$$

From equation (A7.3) decreasing the argument of c by b_n leads to an addition of $\ln(n)$ which is the same as b_n/θ, from equation (A7.2). Therefore, decreasing the argument by x gives

$$c(x) = c(0) - x/\theta$$

Finally, take the exponential of equation (A7.4) twice to obtain

$$G(x) = \exp(-\exp(c(x)))$$
$$= \exp(-\exp(-(x - \xi)/\theta))$$

where ξ is the product of the constant θ and $c(0)$.

Appendix 8 Fitting the Wakeby Distribution (Landwehr *et al.*, 1979)

Estimating all five parameters from n data, $\{x_i\}$.

Step 1. Calculate the PWM for $k = 0, \ldots, 4$

$$a_k = \sum x_{i:n}[(n-i+0.35)/n]^k/n$$

Step 2. For $j = 1, 2, 3$ calculate

$$N_{4-j} = (4)^j a_3 - (3)^{1+j} a_2 + 3(2)^j a_1 - a_0$$
$$C_{4-j} = (5)^j a_4 - 3(4)^j a_3 + (3)^{1+j} a_2 - (2)^j a_1$$

Step 3.

$$\hat{\beta} = \frac{(N_3 C_1 - N_1 C_3) \pm [(N_1 C_3 - N_3 C_1)^2 - 4(N_1 C_2 - N_2 C_1)(N_2 C_3 - N_3 C_2)]^{1/2}}{2(N_2 C_3 - N_3 C_2)}$$

and choose the alternative that satisfies $0.3 < \hat{\beta} < 50$

$$\hat{\delta} = -(N_1 + \hat{\beta} N_2)/(N_2 + \hat{\beta} N_3)$$

Step 4. For $k = 0, 1, 2, 3$

$$\{k\} = (k+1)(k+1+\hat{\beta})(k+1+\hat{\delta})a_k$$

Step 5. $\hat{\xi} = [\{3\} - \{2\} - \{1\} + \{0\}]/4.$

Step 6. $\hat{\theta} = (1+\hat{\beta})(2+\hat{\beta})[\{1\}/(2+\hat{\beta}) - \{0\}/(1+\hat{\beta}) - \hat{\xi}]/(\hat{\beta} - \hat{\delta}).$

Step 7. $\hat{\phi} = -(1+\hat{\delta})(2+\hat{\delta})[-\{1\}/(2+\hat{\delta}) + \{0\}/(1+\hat{\delta}) + \hat{\xi}]/(\hat{\beta} - \hat{\delta}).$

Step 8. Check the parameter estimates correspond to a proper distribution by plotting F against x, where F is from $[0, 1)$ and

$$x = \hat{\xi} + \hat{\theta}\hat{\beta}^{-1}[1 - (1-F)^{\hat{\beta}}] - \hat{\phi}\hat{\delta}^{-1}[1 - (1-F)^{\hat{\delta}}]$$

They will not necessarily do so. If they do not, the options are: try the four-parameter Wakeby distribution; try ad hoc constrained estimation procedures; or conclude that the Wakeby distribution is not plausible for the application.

Estimating four parameters

Step 0. Assume $\xi = 0$.

Step 1. Calculate the PWM, a_k as before.

Step 2. For $j = 1, 2, 3$ calculate

$$N_{4-j} = -(3)^j a_2 + (2)^{1+j} a_1 - a_0$$
$$C_{4-j} = -(4)^j a_3 + 2(3)^j a_2 - (2)^j a_1$$

Steps 3 and 4 as before, omit Step 5, Steps 6 and 7 with $\hat{\xi} = 0$, and finally the check in step 8.

Example AP.1

The results of applying the algorithms to the Thames data follow.

Five parameter algorithm:

$$\hat{\xi} = 122.19, \qquad \hat{\beta} = 5.4195, \qquad \hat{\delta} = -0.0126, \qquad \hat{\theta} = 647.29,$$
$$\hat{\phi} = -100.18$$

But $\hat{\xi}$ exceeds the minimum sample value, which is 94.

Four parameter algorithm:

$$\hat{\beta} = 30.379, \qquad \hat{\delta} = 0.1601, \qquad \hat{\theta} = 6215.1, \qquad \hat{\phi} = -146.67$$

The upper 1% points of the distribution are estimated as 717 and 682 respectively.

Another possibility is to subtract the sample minimum from all the data, and then fit the four-parameter distribution. This gives an estimated upper 1% point of 691.

Appendix 9 Expected Value of the Residual Sum of Squares

The multiple regression model can be written

$$Y = XB + E$$

where the errors are assumed to be independently distributed with mean 0 and variance σ^2.

The residual sum of squares (RSS) is given by the equation

$$RSS = (Y - X\hat{B})^{\mathsf{T}}(Y - X\hat{B})$$

If we remember, from the derivation of $\hat{B}$, that

$$(X^{\mathsf{T}}X)\hat{B} = X^{\mathsf{T}}Y$$

we can rewrite RSS as

$$RSS = Y^{\mathsf{T}}Y - \hat{B}^{\mathsf{T}}X^{\mathsf{T}}X\hat{B}$$

If we use result (iv) of exercise 3 of Chapter 5

$$E[RSS\} = E[Y^{\mathsf{T}}Y] - [B^{\mathsf{T}}X^{\mathsf{T}}XB + \mathrm{tr}((X^{\mathsf{T}}X)(X^{\mathsf{T}}X)^{-1}\sigma^2)]$$

$$= E[Y^{\mathsf{T}}Y] - B^{\mathsf{T}}X^{\mathsf{T}}XB - (k+1)\sigma^2$$

Finally, remembering the scalar version of the same result

$$E[Y_i^2] = \mathrm{var}(Y_i) + E[Y_i]^2$$

the first term on the right-hand side

$$E[Y^{\mathsf{T}}Y] = n\sigma^2 + \sum_{i=1}^{n}(\beta_0 + \beta_1 x_{1i} + \cdots + \beta_k x_{ki})^2$$

$$= n\sigma^2 + B^{\mathsf{T}}X^{\mathsf{T}}XB$$

so

$$E[RSS] = (n - k - 1)\sigma^2$$

The estimator S^2 of σ^2 is $RSS/(n-k-1)$, so it follows that

$$E[S^2] = \sigma^2$$

The expression $\boldsymbol{B}^{\mathrm{T}}(\boldsymbol{X}^{\mathrm{T}}\boldsymbol{X})\boldsymbol{B} \geq 0$, because $\boldsymbol{X}^{\mathrm{T}}\boldsymbol{X}$ is the inverse of a covariance matrix multiplied by a positive constant. In later sections we will use the fact that S^2 is not correlated with $\hat{\boldsymbol{B}}$. A formal proof will not be given, but there is no reason to suppose that $\hat{\beta}_k$ being above its mean β_k has any effect on the relationship of S^2 to its mean σ^2.

Appendix 10 Bayes' Linear Estimator

This account is based on O'Hagan (1994). y and x are scalar and vector random variables respectively.

We wish to minimize the expected value of the squared error

$$D = E[((a + b^T x) - y)^2]$$
$$= E[a^2 + 2ab^T x + b^T xx^T b - 2ay - 2b^T xy + y^2]$$
$$= a^2 + 2ab^T E[x] + b^T E[xx^T]b - 2aE[y] - 2b^T E[xy] + E[y^2]$$

Now

$$\text{var}(x) = E[xx^T] - E[x]E[x^T]$$

and

$$\text{cov}(x, y) = E[xy] - E[x]E[y]$$

and so

$$D = (a + b^T E[x] - E[y])^2 + b^T \text{var}(x)b - 2b^T \text{cov}(x, y) + \text{var}(y)$$

This can be written as

$$D = (a + b^T E[x] - E[y])^2 + (b - b^*)^T \text{var}(x)(b - b^*) + D^*$$

where

$$b^* = \text{var}(x)^{-1} \text{cov}(x, y)$$
$$D^* = \text{var}(y) - \text{cov}(x, y)\text{var}(x)^{-1} \text{cov}(x, y)$$

The two leading terms on the right-hand side of the expression for D are non-negative. Therefore, D will be minimized if they are made zero by setting

$$b = b^*$$
$$a = -b^{*T} E[x] + E[y]$$

We will now show that

$$E[y] + b^{*T}(x - E[x])$$

is the posterior mean for y, given x. It is a particular case of the following result. We let w be an estimator of θ, and find the value of w that minimizes

$$D = E[(w - \theta)^2] = E[w^2 - 2\theta w + \theta^2]$$
$$= w^2 - 2wE[\theta] + E[\theta^2]$$

We set dD/dw equal to zero

$$2w - 2E[\theta] = 0$$

to find that $w = E[\theta]$.

Appendix 11   Simulating From ARIMA Processes

Background

Hosking (1984) gave the following algorithm. For any stationary linear stochastic process $\{X_t\}$ with zero mean, the conditional mean and variance of X_t given the past values can be expressed in terms of the partial linear regression coefficients, ϕ_{tj}. They are defined as the constants which minimize

$$E[(X_t - \phi_{t1}X_{t-1} - \cdots - \phi_{tt}X_0)^2]$$

The conditional mean and variance are given by

$$m_t = E[X_t | x_{t-1}, ..., x_0] = \sum_{j=1}^{t} \phi_{tj}x_{t-j}$$

$$v_t = \mathrm{var}[X_t | x_{t-1}, ..., x_0] = \gamma_0 \prod_{j=1}^{t} (1 - \phi_{jj}^2)$$

where γ_0 is the unconditional variance of $\{X_t\}$.

Algorithm

This algorithm simulates a realization of length $n\{x_0, ..., x_{n-1}\}$ from a stationary process with a normal marginal distribution and correlation function ρ_k.

Step 1. Generate a starting value from $N(0, \gamma_0)$ where γ_0 is the required variance of $\{X_t\}$. Set $N_0 = 0$, $D_0 = 1$.

Step 2. For $t = 1, ..., n-1$ calculate ϕ_{tj} for j from 1 up to t recursively using

$$N_t = \rho_t - \sum_{j=1}^{t-1} \phi_{t-1,j}\rho_{t-j}$$

$$D_t = D_{t-1} - N_{t-1}^2/D_{t-1}, \qquad \phi_{tt} = N_t/D_t$$

$$\phi_{tj} = \tau_{t-1,j} - \phi_{tt}\phi_{t-1,t-j} \quad \text{for} \quad j = 1, ..., t-1$$

Step 3. Calculate

$$m_t = \sum_{j=1}^{t} \phi_{tj} x_{t-j}$$

and

$$v_t = (1 - \phi_{tt}^2) v_{t-1}$$

and generate x_t from $N(m_t, v_t)$.

Notes

(a) The algorithm is simplified for ARIMA$(0, d, 0)$ because ϕ_{tt} in Step 2 is known to be

$$\phi_{tt} = d/(t - d)$$

(b) If a realization $\{y_t\}$ given by

$$y_t = \alpha_1 y_{t-1} + \cdots + \alpha_p y_{t-p} + x_t + \beta_1 x_{t-1} + \cdots + \beta_q x_{t-q}$$

is calculated (Section 7.3.11), it is advisable to set $y_t = 0$ for $t \leqslant -L$, generate y_t for $-L < t \leqslant n$ and discard the initial L values. A value of L of about 50 should suffice. The relationship between the variance of $\{X_t\}$ and $\{Y_t\}$ is quite complex. Hosking (1984) gives the details but it might be quicker to set γ_0 equal to 1 and estimate var$\{Y_t\}$ from a long simulation. It would only be necessary to accumulate the y_t^2. The value of γ_0 could then be found by scaling.

Appendix 12   Neyman-Scott Streamflow Model

Background

The mathematical basis of the Neyman-Scott rainfall and streamflow models is quite specialized. The book by Cox and Isham (1980) is a good starting point. The following results give an indication of the ideas involved, and use their notation.

Suppose we have a stationary point process with a rate of occurrences λ also known as the *intensity*. Let $N(t)$ be the number of points between times 0 and t, and $N(z, z + \delta)$ be the number of occurrences between times z and $z + \delta$. Formally define $dN(z)$ as $N(z, z + \delta)$. Then it is convenient to write

$$N(t) = \int_0^t dN(z)$$

as a limit of sums of counts in small intervals. For small δ, $\Pr(N(z, z + \delta) = 1) = \lambda\delta$ and the probability of more than one occurrence is of order δ^2 and hence negligible. It follows that

$$E[N(z, z + \delta)] = E[dN(z)] = \lambda\delta$$

and hence

$$E[N(t)] = \lambda t$$

Before finding an expression for $\text{var}[N(t)]$ we need to define the *conditional intensity function*

$$h(t) = \underset{\delta \to 0}{\text{Limit}} \; \delta^{-1} \Pr(N(t, t + \delta) = 1 \,|\, \text{a point at the origin})$$

Now use the result for the variance of a linear combination of random variables, and write

$$\text{var}[N(t)] = \int_0^t \text{var}[dN(z)] + 2 \iint_{\substack{0 < z < t \\ 0 < u + z \leqslant t}} \text{cov}[dN(z), dN(z + u)]$$

First consider

$$\text{var}[N(z, z + \delta)] = \text{E}[(N(z, z + \delta))^2] - (\text{E}[N(z, z + \delta)])^2$$
$$= 1^2\lambda\delta + (\lambda\delta)^2 = \lambda\delta + 0(\delta)$$

where the notation $0(\delta)$ means terms in δ^2 and, in general any higher powers. Next, for $u > 0$

$$\text{cov}[N(z, z + \delta_1), N(z + u, z + u + \delta_2)]$$
$$= \text{E}[N(z, z + \delta_1)\text{E}[N(z + u, z + u + \delta_2) | N(z, z + \delta_1)]]$$
$$- \text{E}[N(z, z + \delta_1)\text{E}[N(z + u, z + u + \delta_2)]$$
$$= \lambda h(u)\delta_1\delta_2 - \lambda^2\delta_1\delta_2 + 0(\delta_1\delta_2)$$

Where we have used the general result that:

$$\text{E}[XY] = \iint xy f(x, y)\, dx\, dy = \iint x f(x) y f(y|x)\, dx\, dy = \int x f(x)\, dx \int y f(y|x)\, dy$$

Letting δ_1, δ_2 tend to zero when substituting in the original expression for $\text{var}[N(t)]$ gives

$$\text{var}[N(t)] = \lambda t + 2\lambda \int_0^t (t - u)h(u)\, du - \lambda^2 t^2$$

The integral in the second term on the right follows from writing dz and du for the limits of δ_1 and δ_2 and noticing that

$$\int_{z=0}^t \int_{u=0}^{t-z} h(u)\, dz\, du = \int_{z=0}^{t-u} \int_{u=0}^t h(u)\, du\, dz = \int_0^t (t - u)h(u)\, du$$

The Poisson process is a special case in which $h(u)$ equals λ, and it is easy to verify that the expression for $\text{var}[N(t)]$ simplifies to give λt. A more interesting special case is the Neyman–Scott clustered point process model which, for example, locates the leading edge of raincells in the Neyman–Scott rainfall model (see Section 2.2.2). Let λ_s be the rate of storm arrival and μ_c be the mean number of cell origins. Then the intensity (λ throughout this section) is $\lambda_s\mu_c$. It can be shown (Cox and Isham, 1980) that the conditional intensity is

$$h(u) = \lambda_s\mu_c + (2\mu_c)^{-1}\text{E}[C(C - 1)]\beta e^{-\beta u}$$

where C is the number of cell origins and the mean time from the storm origin to the cell origin is $(1/\beta)$.

Poisson model with rectangular pulses

Rodriguez–Iturbe *et al.* (1987) begin with a Poisson process of rate λ. At each point a rectangular pulse of random duration L and random depth X represents a rainfall intensity. The durations and depths are independent of each other and the Poisson process. Let $Y(t)$ be the total intensity at time t. Then

$$Y(t) = \int_{u=0}^{\infty} X_{t-u}(u)\, dN(t - u)$$

where $X_s(\tau)$ is the depth of a pulse originating at a time s measured a time τ later and dN is as defined in the previous section. So, if R is the survivor function for the duration of the pulse of height X

$$X_{t-u}(u) = \begin{cases} X & \text{with probability } R(u) \\ 0 & \text{with probability } 1 - R(u) \end{cases}$$

Then

$$E[Y(t)] = \lambda \int_0^\infty E[X]\, R(u)\, du = \lambda \mu_X \mu_L$$

$$\text{var}[Y(t)] = \int_0^\infty E[X^2] R(u)\, \text{var}\,[dN(t-u)] = \lambda E[X^2] \int_0^\infty R(u)\, du$$

$$= \lambda(\mu_X^2 + \sigma_X^2) \int_0^\infty R(u)\, du = \lambda(\mu_X^2 + \sigma_X^2)\mu_L$$

using the result of Exercise 4 in Chapter 3, the fact that the variance of a Poisson process is the same as the mean, and the relationship that $E[dN(t-u)] = E[dN(u)]$ for a stationary process. Furthermore,

$$\text{cov}\,[Y(t), Y(t+\tau)] = \lambda(\mu_X^2 + \sigma_X^2) \int_\tau^\infty R(v)\, dv$$

Rodriguez, *et al.* (1987) use similar techniques to develop the Neyman–Scott rainfall model from the Neyman–Scott clustered point process model.

Aggregated process

Rainfall data are usually observed in aggregated form so Rodriguez, et al (1987) consider cumulative totals

$$Y_i^{(h)} = \int_{(i-1)h}^{ih} Y(t)\, dt$$

The second order properties are obtained as

$$E[Y_i^{(h)}] = hE[Y(t)]$$

$$\text{var}\,[Y_i^{(h)}] = 2\int_0^h (h-u)c_y(u)\, du$$

$$\text{cov}\,[Y_i^{(h)}, Y_{i+k}^{(h)}] = \int_{-h}^h (h-(u))c_y(kh+u)\, du$$

where $c_Y(u)$ is the covariance of $Y(t)$. These expressions can be verified by discretizing $Y(t)$ over h and applying the general result for a sum of random variables.

Neyman–Scott streamflow model

Let λ be the rate of storm arrivals, μ_c be the mean number of cells per storm and μ_z be the mean of the pulses. Then

$$E[Y(t)] = \lambda \mu_c \mu_z \left(\frac{\alpha}{\tau_1} + \frac{(1-\alpha)}{\tau_2} \right)$$

Now let β^{-1} be the mean of the exponential distribution of times between cell origins and storm origins, and $Y_i^{(h)}$ denote the average flow for the ith interval of length h. Then

$$\mathrm{var}[Y_i^{(h)}] = \frac{2}{h^2 \tau_1^2} [\tau_1 h - 1 + e^{-\tau_1 h}][A_1 + BC_1]$$

$$+ \frac{2}{h^2 \tau_2^2} [\tau_2 h - 1 + e^{-\tau_2 h}][A_2 + BC_2] + \frac{2BD}{h^2 \beta^2} [\beta h - 1 + e^{-\beta h}]$$

$$\mathrm{cov}[Y_i^{(h)}, Y_{i+k}^{(h)}] = \frac{1}{h^2 \tau_1^2} e^{-\tau_1 h(k-1)} (1 - e^{-\tau_1 h})^2 [A_1 + BC_1]$$

$$+ \frac{1}{h^2 \tau_2^2} e^{-\tau_2 h(k-1)} (1 - e^{-\tau_2 h})^2 [A_2 + BC_2] + \frac{BD}{h^2 \beta^2} e^{-\beta h(k-1)} (1 - e^{-\beta h})^2$$

where

$$A_1 = \frac{\lambda \mu_c E[Z^2]}{2 \tau_1 \tau_2 (\tau_1 + \tau_2)} \; \alpha \tau_2 [\alpha \tau_2 + (2 - \alpha)\tau_1]$$

$$A_2 = \frac{\lambda \mu_c E[Z^2]}{2 \tau_1 \tau_2 (\tau_1 + \tau_2)} \; (1 - \alpha)\tau_1 [(1 - \alpha)\tau_1 + (1 + \alpha)\tau_2]$$

$$B = \frac{\lambda \beta \mu_z^2 E[C^2 - C]}{2 \tau_1 \tau_2 (\tau_1 + \tau_2)(\beta^2 - \tau_1^2)(\beta^2 - \tau_2^2)}$$

$$C_1 = \alpha \beta \tau_2 (\beta^2 - \tau_2^2)[\alpha(\tau_2 - \tau_1) + 2\tau_1]$$

$$C_2 = (1 - \alpha)\beta \tau_1 (\beta^2 - \tau_1^2)[\alpha(\tau_2 - \tau_1) + \tau_1 + \tau_2]$$

$$D = \tau_1 \tau_2 (\tau_1 + \tau_2)[\alpha(\tau_2 - \tau_1) + \tau_1 - \beta][\alpha(\tau_2 - \tau_1) + \tau_1 + \beta]$$

Cowpertwait and O'Connell (1993) continued by taking Z as an exponential variable and $(C - 1)$ as a Poisson variable.

Appendix 13   Fourier Series and Transforms

The description in this section is taken from Hearn and Metcalfe (1995), which is a more detailed account

Finite Fourier series

We start with a sequence of observations

$$\{x_t\} \qquad \text{for} \quad t = -m, \ldots, -1, \, , 0, \, ,1, \ldots, m - 1$$

where $2m = n$. We wish to find X_r so that

$$x_t = \sum_{r=-m}^{m-1} X_r \exp(i2\pi r t/n)$$

Multiplying both sides by $\exp(-i2\pi kt/n)$, where k is an integer between $-m$ and $(m-1)$, and summing over t leads to

$$\sum_{t=-m}^{m-1} x_t \exp(-i2\pi kt/n) = \sum_{t=-m}^{m-1} \sum_{r=-m}^{m-1} X_r \exp(i2\pi(r-k)t/n)$$

$$= \sum_{r=-m}^{m-1} X_r \left\{ \sum_{t=-m}^{m} \exp(i2\pi(r-k)t/n) \right\}$$

If r equals k, all the terms in the bracketed sum are one, and the sum is therefore $2m$. If r is not equal to k the following argument shows that the sum is zero, that is

$$\sum_{t=-m}^{m-1} \exp(i2\pi(r-k)t/n) = \exp(-i2\pi(r-k)m/n) \sum_{t=0}^{2m-1} \exp(i2\pi(r-k)t/n)$$

and so, using the standard result for the sum of a geometric progression, this can be written as

$$= \exp(-i2\pi(r-k)m/n)\{1 - \exp(i2\pi(r-k)2m/n)\}/\{1 - \exp(i2\pi(r-k)/n)\}$$

Since $2m$ equals n and $(r-k)$ is an integer, $\exp(i2\pi(r-k)2m/n)$ equals 1, and the numerator of the expression equals zero. The denominator is not zero because $(r-k)$ is assumed to be non-zero and the range of r and k values restricts it to be at most $2m-1$, which is less than n. This simplification of the bracket leads to

$$X_k = \frac{1}{n} \sum_{r=-n}^{n-1} x_t \exp(-i2\pi kt/n)$$

for $-m \leqslant k \leqslant m-1$. The subscript k is just a 'dummy' and can be replaced by r, or any other letter.

Parseval's theorem

Parseval's theorem for a finite Fourier series can be stated as

$$\sum_{t=-m}^{m-1} x_t^2/n = \sum_{r=-n}^{n-1} |X_r|^2$$

To prove this we start from

$$x_t = \sum_{r=-m}^{m-1} X_r \exp(i2\pi rt/n)$$

The complex conjugate of this result is

$$x_t^* = \sum_{p=-m}^{m-1} X_p^* \exp(-i2\pi pt/n)$$

Now, take the products $x_t x_t^*$ and the sum over t, that is

$$\sum_{t=-m}^{m-1} x_t x_t^* = \sum_{t=-m}^{m-1} \sum_{r=-m}^{m-1} \sum_{p=-m}^{m-1} X_r X_p^* \exp(i2\pi(r-p)t/n)$$

$$= \sum_{r=-m}^{m-1} \sum_{p=-m}^{m-1} X_r X_p^* \left(\sum_{t=-m}^{m-1} \exp(i2\pi(r-p)t/n) \right)$$

It was explained earlier in this section that the sum in brackets is zero unless m equals p, in which case it equals n. Hence, since the data x_t are real

$$\sum_{t=-m}^{m-1} x_t^2 = n \sum_{r=-m}^{m-1} |X_r|^2$$

Fourier series

Now assume $x(t)$ is a continuous signal of length T, defined from $-T/2$ up to $T/2$. A similar argument to that for the finite Fourier series gives

$$x(t) = \sum_{r=-\infty}^{\infty} X_r \exp(i2\pi rt/T)$$

where

$$X_r = \frac{1}{T} \int_{-T/2}^{T/2} x(t)\exp(-i2\pi rt/T)\,dt$$

Fourier transform

Suppose the signal $x(t)$ is defined for $-\infty < t < \infty$, and $\int_{-\infty}^{\infty} |x(t)|\,dt < \infty$. (Note that this condition can be relaxed by the use of generalized functions. Then the Fourier transform of a sine wave is defined using a Dirac delta function.) We start by writing the Fourier series for $x(t)$ defined for $-T/2 \leqslant t \leqslant T/2$ as

$$x(t) = \sum_{r=-\infty}^{\infty} (TX_r)\exp(i2\pi rt/T)(1/T)$$

that is

$$(TX_r) = \int_{-T/2}^{T/2} x(t)\exp(-i2\pi rt/T)\,dt$$

Now write ω for $2\pi r/T$, $d\omega$ for $(2\pi/T)$, define $X(\omega)$ as TX_r and let T tend to infinity. Then

$$x(t) = \frac{1}{2\pi} \int_{-\infty}^{\infty} X(\omega)e^{i\omega t}\,d\omega$$

and

$$X(\omega) = \int_{-\infty}^{\infty} x(t)\exp(-i\omega t)\,dt$$

$X(\omega)$ is the Fourier tranform of $x(t)$, and $x(t)$ is the inverse Fourier transform of $X(\omega)$.

Discrete Fourier transform

Consider a discrete signal x_t defined for $-\infty < t < \infty$. The integral for $X(\omega)$ becomes the sum

$$X(\omega) = \sum_{t=-\infty}^{\infty} x_t \exp(-i\omega t) \qquad -\pi < \omega < \pi$$

and the other half of the transform pair is

$$x_t = \frac{1}{2\pi} \int_{-\pi}^{\pi} X(\omega) e^{i\omega t} \, d\omega$$

The factor of $(1/2\pi)$ could be applied to the sum defining $X(\omega)$ instead of the integral giving x_t. This convention is used in the definition of the spectrum.

Appendix 14   Wavelet Transforms

Background

This summary of wavelet analysis is based on the account given by Newland (1993) and uses the same notation. Wavelets can be defined in terms of scaling functions, $\phi(x)$, which satisfy dilation equations. That is, the scaling function is expressed as a linear combination of stretched versions of itself. For the D4 scaling function:

$$\phi(x) = c_0\phi(2x) + c_1\phi(2x-1) + c_2\phi(2x-2) + c_3\phi(2x-3)$$

The D stands for Ingrid Daubechies who developed the theory in the 1980s. The numeral, which must be even, is the number of terms and the larger it is the smoother the wavelet becomes, although it still has a fractal nature. The scaling function has to be calculated iteratively, except for the D2 Haar wavelet, from a unit square with a base between 0 and 1. The coefficients need to be chosen so that the area remains constant during the iterations. Two other sets of conditions, accuracy and orthogonality, lead to unique values for the coefficients. Daubechies (1988) gives tables of these.

For the D4 wavelet

$$c_0 = (1 + \sqrt{3})/4 \qquad c_1 = (3 + \sqrt{3})/4$$
$$c_2 = (3 - \sqrt{3})/4 \qquad c_3 = (1 - \sqrt{3})/4$$

The wavelet, $W(x)$, is itself defined by

$$W(x) = -c_3\phi(2x) + c_2\phi(2x-1) - c_1\phi(2x-2) + c_0\phi(2x-3)$$

The shape of the D4 wavelet is shown in Figure A14.1 It has been calculated for 12 288 points using a program written by Gordon Sutcliffe in J. It extends over an interval from 0 to almost 3, and is the basic shape which is used to reconstruct a signal.

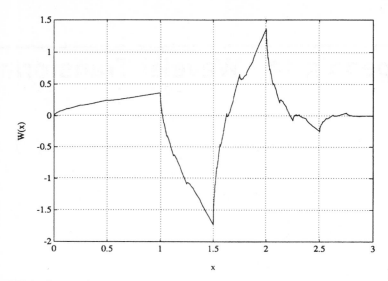

Figure A14.1   D4 wavelet

Now suppose $f(x)$ is a continuous signal which has been scaled so that it extends over the interval [0, 1]. The D4 wavelet expansion of $f(x)$ is

$$f(x) = a_0\phi(x) + a_1 W(x) + [a_2 \ a_3] \begin{bmatrix} W(2x) \\ W(2x - 1) \end{bmatrix} + [a_4 \ a_5 \ a_6 \ a_7] \begin{bmatrix} W(4x) \\ W(4x - 1) \\ W(4x - 2) \\ W(4x - 3) \end{bmatrix} + \dots$$

where the $W(\ )$ are wrapped around [0, 1] as necessary. For example, $W(x)$ extends over [0, 3] so $f(x)$ receives contributions from $W(x)$, $W(x+1)$ and $W(x+2)$ for $0 \leqslant x \leqslant 1$. The $W(x+1)$, accounts for the middle third of $W(x)$ with $0 \leqslant x \leqslant 3$, and $W(x+2)$ accounts for the last third. A consequence of the wrap around is that $a_0\phi(x)$ is the mean value of the signal. The main result is that the mean squared value of the signal can be expressed as

$$\int_0^1 f^2(x)\,\mathrm{d}x = a_0^2 + a_1^2 + \frac{1}{2}(a_2^2 + a_3^2) + \frac{1}{4}(a_4^2 + a_5^2 + a_6^2 + a_7^2) + \cdots$$

If the squared wavelet amplitudes are plotted on the grid base shown in Figure A14.2 (after Newland, 1993) they will enclose a volume equal to the mean square of the signal.

In practice the signal will be sampled at n points, where n is some power of 2, i.e. $n = 2^m$, so the series is truncated after n points and the integral on the left is replaced by

$$\frac{1}{n}\sum_{r=0}^{n-1} f_r^2$$

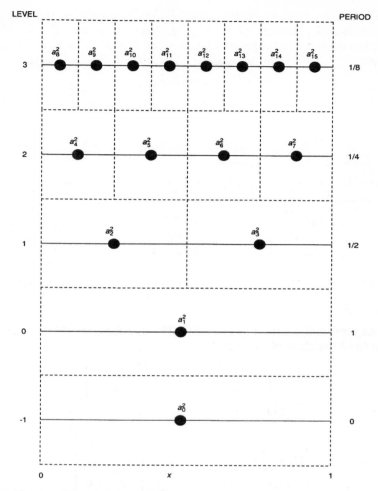

Figure A14.2 Basis for mean square maps

The wavelet coefficients are given by the integral

$$a_{2^j+k} = \int_0^t f(x)W(2^j x - k)\,\mathrm{d}x$$

The orthogonality conditions guarantee that all other terms are zero. Even so, we would need discrete approximations to all the wavelets to use this formula. Fortunately the discrete wavelet transform (DWT) invented by Mallat (1989) removes the need to calculate the wavelets explicitly.

Mallat's pyramid algorithm

The DWT takes a discrete signal to the wavelet coefficients, and the inverse transform (IDWT) reconstructs the signal from the wavelet coefficients. The algorithm is shown schematically. The pattern continues as the number of points increases by factors of 2, e.g. M_4 has dimensions $2^4 \times 2^3$. Gordon Sutcliffe has contributed programs written in the subset of J which is available, free, from

http://www.jsoftware.com

IDWT

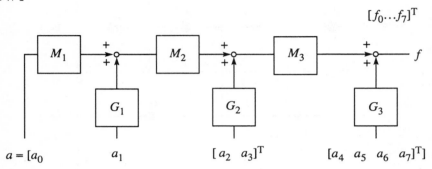

The boxes represent matrix multiplication. The matrices, for D4 wavelets, have the form (missing elements are zero):

$$M_1 = \begin{bmatrix} c_0 + c_2 \\ c_1 + c_3 \end{bmatrix} \qquad G_1 = \begin{bmatrix} -c_3 - c_1 \\ c_2 + c_0 \end{bmatrix}$$
$$(2 \times 1)$$

$$M_2 = \begin{bmatrix} c_0 & c_2 \\ c_1 & c_3 \\ c_2 & c_0 \\ c_3 & c_1 \end{bmatrix} \qquad G_2 = \begin{bmatrix} -c_3 & -c_1 \\ c_2 & c_0 \\ -c_1 & -c_3 \\ c_0 & c_2 \end{bmatrix}$$
$$(2^2 \times 2)$$

$$M_3 = \begin{bmatrix} c_0 & & & c_2 \\ c_1 & & & c_3 \\ c_2 & c_0 & & \\ c_3 & c_1 & & \\ & c_2 & c_0 & \\ & c_3 & c_1 & \end{bmatrix} \qquad G_3 = \begin{bmatrix} -c_3 & & & -c_1 \\ c_2 & & & c_0 \\ -c_1 & -c_3 & & \\ c_0 & c_2 & & \\ & -c_1 & -c_3 & \\ & c_0 & c_2 & \end{bmatrix}$$
$$(2^3 \times 2^2)$$

DWT

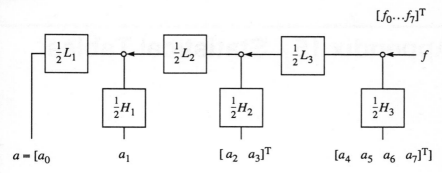

$[f_0 \ldots f_7]^T$

$$a = [a_0 \qquad a_1 \qquad [a_2 \quad a_3]^T \qquad [a_4 \quad a_5 \quad a_6 \quad a_7]^T]$$

The matrices H and L are the transposes of G and M

$$H_k = G_k^T \qquad L_k = M_k^T$$

Haar wavelets

The matrices have a particularly simple form

$$L_1 = [1 \quad 1] \qquad H_1 = [-1 \quad 1]$$

$$L_2 = \begin{bmatrix} 1 & 1 & & \\ & & 1 & 1 \end{bmatrix} \qquad H_2 = \begin{bmatrix} -1 & 1 & & \\ & & -1 & 1 \end{bmatrix}$$

$$L_3 = \begin{bmatrix} 1 & 1 & & & & & & \\ & & 1 & 1 & & & & \\ & & & & 1 & 1 & & \\ & & & & & & 1 & 1 \end{bmatrix} \qquad H_3 = \begin{bmatrix} -1 & 1 & & & & & & \\ & & -1 & 1 & & & & \\ & & & & -1 & 1 & & \\ & & & & & & -1 & 1 \end{bmatrix}$$

Appendix 15 Statistical Tables

Table A15.1 *Values of z, the standard normal variable, from 0.0 by steps of 0.01 to 3.9, showing the cumulative probability up to z. (Probability correct to 4 decimal places).*

z	0.00	0.01	0.02	0.03	0.04	0.05	0.06	0.07	0.08	0.09
0.0	0.5000	0.5040	0.5080	0.5120	0.5160	0.5199	0.5239	0.5279	0.5319	0.5359
0.1	0.5398	0.5438	0.5478	0.5517	0.5557	0.5596	0.5636	0.5675	0.5714	0.5753
0.2	0.5793	0.5832	0.5871	0.5910	0.5948	0.5987	0.6026	0.6064	0.6103	0.6141
0.3	0.6179	0.6217	0.6255	0.6293	0.6331	0.6368	0.6406	0.6443	0.6480	0.6517
0.4	0.6554	0.6591	0.6628	0.6664	0.6700	0.6736	0.6772	0.6808	0.6844	0.6879
0.5	0.6915	0.6950	0.6985	0.7019	0.7054	0.7088	0.7123	0.7157	0.7190	0.7224
0.6	0.7257	0.7291	0.7324	0.7357	0.7389	0.7422	0.7454	0.7486	0.7517	0.7549
0.7	0.7580	0.7611	0.7642	0.7673	0.7704	0.7734	0.7764	0.7794	0.7823	0.7852
0.8	0.7881	0.7910	0.7939	0.7967	0.7995	0.8023	0.8051	0.8078	0.8106	0.8133
0.9	0.8159	0.8186	0.8212	0.8238	0.8264	0.8289	0.8315	0.8340	0.8365	0.8389
1.0	0.8413	0.8438	0.8461	0.8485	0.8508	0.8531	0.8554	0.8577	0.8599	0.8621
0.1	0.8643	0.8665	0.8686	0.8708	0.8729	0.8749	0.8770	0.8790	0.8810	0.8830
0.2	0.8849	0.8869	0.8888	0.8907	0.8925	0.8944	0.8962	0.8980	0.8997	0.9015
0.3	0.9032	0.9049	0.9066	0.9082	0.9099	0.9115	0.9131	0.9147	0.9162	0.9177
0.4	0.9192	0.9207	0.9222	0.9236	0.9251	0.9265	0.9279	0.9292	0.9306	0.9319
0.5	0.9332	0.9345	0.9357	0.9370	0.9382	0.9394	0.9406	0.9418	0.9429	0.9441
0.6	0.9452	0.9463	0.9474	0.9484	0.9495	0.9505	0.9515	0.9525	0.9535	0.9545
0.7	0.9554	0.9564	0.9573	0.9582	0.9591	0.9599	0.9608	0.9616	0.9625	0.9633
0.8	0.9641	0.9649	0.9656	0.9664	0.9671	0.9678	0.9686	0.9693	0.9699	0.9706
0.9	0.9713	0.9719	0.9726	0.9732	0.9738	0.9744	0.9750	0.9756	0.9761	0.9767
2.0	0.9772	0.9778	0.9783	0.9788	0.9793	0.9798	0.9803	0.9808	0.9812	0.9817
0.1	0.9821	0.9826	0.9830	0.9834	0.9838	0.9842	0.9846	0.9850	0.9854	0.9857
0.2	0.9861	0.9864	0.9868	0.9871	0.9875	0.9878	0.9881	0.9884	0.9887	0.9890
0.3	0.9893	0.9896	0.9898	0.9901	0.9904	0.9906	0.9909	0.9911	0.9913	0.9916
0.4	0.9918	0.9920	0.9922	0.9925	0.9927	0.9929	0.9931	0.9932	0.9934	0.9936
0.5	0.9938	0.9940	0.9941	0.9943	0.9945	0.9946	0.9948	0.9949	0.9951	0.9952
0.6	0.9953	0.9955	0.9956	0.9957	0.9959	0.9960	0.9961	0.9962	0.9963	0.9964
0.7	0.9965	0.9966	0.9967	0.9968	0.9969	0.9970	0.9971	0.9972	0.9973	0.9974
0.8	0.9974	0.9975	0.9976	0.9977	0.9977	0.9978	0.9979	0.9979	0.9980	0.9981
0.9	0.9981	0.9982	0.9982	0.9983	0.9984	0.9984	0.9985	0.9985	0.9986	0.9986
3.0	0.9987	0.9987	0.9987	0.9988	0.9988	0.9989	0.9989	0.9989	0.9990	0.9990
0.1	0.9990	0.9991	0.9991	0.9991	0.9992	0.9992	0.9992	0.9992	0.9993	0.9993
0.2	0.9993	0.9993	0.9994	0.9994	0.9994	0.9994	0.9994	0.9995	0.9995	0.9995
0.3	0.9995	0.9995	0.9995	0.9996	0.9996	0.9996	0.9996	0.9996	0.9996	0.9997
0.4	0.9997	0.9997	0.9997	0.9997	0.9997	0.9997	0.9997	0.9997	0.9997	0.9998
0.5	0.9998	0.9998	0.9998	0.9998	0.9998	0.9998	0.9998	0.9998	0.9998	0.9998
0.6	0.9998	0.9998	0.9999	0.9999	0.9999	0.9999	0.9999	0.9999	0.9999	0.9999
0.7	0.9999	0.9999	0.9999	0.9999	0.9999	0.9999	0.9999	0.9999	0.9999	0.9999
0.8	0.9999	0.9999	0.9999	0.9999	0.9999	0.9999	0.9999	0.9999	0.9999	0.9999
0.9	1.0000									

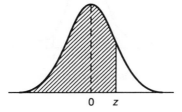

The curve is $N(0, 1)$, the standard normal variable. The table entry is the shaded area $\Phi(z) = \Pr(Z < z)$. For example when $z = 1.96$ the shaded area is 0.9750. Critical values of the standard normal distribution will be found in the bottom row of Table A15.2

Table A15.2 *Percentage points of Student's t-distribution*

d.f.	P = 0.05	0.025	0.01	0.005	0.001	0.0005
1	6.314	12.706	31.821	63.657	318.31	636.62
2	2.920	4.303	6.965	9.925	22.327	31.598
3	2.353	3.182	4.541	5.841	10.214	12.924
4	2.132	2.776	3.747	4.604	7.173	8.610
5	2.015	2.571	3.365	4.032	5.893	6.869
6	1.943	2.447	3.143	3.707	5.208	5.959
7	1.895	2.365	2.998	3.499	4.785	5.408
8	1.860	2.306	2.896	3.355	4.501	5.041
9	1.833	2.262	2.821	3.250	4.297	4.781
10	1.812	2.228	2.764	3.169	4.144	4.587
11	1.796	2.201	2.718	3.106	4.025	4.437
12	1.782	2.179	2.681	3.055	3.930	4.318
13	1.771	2.160	2.650	3.012	3.852	4.221
14	1.761	2.145	2.624	2.977	3.787	4.140
15	1.753	2.131	2.602	2.947	3.733	4.073
16	1.746	2.210	2.583	2.921	3.686	4.015
17	1.740	2.110	2.567	2.898	3.646	3.965
18	1.734	2.101	2.552	2.878	3.610	3.922
19	1.729	2.093	2.539	2.861	3.579	3.883
20	1.725	2.086	2.528	2.845	3.552	3.850
21	1.721	2.080	2.518	2.831	3.527	3.819
22	1.717	2.074	2.508	2.819	3.505	3.792
23	1.714	2.069	2.500	2.807	3.485	3.767
24	1.711	2.064	2.492	2.797	3.467	3.745
25	1.708	2.060	2.485	2.787	3.450	3.725
26	1.706	2.056	2.479	2.779	3.435	3.707
27	1.703	2.052	2.473	2.771	3.421	3.690
28	1.701	2.048	2.467	2.763	3.408	3.674
29	1.699	2.045	2.462	2.756	3.396	3.659
30	1.697	2.042	2.457	2.750	3.385	3.646
40	1.684	2.021	2.423	2.704	3.307	3.551
60	1.671	2.000	2.390	2.660	3.232	3.460
120	1.658	1.980	2.358	2.617	3.160	3.373
∞	1.645	1.960	2.326	2.576	3.090	3.291

The last row of the table (∞) gives values of z, the standard normal variable.

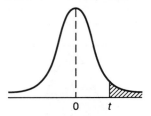

P is the shaded area

Table A15.3 *Percentage points of the χ^2 distribution exceeded with probability P.*

d.f	0.995	0.975	0.050	0.025	0.010	0.005	0.001
			P				
1	3.9×10^{-5}	9.8×10^{-4}	3.84	5.02	6.63	7.88	10.83
2	0.010	0.051	5.99	7.38	9.21	10.60	13.81
3	0.071	0.22	7.81	9.35	11.34	12.84	16.27
4	0.21	0.48	9.49	11.14	13.28	14.86	18.47
5	0.41	0.83	11.07	12.83	15.09	16.75	20.52
6	0.68	1.24	12.59	14.45	16.81	18.55	22.46
7	0.99	1.69	14.07	16.01	18.48	20.28	24.32
8	1.34	2.18	15.51	17.53	20.09	21.96	26.13
9	1.73	2.70	16.92	19.02	21.67	23.59	27.88
10	2.16	3.25	18.31	20.48	23.21	25.19	29.59
11	2.60	3.82	19.68	21.92	24.73	26.76	31.26
12	3.07	4.40	21.03	23.34	26.22	28.30	32.91
13	3.57	5.01	22.36	24.74	27.69	29.82	34.53
14	4.07	5.63	23.68	26.12	29.14	31.32	36.12
15	4.60	6.26	25.00	27.49	30.58	32.80	37.70
16	5.14	6.91	26.30	28.85	32.00	34.27	39.25
17	5.70	7.56	27.59	30.19	33.41	35.72	40.79
18	6.26	8.23	28.87	31.53	34.81	37.16	42.31
19	6.84	8.91	30.14	32.85	36.19	38.58	43.82
20	7.43	9.59	31.41	34.17	37.57	40.00	45.32
21	8.03	10.28	32.67	35.48	38.93	41.40	46.80
22	8.64	10.98	33.92	36.78	40.29	42.80	48.27
23	9.26	11.69	35.17	38.08	41.64	44.18	49.73
24	9.89	12.40	36.42	39.36	42.98	45.56	51.18
25	10.52	13.12	37.65	40.65	44.31	46.93	52.62
26	11.16	13.84	38.89	41.92	45.64	48.29	54.05
27	11.81	14.57	40.11	43.19	46.96	49.64	55.48
28	12.46	15.31	41.34	44.46	48.28	50.99	56.89
29	13.12	16.05	42.56	45.72	49.59	52.34	58.30
30	13.79	16.79	43.77	46.98	50.89	53.67	59.70
40	20.71	24.43	55.76	59.34	63.69	66.77	73.40
50	27.99	32.36	67.50	71.42	76.16	79.49	86.66
60	35.53	40.48	79.08	83.30	88.38	91.95	99.61
70	43.28	48.76	90.53	95.02	100.43	104.22	112.32
80	51.17	57.15	101.88	106.63	112.33	116.32	124.84
90	59.20	65.65	113.15	118.14	124.12	128.30	137.21
100	67.33	74.22	124.34	129.56	135.81	140.17	149.44

For degrees of freedom $\nu > 100$, test $\sqrt{2\chi_\nu^2}$ as $N(\sqrt{2\nu - 1}, 1)$.

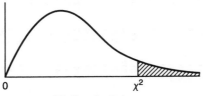

P is the shaded area

Table A15.4   *Table of F-distribution. Upper 5% points.*

v_1

v_2	1	2	3	4	5	6	7	8	9	10	12	15	20	24	30	40	60	120	∞
1	161.4	199.5	215.7	224.6	230.2	234.0	236.8	238.9	240.5	241.9	243.9	245.9	248.0	249.1	250.1	251.1	252.2	253.3	254.3
2	18.51	19.00	19.16	19.25	19.30	19.33	19.35	19.37	19.38	19.40	19.41	19.43	19.45	19.45	19.46	19.47	19.48	19.49	19.50
3	10.13	9.55	9.28	9.12	9.01	8.94	8.89	8.85	8.81	8.79	8.74	8.70	8.66	8.64	8.62	8.59	8.57	8.55	8.53
4	7.71	6.94	6.59	6.39	6.26	6.16	6.09	6.04	6.00	5.96	5.91	5.86	5.80	5.77	5.75	5.72	5.69	5.66	5.63
5	6.61	5.79	5.41	5.19	5.05	4.95	4.88	4.82	4.77	4.74	4.68	4.62	4.56	4.53	4.50	4.46	4.43	4.40	4.36
6	5.99	5.14	4.76	4.53	4.39	4.28	4.21	4.15	4.10	4.06	4.00	3.94	3.87	3.84	3.81	3.77	3.74	3.70	3.67
7	5.59	4.74	4.35	4.12	3.97	3.87	3.79	3.73	3.68	3.64	3.57	3.51	3.44	3.41	3.38	3.34	3.30	3.27	3.23
8	5.32	4.46	4.07	3.84	3.69	3.58	3.50	3.44	3.39	3.35	3.28	3.22	3.15	3.12	3.08	3.04	3.01	2.97	2.93
9	5.12	4.26	3.86	3.63	3.48	3.37	3.29	3.23	3.18	3.14	3.07	3.01	2.94	2.90	2.86	2.83	2.79	2.75	2.71
10	4.96	4.10	3.71	3.48	3.33	3.22	3.14	3.07	3.02	2.98	2.91	2.85	2.77	2.74	2.70	2.66	2.62	2.58	2.54
11	4.84	3.98	3.59	3.36	3.20	3.09	3.01	2.95	2.90	2.85	2.79	2.72	2.65	2.61	2.57	2.53	2.49	2.45	2.40
12	4.75	3.89	3.49	3.26	3.11	3.00	2.91	2.85	2.80	2.75	2.69	2.62	2.54	2.51	2.47	2.43	2.38	2.34	2.30
13	4.67	3.81	3.41	3.18	3.03	2.92	2.83	2.77	2.71	2.67	2.60	2.53	2.46	2.42	2.38	2.34	2.30	2.25	2.21
14	4.60	3.74	3.34	3.11	2.96	2.85	2.76	2.70	2.65	2.60	2.53	2.46	2.39	2.35	2.31	2.27	2.22	2.18	2.13
15	4.54	3.68	3.29	3.06	2.90	2.79	2.71	2.64	2.59	2.54	2.48	2.40	2.33	2.29	2.25	2.20	2.16	2.11	2.07
16	4.49	3.63	3.24	3.01	2.85	2.74	2.66	2.59	2.54	2.49	2.42	2.35	2.28	2.24	2.19	2.15	2.11	2.06	2.01
17	4.45	3.59	3.20	2.96	2.81	2.70	2.61	2.55	2.49	2.45	2.38	2.31	2.23	2.19	2.15	2.10	2.06	2.01	1.96
18	4.41	3.55	3.16	2.93	2.77	2.66	2.58	2.51	2.46	2.41	2.34	2.27	2.19	2.15	2.11	2.06	2.02	1.97	1.92
19	4.38	3.52	3.13	2.90	2.74	2.63	2.54	2.48	2.42	2.38	2.31	2.23	2.16	2.11	2.07	2.03	1.98	1.93	1.88
20	4.35	3.49	3.10	2.87	2.71	2.60	2.51	2.45	2.39	2.35	2.28	2.20	2.12	2.08	2.04	1.99	1.95	1.90	1.84
21	4.32	3.47	3.07	2.84	2.68	2.57	2.49	2.42	2.37	2.32	2.25	2.18	2.10	2.05	2.01	1.96	1.92	1.87	1.81
22	4.30	3.44	3.05	2.82	2.66	2.55	2.46	2.40	2.34	2.30	2.23	2.15	2.07	2.03	1.98	1.94	1.89	1.84	1.78
23	4.28	3.42	3.03	2.80	2.64	2.53	2.44	2.37	2.32	2.27	2.20	2.13	2.05	2.01	1.96	1.91	1.86	1.81	1.76
24	4.26	3.40	3.01	2.78	2.62	2.51	2.42	2.36	2.30	2.25	2.18	2.11	2.03	1.98	1.94	1.89	1.84	1.79	1.73
25	4.24	3.39	2.99	2.76	2.60	2.49	2.40	2.34	2.28	2.24	2.16	2.09	2.01	1.96	1.92	1.87	1.82	1.77	1.71
26	4.23	3.37	2.98	2.74	2.59	2.47	2.39	2.32	2.27	2.22	2.15	2.07	1.99	1.95	1.90	1.85	1.80	1.75	1.69
27	4.21	3.35	2.96	2.73	2.57	2.46	2.37	2.31	2.25	2.20	2.13	2.06	1.97	1.93	1.88	1.84	1.79	1.73	1.67
28	4.20	3.34	2.95	2.71	2.56	2.45	2.36	2.29	2.24	2.19	2.12	2.04	1.96	1.91	1.87	1.82	1.77	1.71	1.65
29	4.18	3.33	2.93	2.70	2.55	2.43	2.35	2.28	2.22	2.18	2.10	2.03	1.94	1.90	1.85	1.81	1.75	1.70	1.64
30	4.17	3.32	2.92	2.69	2.53	2.42	2.33	2.27	2.21	2.16	2.09	2.01	1.93	1.89	1.84	1.79	1.74	1.68	1.62
40	4.08	3.23	2.84	2.61	2.45	2.34	2.25	2.18	2.12	2.08	2.00	1.92	1.84	1.79	1.74	1.69	1.64	1.58	1.51
60	4.00	3.15	2.76	2.53	2.37	2.25	2.17	2.10	2.04	1.99	1.92	1.84	1.75	1.70	1.65	1.59	1.53	1.47	1.39
120	3.92	3.07	2.68	2.45	2.29	2.17	2.09	2.02	1.96	1.91	1.83	1.75	1.66	1.61	1.55	1.50	1.43	1.35	1.25
∞	3.84	3.00	2.60	2.37	2.21	2.10	2.01	1.94	1.88	1.83	1.75	1.67	1.57	1.52	1.46	1.39	1.32	1.22	1.00

v_1, v_2 are numerator, denominator d.f. respectively.
Tabulated values are those exceeded with probability 0.05.

Table 15.4 (cont.)   *Table of F-distribution. Upper 1% points.*

ν_1

ν_2	1	2	3	4	5	6	7	8	9	10	12	15	20	24	30	40	60	120	∞
1	4052	4999.5	5403	5625	5764	5859	5928	5981	6022	6056	6106	6157	6209	6235	6261	6287	6313	6339	6366
2	98.50	99.00	99.17	99.25	99.30	99.33	99.36	99.37	99.39	99.40	99.42	99.43	99.45	99.46	99.47	99.47	99.48	99.49	99.50
3	34.12	30.82	29.46	28.71	28.24	27.91	27.67	27.49	27.35	27.23	27.05	26.87	26.69	26.60	26.50	26.41	26.32	26.22	26.13
4	21.20	18.00	16.69	15.98	15.52	15.21	14.98	14.80	14.66	14.55	14.37	14.20	14.02	13.93	13.84	13.75	13.65	13.56	13.46
5	16.26	13.27	12.06	11.39	10.97	10.67	10.46	10.29	10.16	10.05	9.89	9.72	9.55	9.47	9.38	9.29	9.20	9.11	9.02
6	13.75	10.92	9.78	9.15	8.75	8.47	8.26	8.10	7.98	7.87	7.72	7.56	7.40	7.31	7.23	7.14	7.06	6.97	6.88
7	12.25	9.55	8.45	7.85	7.46	7.19	6.99	6.84	6.72	6.62	6.47	6.31	6.16	6.07	5.99	5.91	5.82	5.74	5.65
8	11.26	8.65	7.59	7.01	6.63	6.37	6.18	6.03	5.91	5.81	5.67	5.52	5.36	5.28	5.20	5.12	5.03	4.95	4.86
9	10.56	8.02	6.99	6.42	6.06	5.80	5.61	5.47	5.35	5.26	5.11	4.96	4.81	4.73	4.65	4.57	4.48	4.40	4.31
10	10.04	7.56	6.55	5.99	5.64	5.39	5.20	5.06	4.94	4.85	4.71	4.56	4.41	4.33	4.25	4.17	4.08	4.00	3.91
11	9.65	7.21	6.22	5.67	5.32	5.07	4.89	4.74	4.63	4.54	4.40	4.25	4.10	4.02	3.94	3.86	3.78	3.69	3.60
12	9.33	6.93	5.95	5.41	5.06	4.82	4.64	4.50	4.39	4.30	4.16	4.01	3.86	3.78	3.70	3.62	3.54	3.45	3.36
13	9.07	6.70	5.74	5.21	4.86	4.62	4.44	4.30	4.19	4.10	3.96	3.82	3.66	3.59	3.51	3.43	3.34	3.25	3.17
14	8.86	6.51	5.56	5.04	4.69	4.46	4.28	4.14	4.03	3.94	3.80	3.66	3.51	3.43	3.35	3.27	3.18	3.09	3.00
15	8.68	6.36	5.42	4.89	4.56	4.32	4.14	4.00	3.89	3.80	3.67	3.52	3.37	3.29	3.21	3.13	3.05	2.96	2.87
16	8.53	6.23	5.29	4.77	4.44	4.20	4.03	3.89	3.78	3.69	3.55	3.41	3.26	3.18	3.10	3.02	2.93	2.84	2.75
17	8.40	6.11	5.18	4.67	4.34	4.10	3.93	3.79	3.68	3.59	3.46	3.31	3.16	3.08	3.00	2.92	2.83	2.75	2.65
18	8.29	6.01	5.09	4.58	4.25	4.01	3.84	3.71	3.60	3.51	3.37	3.23	3.08	3.00	2.92	2.84	2.75	2.66	2.57
19	8.18	5.93	5.01	4.50	4.17	3.94	3.77	3.63	3.52	3.43	3.30	3.15	3.00	2.92	2.84	2.76	2.67	2.58	2.49
20	8.10	5.85	4.94	4.43	4.10	3.87	3.70	3.56	3.46	3.37	3.23	3.09	2.94	2.86	2.78	2.69	2.61	2.52	2.42
21	8.02	5.78	4.87	4.37	4.04	3.81	3.64	3.51	3.40	3.31	3.17	3.03	2.88	2.80	2.72	2.64	2.55	2.46	2.36
22	7.95	5.72	4.82	4.31	3.99	3.76	3.59	3.45	3.35	3.26	3.12	2.98	2.83	2.75	2.67	2.58	2.50	2.40	2.31
23	7.88	5.66	4.76	4.26	3.94	3.71	3.54	3.41	3.30	3.21	3.07	2.93	2.78	2.70	2.62	2.54	2.45	2.35	2.26
24	7.82	5.61	4.72	4.22	3.90	3.67	3.50	3.36	3.26	3.17	3.03	2.89	2.74	2.66	2.58	2.49	2.40	2.31	2.21
25	7.77	5.57	4.68	4.18	3.85	3.63	3.46	3.32	3.22	3.13	2.99	2.85	2.70	2.62	2.54	2.45	2.36	2.27	2.17
26	7.72	5.53	4.64	4.14	3.82	3.59	3.42	3.29	3.18	3.09	2.96	2.81	2.66	2.58	2.50	2.42	2.33	2.23	2.13
27	7.68	5.49	4.60	4.11	3.78	3.56	3.39	3.26	3.15	3.06	2.93	2.78	2.63	2.55	2.47	2.38	2.29	2.20	2.10
28	7.64	5.45	4.57	4.07	3.75	3.53	3.36	3.23	3.12	3.03	2.90	2.75	2.60	2.52	2.44	2.35	2.26	2.17	2.06
29	7.60	5.42	4.54	4.04	3.73	3.50	3.33	3.20	3.09	3.00	2.87	2.73	2.57	2.49	2.41	2.33	2.23	2.14	2.03
30	7.56	5.39	4.51	4.02	3.70	3.47	3.30	3.17	3.07	2.98	2.84	2.70	2.55	2.47	2.39	2.30	2.21	2.11	2.01
40	7.31	5.18	4.31	3.83	3.51	3.29	3.12	2.99	2.89	2.80	2.66	2.52	2.37	2.29	2.20	2.11	2.02	1.92	1.80
60	7.08	4.98	4.13	3.65	3.34	3.12	2.95	2.82	2.72	2.63	2.50	2.35	2.20	2.12	2.03	1.94	1.84	1.73	1.60
120	6.85	4.79	3.95	3.48	3.17	2.96	2.79	2.66	2.56	2.47	2.34	2.19	2.03	1.95	1.86	1.76	1.66	1.53	1.38
∞	6.63	4.61	3.78	3.32	3.02	2.80	2.64	2.51	2.41	2.32	2.18	2.04	1.88	1.79	1.70	1.59	1.47	1.32	1.00

ν_1, ν_2 are numerator, denominator d.f. respectively
Tabulated values are those exceeded with probability 0.01.

Table A15.5   *Random Digits.*

12005	84000	51051	92674	76575	35789	04180	75029	32490	39949
98859	09884	45275	09467	93026	32912	13941	23206	62419	67776
26604	95099	93751	00590	93060	64776	83565	69919	51623	27483
82984	65780	94428	30160	86023	52284	62463	70712	40687	92630
70888	14063	96700	83008	17579	71321	62664	51514	92195	46722
77803	61872	86245	68220	66267	01379	11304	01658	82404	46728
35228	49673	53552	51215	45611	83927	00772	99295	72154	24126
69965	74926	63366	47688	14279	42943	98863	86630	53925	22310
89716	61713	30650	49028	20285	37791	69149	41701	42403	64009
68348	85228	97590	90997	83339	95822	72969	14037	32379	96225
33821	41538	86376	71823	16285	92630	89531	59337	05421	17043
63162	18167	32088	41917	60942	63252	83886	54130	31841	04502
03431	44528	41760	68035	33731	43262	12789	40348	15532	95309
99198	35092	63655	23987	31112	88069	58720	41729	18757	96096
75535	45156	49477	10673	48262	78240	94031	06192	75221	13363
98554	52502	11780	04060	56634	58077	02005	80217	65893	78381
89725	00679	28401	79434	00909	22989	31446	76251	17061	66680
49221	37750	26367	44817	09214	82674	65641	14332	58211	49564
31783	96028	69352	78426	94411	38335	22540	37881	10784	84658
61025	72770	13689	21456	48391	00157	61957	11262	12640	17228
10581	30143	89214	52134	76280	77823	61674	96898	90487	43998
51753	56087	71524	64913	81706	33984	90919	86969	75553	87375
96050	08123	28557	04240	33606	10776	64239	81900	74830	92654
93998	95705	73353	26933	66089	25177	62387	34932	62021	34044
70974	45757	31830	09589	31037	91886	51780	21912	16444	52881
25833	71286	76375	43640	92551	46510	68950	60168	26349	04599
55060	28982	92650	71622	36740	05869	17828	29937	01020	90851
29436	79967	34383	85646	04715	80695	39283	5O543	26875	94047
80180	08706	17875	72123	69723	52846	71310	72507	25702	33449
40842	32742	44671	72953	54811	39495	05023	61569	60805	26580
31481	16208	60372	94367	88977	35393	08681	53325	92547	31622
06045	35097	38319	17264	40640	63022	01496	28439	04197	63858
41446	12336	54072	47198	56085	25215	89943	41153	18496	76869
22301	07404	60943	75921	02932	50090	51949	86415	51919	98125
38199	09042	26771	15881	80204	61281	61610	24501	01935	33256
06273	93282	55034	79777	75241	1l762	11274	41685	24117	98311
92201	02587	31599	27987	25678	69736	94487	41653	79550	92949
70782	80894	95413	36338	04237	19954	71137	23584	87069	10407
05245	40934	96832	33415	62058	87179	31542	18174	54711	21882
85607	45719	65640	33241	04852	87636	43840	42242	22092	28975
61175	56493	93453	90267	99471	04519	78694	17115	00371	64703
36079	22448	22686	31272	01245	66265	12670	29560	49346	20049
94688	39732	02785	73373	44876	39888	69352	40488	43849	95406
54047	85793	53994	28605	46114	91174	49646	85123	66246	72392
24997	69553	468O2	24331	88523	89026	69776	55460	21984	76677

Table A15.6 *Control chart factors for the sample range*

	Lower percentage factors		Upper percentage factors	
Sample size	0.1%	2.5%	2.5%	0.1%
2	0.00	0.04	3.17	4.65
3	0.06	0.30	3.68	5.06
4	0.20	0.59	3.98	5.31
5	0.37	0.85	4.20	5.48
6	0.53	1.07	4.36	5.62
7	0.69	1.25	4.49	5.73
8	0.83	1.41	4.60	5.82
9	0.97	1.55	4.70	5.90
10	1.08	1.67	4.78	5.97

Multiply the estimate of the standard deviation by the tabled factors, which are based on a normal distribution.

Table A15.7 *Gamma function (from Kreyszig, 1993)*

α	$\Gamma(\alpha)$	α	$\Gamma(\alpha)$	α	$\Gamma(\alpha)$	α	$\Gamma(\alpha)$	α	$\Gamma(\alpha)$
1.00	1.000 000	1.20	0.918 169	1.40	0.887 264	1.60	0.893 515	1.80	0.931 384
1.02	0.988 844	1.22	0.913 106	1.42	0.886 356	1.62	0.895 924	1.82	0.936 845
1.04	0.978 438	1.24	0.908 521	1.44	0.885 805	1.64	0.898 642	1.84	0.942 612
1.06	0.968 744	1.26	0.904 397	1.46	0.885 604	1.66	0.901 668	1.86	0.948 687
1.08	0.959 725	1.28	0.900 718	1.48	0.885 747	1.68	0.905 001	1.88	0.955 071
1.10	0.951 351	1.30	0.897 471	1.50	0.886 227	1.70	0.908 639	1.90	0.961 766
1.12	0.943 590	1.32	0.894 640	1.52	0.887 039	1.72	0.912 581	1.92	0.968 774
1.14	0.936 416	1.34	0.892 216	1.54	0.888 178	1.74	0.916 826	1.94	0.976 099
1.16	0.929 803	1.36	0.890 185	1.56	0.889 639	1.76	0.921 375	1.96	0.983 743
1.18	0.923 728	1.38	0.888 537	1.58	0.891 420	1.78	0.926 227	1.98	0.991 708
1.20	0.918 169	1.40	0.887 264	1.60	0.893 515	1.80	0.931 384	2.00	1.000 000

Appendix 16   Selected Data Tables

Table A16.1 *Annual maximum daily mean discharges* (m^3/s), *naturalized, for the River Thames at Kingston (data courtesy of the NRA Thames Region). For the water years 1983/84 1994/95 (water years run from 1 October until 30 September)*

Water year	Flow	Month	Rank	Water year	Flow	Month	Rank
1882/83	292	*	65.0	1939/40	410	2	19.0
1883/84	231	2	88.0	1940/41	384	11	27.5
1884/85	229	2	90.5	1941/42	298	1	63.5
1885/86	244	12	83.0	1942/43	457	2	13.0
1886/87	284	1	69.0	1943/44	115	1	112.0
1887/88	207	3	99.0	1944/45	261	12	76.0
1888/89	237	3	85.0	1945/46	256	12	77.5
1889/90	204	1	100.5	1946/47	714	3	2.0
1890/91	171	2	106.5	1947/48	227	1	93.5
1891/92	339	10	45.0	1948/49	299	1	61.5
1892/93	299	2	61.5	1949/50	324	2	52.0
1893/94	173	2	105.0	1950/51	384	2	27.5
1894/95	789	11	1.0	1951/52	376	11	30.0
1895/96	201	3	102.0	1952/53	263	12	73.0
1896/97	351	2	41.5	1953/54	222	6	96.0
1897/98	171	1	106.5	1954/55	452	12	15.0
1898/99	262	2	74.5	1955/56	315	2	55.0
1899/00	533	2	7.0	1956/57	314	2	56.5
1900/01	200	4	103.0	1957/58	316	2	54.0
1901/02	162	12	109.0	1958/59	375	1	31.5
1902/03	386	6	25.0	1959/60	308	1	59.0
1903/04	517	2	10.0	1960/61	456	11	14.0
1904/05	229	3	90.5	1961/62	344	1	44.0
1905/06	249	1	81.0	1962/63	285	3	68.0
1906/07	220	1	97.5	1963/64	369	3	37.0
1907/08	375	12	31.5	1964/65	122	9	111.0
1908/09	204	3	100.5	1965/66	324	2	52.0
1909/10	231	2	88.0	1966/67	313	3	58.0
1910/11	428	12	17.0	1967/68	600	9	3.0
1911/12	366	1	39.0	1968/69	369	12	37.0
1912/13	255	1	79.0	1969/70	224	1	95.0
1913/14	256	3	77.5	1970/71	362	1	40.0
1914/15	585	1	4.0	1971/72	330	3	48.0
1915/16	373	3	34.0	1972/73	266	12	72.0
1916/17	327	12	49.0	1973/74	396	2	23.0
1917/18	351	1	41.5	1974/75	559	11	5.0
1918/19	334	3	46.5	1975/76	152	12	110.0
1919/20	251	2	80.0	1976/77	334	1	46.5
1920/21	240	1	84.0	1977/78	326	1	50.0
1921/22	198	3	104.0	1978/79	324	4	52.0
1922/23	231	4	88.0	1979/80	393	12	24.0
1923/24	298	1	63.5	1980/81	289	3	66.0
1924/25	522	1	9.0	1981/82	314	12	56.5
1925/26	370	1	35.0	1982/83	345	12	43.0
1926/27	374	3	33.0	1983/84	286	3	67.0
1927/28	526	1	8.0	1984/85	270	11	71.0
1928/29	235	12	86.0	1985/86	408	12	20.0
1929/30	551	12	6.0	1986/87	304	4	60.0
1930/31	228	12	92.0	1987/88	402	1	21.0
1931/32	274	5	70.0	1988/89	262	2	74.5
1932/33	479	2	11.0	1989/90	427	2	18.0
1933/34	94	3	113.0	1990/91	220	1	97.5
1934/35	227	3	93.5	1991/92	165	9	108.0
1935/36	478	1	12.0	1992/93	378	12	29.0
1936/37	437	1	16.0	1993/94	400	1	22.0
1937/38	247	12	82.0	1994/95	385	2	26.0
1938/39	369	2	37.0				

Table A16.2 *The following data are the annual maximum sea levels at Lowestoft from 1952–1994, and for Sheerness for 27 years between 1952 and 1994, in metres relative to Ordnance Datum Newlyn. (Data courtesy of Proudman Oceanographic Laboratory, Bidston Observatory, Birkenhead, UK)*

	Lowestoft	Sheerness		Lowestoft	Sheerness		Lowestoft	Sheerness
1952	3.508		1967	1.948	3.661	1982	1.952	3.402
1953	3.367		1968	1.856	3.539	1983	2.688	3.732
1954	2.508		1969	2.709	3.688	1984	2.053	3.373
1955	1.954		1970	1.996	3.487	1985	1.869	3.489
1956	1.831		1971	2.262	3.624	1986	1.721	3.608
1957	1.865		1972	1.832	3.573	1987	1.913	3.427
1958	1.941	3.447	1973	2.47	3.903	1988	1.871	3.818
1959	1.903		1974	1.81	3.313	1989	2.305	3.514
1960	1.929		1975	1.85	3.500	1990	2.236	3.726
1961	2.381		1976	2.68		1991	1.88	3.318
1962	2.365		1977	1.961		1992	1.645	3.481
1963	1.612		1978	2.346		1993	2.641	3.866
1964	1.856		1979	1.659		1994	2.413	3.735
1965	2.131	4.023	1980	1.932	3.525			
1966	1.917	3.614	1981	2.02	3.613			

Table A16.3 *Monthly maximum wind speeds (knots) at Weston Park, Sheffield, from January 1990 until December 1995. The direction (degrees clockwise from N) and month are also shown. (Data courtesy of Sheffield City Museums.)*

Speed	Direction	Month	Speed	Direction	Month	Speed	Direction	Month
66	292.5	1	44	247.5	1	53	225	1
59	270	2	50	292.5	2	39	292.5	2
49	270	3	51	292.5	3	60	270	3
35	270	4	50	22.5	4	44	0	4
31	292.5	5	46	270	5	35	112.5	5
28	292.5	6	30	22.5	6	39	247.5	6
31	270	7	31	247.5	7	25	270	7
32	292.5	8	43	247.5	8	30	337.5	8
40	292.5	9	40	225	9	33	270	9
42	247.5	10	32	270	10	38	270	10
39	*	11	46	270	11	36	225	11
47	270	12	41	157.5	12	49	270	12
56	270	1	56	225	1	61	315	1
35	225	2	52	315	2	49	247.5	2
41	247.5	3	41	315	3	57	270	3
39	157.5	4	43	292.5	4	43	315	4
31	0	5	44	*	5	33	292.5	5
31	270	6	35	315	6	33	337.5	6
27	225	7	35	247.5	7	27	315	7
31	292.5	8	43	247.5	8	36	315	8
37	225	9	31	90	9	47	292.5	9
45	270	10	31	22.5	10	37	247.5	10
46	247.5	11	47	337.5	11	33	247.5	11
52	292.5	12	67	292.5	12	29	90	12

Table A16.4 *River Coquet peak flows (cumecs) extracted from a 21 year period, 1973–1993. The following independence criteria were used: all peaks above mean level, peaks separated by at least 3 days, trough between peaks less than 2/3 of earlier peak (courtesy Environment Agency)*

23.9	5.7	13.1	30.8	29.7	7.2	10.3	29.9	6.9	48.5
5.8	10.5	18.9	25.3	10.3	28.0	11.7	57.0	16.2	10.0
6.4	12.8	27.0	19.4	46.7	6.1	29.9	15.1	23.0	6.4
17.1	7.6	7.1	17.9	54.7	12.7	7.4	8.0	23.2	8.7
7.2	6.1	6.8	17.0	37.8	15.0	23.6	5.9	15.3	190.9
5.9	20.2	11.4	39.4	12.1	23.0	8.0	7.6	9.0	26.8
5.7	15.8	66.2	13.4	44.6	48.1	8.1	25.1	6.7	12.0
10.5	8.4	16.7	8.5	16.9	17.9	29.0	5.7	9.7	9.6
8.7	6.0	89.6	10.1	36.3	8.5	20.3	24.4	9.2	8.5
9.9	6.8	25.4	17.5	15.7	21.6	11.4	55.7	17.4	8.1
22.0	14.8	8.5	18.5	21.4	8.9	33.1	5.7	6.4	7.0
19.9	16.1	10.0	66.3	180.7	19.3	16.4	5.9	18.7	14.2
6.4	27.0	8.4	34.2	8.0	86.6	32.1	42.1	8.0	5.7
23.0	25.4	12.5	9.1	39.2	19.5	13.1	49.8	6.7	11.5
20.4	10.0	15.6	21.5	19.1	9.5	14.8	13.5	9.0	35.1
19.0	18.2	11.6	9.1	10.1	13.6	13.8	21.0	20.4	7.4
17.1	6.2	6.7	37.8	8.4	7.4	10.2	9.0	14.4	8.4
18.0	13.7	8.6	43.5	9.3	76.4	66.3	46.2	38.8	15.6
25.4	10.4	9.4	9.5	10.1	6.3	10.8	8.3	23.7	14.5
13.9	23.3	6.5	6.5	11.5	7.1	11.6	10.3	30.0	30.2
23.8	6.6	6.4	13.9	9.1	7.2	7.3	8.9	24.3	15.0
6.7	31.5	7.4	11.2	7.7	81.3	63.1	39.3	9.4	32.3
10.3	47.5	24.6	9.9	33.7	19.6	32.3	17.4	11.4	6.0
14.9	9.9	9.4	17.3	18.9	35.8	21.0	91.4	6.4	9.0
7.5	63.9	7.7	21.3	10.2	25.9	7.4	15.2	7.7	13.0
7.3	27.8	14.3	7.6	11.4	11.6	5.8	8.6	14.7	53.1
9.4	17.7	40.2	8.9	17.7	12.1	7.2	14.3	13.4	29.4
15.4	13.7	16.2	16.3	28.3	5.8	6.7	11.8	7.8	10.1
7.8	8.5	12.7	19.1	30.2	11.0	142.5	21.1	13.7	7.1
14.6	37.5	57.4	21.8	16.9	6.5	23.6	24.5	25.2	13.1
12.2	30.9	77.4	8.3	25.5	12.0	6.3	9.5	8.4	63.8
54.7	22.0	72.8	32.6	21.1	47.2	8.6	6.9	17.0	21.4
25.7	29.9	26.2	15.4	9.6	6.1	10.2	6.4	27.1	13.8
7.3	60.2	10.9	6.1	17.8	8.2	9.8	21.0	62.5	58.4
8.6	34.9	9.1	9.1	8.5	26.4	7.5	9.5	30.4	20.2
9.0	25.8	113.2	13.2	12.2	32.1	27.2	6.3	15.9	6.3
17.4	77.5	101.7	5.8	13.1	21.2	28.8	23.0	28.7	25.8
20.3	19.0	21.1	11.3	13.1	12.0	21.4	25.8	7.0	71.7
10.0	10.6	56.0	117.1	23.2	5.8	11.3	6.0	106.8	17.2
41.5	13.6	85.2	49.4	17.5	8.2	44.3	5.7	15.9	13.3
33.8	13.1	28.1	19.8	8.0	5.8	37.1	15.9	31.6	29.7
15.8	9.5	8.0	12.3	13.9	8.0	51.3	36.2	11.9	58.8
7.8	17.4	12.9	23.7	32.2	10.9	12.4	25.6	27.7	54.2
13.1	6.1	6.7	5.8	22.6	18.1	18.1	9.2	6.4	18.3
8.5	8.3	17.1	24.4	34.6	16.7	22.9	57.9	23.7	
11.8	11.3	9.3	7.5	27.3	9.6	10.6	25.8	7.5	
11.6	13.4	10.3	6.4	9.5	21.0	10.4	7.4	8.4	
23.0	7.6	15.8	7.1	30.0	9.7	37.8	5.8	23.5	
8.2	17.9	10.9	12.9	16.1	16.3	17.5	9.0	10.8	

Table A16.5 *Extreme wind speeds, Jacksonville, FL.*
T indicates a tropical storm

Year	mph	Type	Year	mph	Type
1950	65	T	1965	52	T
1951	38	T	1966	44	T
1952	51		1967	69	
1953	47		1968	47	T
1954	42		1967	53	
1955	42		1970	40	
1956	44		1971	51	
1957	42		1972	48	
1958	38		1973	53	
1959	34		1974	48	
1960	42	T	1975	68	
1961	44		1976	46	
1962	49		1977	36	T
1963	56		1978	43	
1964	74	T	1979	37	

Notes: Taken from Changery (1982)

Table A16.6 *Annual maximum flows (cumecs) at Skelton, near York, on the River Ouse and year of occurrence (courtesy of the Environment Agency). The data are given in descending order of magnitude*

622	82	361	83	296	71
592	91	354	93	277	75
481	92	348	94	268	69
480	95	335	89	265	70
414	84	322	79	257	74
400	85	317	81	247	72
381	86	317	80	244	96
377	90	313	76	221	73
369	88	310	87		
363	78	304	77		

Table A16.7 *Annual flood series for Namibia. The format is:*

1st line – Code, Station name
2nd line – No. of years
Main bit – Water year, Flood value, Water year, Flood, etc ...
Last line of block – Latitude (S), Longitude (E), Catchment area (km²),
*   Mean annual rainfall (mm), Slope*
Then repeat for next station .....

[Data courtesy of the Department of Water Affairs in Namibia]

NM048201Ham at Tsamab
18

	7071	28.003	7172	119.450	7273	519.579	7475	29.238
	7576	123.158	7677	119.450	7778	18.916	7879	26.782
	7980	73.641	8182	0.000	8283	0.000	8384	35.619
	8485	34.966	8586	105.095	8687	0.523	8788	89.328
	8889	268.682	8990	65.635				
28:09	19:15		2470		140	0.00403		

NM049101Fish at Gras
17

	7071	1059.473	7172	1084.019	7273	165.134	7374	1053.348
	7475	562.410	7576	937.812	7677	883.675	7778	441.060
	7879	522.888	7980	556.742	8081	95.230	8182	105.568
	8384	330.518	8485	659.785	8586	428.094	8687	500.466
	8788	545.429						
24:11	17:21		9170		211	0.00270		

NM049102Kam at Draaihoek
11

	7778	819.203	7879	108.904	8081	33.159	8182	64.378
	8485	199.698	8586	121.875	8687	95.356	8788	129.951
	8889	239.882	8990	30.895	9091	0.000		
24:12	17:02		3450		229	0.00325		

NM049103Fish at Dirichas
12

	7778	160.029	7879	87.942	7980	178.365	8081	0.076
	8182	61.723	8283	97.778	8384	193.390	8485	114.356
	8586	209.089	8687	230.274	8990	77.375	9091	189.029
24:16	17:05		2830		180	0.00262		

NM049106Kam at Klein Aub
11

	7374	118.266	7475	5.920	7879	55.503	7980	46.015
	8081	65.637	8182	73.947	8283	107.911	8384	197.818
	8687	64.480	8788	123.673	8990	90.335		
23:49	16:37		1080		241	0.00417		

NM049202Packriem at Karris
11

	7980	68.616	8081	0.000	8182	0.000	8283	19.786
	8384	498.747	8485	106.009	8687	125.966	8788	204.793
	8889	440.573	8990	62.779	9091	163.180		
24:21	17:35		1520		205	0.00218		

NM049301Hutup at Rietkuil
13

	7576	395.999	7778	58.164	7879	14.078	7980	123.079
	8081	0.122	8182	0.000	8485	40.486	8586	128.171
	8687	39.660	8788	138.713	8889	78.572	8990	1.908
	9091	87.888						
25:07	17:31		5850		179	0.00206		

continued

Table A16.7 *Continued*

10

7879	31.025	7980	55.308	8081	38.718	8182	5.684
8283	24.685	8485	9.345	8586	105.424	8687	35.216
8788	122.248	8889	155.299				

24:50 17:10 4780 177 0.00174

NM049601Fish at Seeheim
21

6162	860.696	6465	356.158	6566	585.374	6667	1205.186
6768	418.487	7071	400.286	7172	8300.113	7273	356.158
7374	6125.126	7475	58.438	7576	2476.661	7778	340.309
7879	78.601	7980	368.169	8081	173.868	8182	0.000
8283	38.444	8485	593.790	8586	178.715	8687	101.400
9091	204.815						

26:49 17:48 46400 181 0.00124

NM049602Fish at Tses
7

7980	721.856	8081	213.819	8182	7.951	8384	399.228
8687	307.077	8990	62.752	9091	148.364		

25:54 17:59 37600 189 0.00161

NM049703Loewen at Altdorn
12

7677	164.438	7778	53.411	7879	9.979	7980	190.446
8182	51.211	8283	92.730	8384	70.270	8485	53.411
8687	37.130	8788	144.374	8889	1260.283	8990	81.068

26:48 18:14 7000 153 0.00391

NM049705Loewen at Geduld
12

7778	86.299	7879	23.812	7980	62.403	8081	66.341
8182	70.398	8283	5.969	8586	134.024	8687	56.107
8788	70.398	8889	255.384	9091	103.808	9192	38.760

26:46 18:29 3200 160 0.00481

NM049802Konkiep at Bethanien
10

7475	57.030	7778	0.000	7879	0.000	7980	39.273
8081	0.401	8182	0.000	8283	96.536	8485	12.206
8586	100.256	8889	164.273				

26:27 17:08 4140 171 0.00407

NM049902Fish at Ai-ais
10

7778	321.362	7879	85.309	7980	365.854	8081	51.936
8182	0.000	8283	15.200	8485	611.939	8586	450.746
8687	96.777	8889	1973.135				

27:55 17:29 63300 167 0.00148

NM049903Gab at Holoog
9

8081	0.000	8182	0.000	8283	0.000	8384	17.180
8485	20.000	8687	0.307	8889	377.827	8990	0.641
9091	17.745						

27:27 17:59 2510 135 0.00402

Table A16.7 *Continued*

NM253101Omatako at Ousema
27

6162	7.045	6263	434.807	6364	13.660	6465	182.210
6566	113.326	6667	43.529	6768	46.925	6869	31.227
6970	59.177	7071	77.246	7172	111.592	7273	45.779
7374	216.573	7475	8.754	7576	169.276	7677	30.294
7778	160.911	7879	74.309	7980	38.153	8081	20.976
8182	28.472	8283	20.215	8384	111.592	8586	30.294
8687	106.270	8788	91.176	8889	64.479		

21:13   17:06   4970   402   0.00232

NM295103Huab at Monte Carlo
11

7677	0.084	7778	0.310	7879	12.546	7980	148.274
8081	14.467	8283	73.867	8384	30.025	8485	32.010
8788	35.112	9091	0.000	9192	0.000		

19:59   14:45   2670   300   0.00481

NM295401Aba-Huab at Rooiberg
14

7576	77.136	7677	78.484	7778	57.008	7879	45.826
7980	38.654	8081	0.000	8182	70.555	8283	258.836
8384	47.974	8485	89.640	8586	67.996	8687	175.920
8788	110.744	9091	91.082				

20:29   14:35   1570   198   0.00505

NM296102Ugab at Petersburg
21

6162	22.263	6263	38.935	6364	3.054	6465	87.027
6566	9.249	6970	31.529	7172	28.203	7273	27.621
7374	42.655	7475	2.745	7576	15.736	7677	4.205
7778	21.718	7879	14.996	7980	86.053	8182	36.495
8384	29.879	8485	119.781	8687	301.795	8788	67.064
8990	14.718						

20:12   16:08   7720   485   0.00364

NM296203Ugab at Vingerklip
17

6869	130.436	6970	42.279	7071	188.215	7172	99.196
7273	50.143	7374	262.889	7475	177.818	7576	127.174
7778	59.712	7879	71.185	8182	97.827	8384	310.206
8485	416.653	8586	153.150	8687	146.032	8889	113.275
9091	167.673						

20:25   15:28   14200   431   0.00317

NM296206Ugab at Ugab Slab
10

7778	2.250	7879	0.000	7980	13.061	8081	0.000
8182	177.130	8384	51.524	8485	257.671	8687	0.879
8889	0.371	8990	5.886				

21:05   13:48   28900   305   0.00338

NM297101Omaruru at Omaruru
15

6465	196.162	6566	360.907	6667	96.464	6869	225.665
7273	0.000	7475	0.000	7576	259.848	7677	37.615
7778	142.566	7879	173.354	7980	156.455	8081	33.705
8182	180.822	8283	299.188	8384	347.502		

21:26   15:57   2520   379   0.00383

continued

Table A16.7   *Continued*

NM297102Omaruru at Etemba
21

6970	63.544	7071	404.277	7172	433.103	7273	66.783
7374	760.214	7475	1.246	7576	380.680	7677	2.092
7778	172.824	7879	180.702	7980	135.991	8081	16.992
8182	153.288	8283	549.994	8384	591.011	8485	702.045
8687	172.059	8788	202.473	8889	76.618	9091	242.689
9192	17.401						

21:26   15:41   3810   350   0.00383

NM297103Omaruru at Omburo
17

7475	5.996	7576	262.034	7677	26.683	7778	101.361
7879	260.302	8081	0.000	8182	176.790	8283	714.037
8384	764.361	8485	684.892	8586	128.673	8687	321.558
8788	275.050	8889	139.353	8990	256.675	9091	211.684
9192	40.293						

21:18   16:12   1320   390   0.00386

NM298205Otjiseva at Duesternbrook
19

6970	32.573	7071	267.893	7172	24.320	7273	66.513
7475	0.000	7576	105.260	7677	12.065	7778	22.446
7879	12.062	7980	9.525	8081	27.167	8182	24.320
8283	113.469	8384	107.365	8485	62.546	8586	124.089
8687	83.178	8788	467.794	8990	48.983		

22:16   16:54   1250   359   0.00862

NM298401Swakop at Westfalenhof
15

6263	482.027	6364	10.400	6465	120.661	6566	126.195
6667	120.661	6768	12.432	6869	0.000	6970	57.741
7071	192.789	7172	349.990	7273	45.249	7374	289.008
7475	25.939	7576	276.943	7677	50.334		

22:17   16:25   8860   389   0.00315

NM298601Khan at Ameib
17

6768	192.940	6970	4.620	7071	282.044	7172	48.011
7273	0.000	7374	642.389	7475	0.000	7576	157.795
7879	155.802	7980	214.697	8384	88.412	8485	987.412
8586	86.774	8687	4.620	8990	255.776	9091	28.283
9192	20.774						

21:50   15:38   4010   319   0.00450

NM298603Dawib at Dawib
7

7980	64.072	8182	0.000	8687	7.497	8788	50.741
8889	1.450	8990	33.985	9192	0.000		

21:53   15:35   554   238   0.00763

NM299101Kuiseb at Schlesien Weir
28

6263	841.395	6364	0.520	6465	308.973	6566	168.825
6667	456.171	6768	20.583	6869	205.704	6970	122.120
7071	98.683	7273	218.828	7374	396.783	7475	133.934
7576	133.934	7677	267.432	7778	71.488	7879	73.035
7980	8.867	8081	73.035	8182	1.801	8283	0.488
8384	58.138	8485	123.619	8586	151.333	8687	29.079
8788	138.782	8889	175.974	9091	13.230	9192	16.247

23:17   15:48   6520   239   0.00568

Table A16.7 *Continued*

NM299103Kuiseb at Us
11

7778	77.571	7879	41.794	7980	36.189	8081	7.110
8182	17.596	8283	16.556	8384	104.766	8586	333.029
8687	78.419	8788	98.098	8889	72.571		

22:58   16:24   1900   290   0.00794

NM299106Bismarck at Stanco
12

8182	8.236	8283	8.236	8384	62.512	8485	26.968
8586	24.326	8687	40.858	8788	60.322	8889	8.236
8990	114.558	9091	1.403	9192	1.000	9293	114.558

22:44   16:36   276   320   0.00877

NM299107Simmenau at Wasservallei
13

7879	29.357	8182	1.199	8283	10.488	8384	97.523
8485	87.423	8586	81.905	8687	51.584	8788	15.129
8889	43.389	8990	13.876	9091	18.539	9192	22.357
9293	29.974						

22:48   16:32   266   295   0.01471

NM299108Heusis at Heusis
12

7879	56.556	7980	2.212	8081	0.000	8182	12.775
8384	34.542	8485	36.140	8586	3.370	8788	18.416
8990	37.779	9091	0.000	9192	1.289	9293	54.360

22:38   16:39   38.4   310   0.0156

NM299109Westende at Westende
10

7879	0.230	7980	1.557	8081	1.388	8182	0.314
8283	0.183	8384	2.360	8485	10.860	8586	7.535
8687	23.414	8788	0.000				

22:53   16:34   17.3   285   0.00243

NM299110Katros at Tweespruit
10

7980	6.125	8081	0.000	8182	0.772	8283	6.621
8384	0.542	8485	136.580	8586	96.380	8687	6.371
8788	25.415	8889	3.354				

22:56   15:56   81.6   243   0.00348

NM299111Huis at Kos weir
10

7980	0.000	8081	0.000	8182	8.652	8283	0.000
8384	0.806	8485	0.718	8586	4.743	8687	2.155
8788	19.247	8889	6.957				

23:13   16:11   20.1   205   0.01887

NM299203Gaub at Greylingshof
10

8081	0.000	8182	18.248	8283	3.177	8384	43.652
8485	41.612	8586	115.606	8687	99.137	8788	67.971
8990	38.715	9091	131.705				

23:29   15:46   2490   181   0.01163

continued

Table A16.7   *Continued*

NM299302Kuiseb at Gobabeb
12

7879	9.283	7980	0.000	8081	0.000	8182	0.000
8283	0.000	8384	19.697	8485	97.240	8586	70.362
8687	24.885	8889	69.860	8990	59.080	9091	9.904

23:30   14:58   11700   190   0.00431

NM299401Kuiseb at Rooibank
15

7677	0.000	7778	0.000	7879	0.000	7980	0.000
8081	0.000	8182	0.000	8283	0.000	8384	0.000
8485	29.475	8586	9.565	8687	0.295	8788	16.680
8889	12.593	8990	10.842	9091	0.000		

23:11   14:39   14700   159   0.00391

NM302201Tsauchab at Sesriem
10

7980	0.000	8081	13.546	8182	0.000	8283	0.000
8384	151.570	8485	29.363	8586	222.243	8687	75.758
8788	30.697	8889	69.562				

24:31   15:46   1480   130

NM311101Black Nossob at Henopsrus
17

1971	24.737	1972	310.367	1973	416.436	1974	50.000
1975	613.447	1976	76.545	1977	887.478	1978	92.943
1979	0.459	1980	10.000	1981	22.342	1983	416.436
1984	50.000	1986	71.787	1987	85.370	1988	92.637
1989	0.000						

22:09   18:50   4530   398   0.00138

NM311102Black Nossob at Mentz
11

7475	8.988	7677	6.586	7980	0.553	8182	1.043
8384	4.162	8485	5.272	8586	8.988	8687	11.979
8788	33.940	8990	0.000	9192	0.000		

23:07   18:42   8160   362   0.00142

NM311202White Nossob at Amasib
15

7475	0.000	7576	50.177	7677	0.558	7778	100.910
7879	0.319	7980	4.049	8081	8.717	8182	1.456
8384	16.964	8485	2.612	8586	8.366	8788	11.401
8990	0.000	9091	13.569	9192	0.000		

23:05   18:39   9250   362   0.00171

NM312402Auob at Stampriet
10

7980	2.253	8081	0.335	8182	2.399	8283	1.337
8384	13.323	8485	6.939	8586	3.597	8687	47.153
8788	45.976	8889	3.345				

24:19   18:27   19200   249

Table A16.8 *Annual maximum peak discharges (10³ cumecs) and associated flood volumes (10⁹ m³) at Concordia on the Rio Uruguay between 1898 and 1993 (Data courtesy of the Government of Uruguay and Sir William Halcrow and Partners Ltd)*

10.0	10.0	13.2	31.0	17.0	54.0	23.8	65.0	17.8	180.0
09.4	11.0	11.8	32.0	18.4	56.0	21.0	80.0	35.6	106.0
08.1	16.0	12.4	38.0	17.8	60.0	22.0	95.0	28.4	124.0
09.2	11.0	13.4	40.0	15.4	63.0	19.8	93.0	27.6	124.0
10.8	14.0	15.8	40.0	14.4	63.0	20.0	99.0	25.2	140.0
09.4	20.0	14.8	45.0	17.0	69.0	18.8	99.0	27.8	165.0
10.8	20.0	12.0	48.0	20.0	70.0	17.4	110.0	25.0	163.0
10.0	24.0	12.6	50.0	17.4	73.0	16.6	112.0	22.8	171.0
14.2	14.0	14.6	49.0	17.4	75.0	20.2	124.0	23.6	181.0
13.8	19.0	14.0	54.0	14.4	71.0	17.2	118.0	21.8	177.0
12.8	18.0	12.8	56.0	19.2	85.0	15.4	120.0	29.4	197.0
15.4	25.0	20.4	35.0	18.2	83.0	16.0	124.0	26.2	221.0
13.6	28.0	18.4	33.0	10.4	81.0	14.4	154.0	22.4	214.0
12.6	27.0	17.6	35.0	13.4	88.0	23.2	93.0	18.8	196.0
08.4	29.0	17.2	37.0	10.6	92.0	23.0	116.0	28.8	248.0
09.4	29.0	17.2	42.0	15.2	92.0	21.6	123.0		
10.8	30.0	18.0	41.0	17.4	100.0	20.4	140.0		
14.4	31.0	17.6	51.0	14.4	89.0	20.4	152.0		
12.4	29.0	19.7	53.0	24.0	56.0	22.4	163.0		

Table A16.9 *Trips per occupied dwelling unit day, average car ownership, average household size, socio-economic index, urbanization index, for 57 traffic analysis zones. Data from the Transportation Centre at Northwestern University, Chicago*

3.18	0.59	3.26	28.32	60.10	5.10	0.75	3.38	43.67	56.64
3.89	0.57	3.13	20.89	65.71	4.70	0.83	3.11	52.74	54.02
3.98	0.61	3.02	25.99	63.19	5.17	0.76	3.20	52.29	58.35
4.16	0.61	3.14	28.52	66.24	5.41	0.87	3.24	43.42	47.78
3.60	0.63	3.75	27.18	58.36	6.46	1.16	3.60	45.94	51.21
4.10	0.66	3.24	27.95	59.58	6.03	0.90	3.02	61.53	54.92
4.36	0.71	2.77	39.91	64.64	4.79	0.53	3.09	49.37	58.63
4.87	0.77	2.74	48.36	67.88	4.83	0.75	2.46	87.38	65.67
5.85	0.84	3.02	42.15	56.86	6.30	0.78	3.36	55.85	59.00
4.97	0.74	2.84	38.14	62.44	4.94	0.69	2.94	50.15	61.09
3.54	0.67	2.93	51.30	68.67	6.01	0.96	3.27	67.01	48.39
4.31	0.64	3.87	43.90	59.49	6.39	0.86	3.32	62.18	50.04
4.54	0.73	3.16	30.27	57.76	5.82	1.09	3.29	45.58	46.47
4.82	0.86	3.42	32.18	63.06	6.25	1.15	3.58	60.85	26.36
4.04	0.66	3.54	34.45	47.73	6.13	0.90	3.09	55.59	43.58
4.60	0.64	3.49	43.32	59.36	6.70	1.02	3.02	75.73	35.89
3.40	0.50	2.76	75.32	75.81	7.10	1.00	3.33	57.84	28.28
4.65	0.58	2.91	62.20	75.26	7.89	1.32	3.58	79.69	25.37
3.02	0.53	1.83	82.53	83.66	7.80	1.06	3.17	57.01	31.97
9.14	1.11	3.00	67.31	38.21	8.02	1.02	3.35	50.93	38.17
4.30	0.70	2.94	64.01	55.51	7.20	0.98	3.43	49.75	34.69
4.24	0.80	3.19	51.16	52.44	5.14	0.82	3.31	36.36	46.98
5.00	0.77	2.61	59.15	59.38	5.56	0.94	3.21	62.27	36.27
5.93	0.96	3.24	48.51	46.51	5.74	0.90	3.52	42.64	26.15
5.11	0.86	2.95	47.44	51.17	6.77	0.62	3.92	21.66	24.08
5.84	0.92	2.95	57.34	58.60	4.94	0.77	3.02	49.18	51.39
4.70	0.80	3.00	62.60	62.40	7.64	0.93	3.37	34.74	44.54
4.54	0.79	2.71	73.00	67.23	7.25	0.75	4.50	26.21	44.80
5.51	0.91	3.46	33.96	41.29					

Table A16.10   *Monthly effective inflows (m³ s) to the Font Reservoir, Northumberland, for the period from January 1909 until December 1980. Rows are years and columns are months. [Data courtesy of Northumbrian Water Plc]*

0.423	0.524	1.375	0.694	0.352	0.281	0.100	0.127	0.144	0.576	0.247	0.949
0.503	1.241	0.313	0.226	0.171	0.034	0.260	0.385	0.089	0.242	0.853	0.846
0.710	0.370	0.674	0.278	0.080	0.596	0.334	0.453	0.030	0.458	0.932	1.070
1.129	1.087	0.579	0.147	0.097	0.929	0.337	0.911	0.400	0.923	0.602	0.822
1.307	0.504	0.784	0.544	0.662	0.110	0.077	0.041	0.089	0.334	0.278	0.157
0.822	0.602	0.721	0.150	0.092	0.293	0.077	0.050	0.046	0.174	1.402	2.158
1.200	1.617	0.932	0.232	0.145	0.095	0.148	0.186	0.116	0.171	0.330	2.102
1.132	1.002	2.484	0.474	0.494	0.247	0.464	0.154	0.296	0.855	1.326	1.094
0.382	1.067	0.597	1.079	0.198	0.049	0.038	0.846	0.223	0.260	0.281	0.568
0.816	0.439	0.287	0.223	0.293	0.266	0.077	0.047	0.522	0.509	0.281	0.494
1.026	0.674	0.627	0.355	0.242	0.027	0.041	0.062	0.055	0.408	0.529	1.141
0.852	0.534	0.497	0.941	0.532	0.070	0.118	0.171	0.046	0.293	0.263	1.070
1.313	0.304	0.145	0.058	0.041	0.024	0.021	0.390	0.278	0.266	0.571	0.316
0.970	0.818	0.423	0.358	0.097	0.247	0.198	0.118	0.675	0.242	0.214	0.822
0.405	1.165	0.526	0.394	0.207	0.058	0.269	0.284	0.229	0.322	0.638	0.763
0.911	0.298	0.180	0.290	0.594	0.562	0.597	0.213	0.370	0.627	0.376	0.568
0.491	1.149	0.585	0.483	0.627	0.070	0.044	0.311	0.391	0.343	0.422	1.129
0.878	0.989	0.311	0.183	0.237	0.541	0.573	0.272	0.287	0.585	1.204	0.231
0.473	0.255	0.665	0.370	0.204	0.284	0.597	0.582	0.715	0.520	0.794	0.656
1.215	0.530	1.807	0.165	0.083	0.574	0.115	0.163	0.076	0.535	0.910	0.464
0.769	0.377	0.180	0.156	0.133	0.027	0.139	0.281	0.034	0.183	0.736	1.183
0.739	0.949	0.609	0.412	0.154	0.040	0.260	0.529	0.587	0.358	0.969	0.671
0.473	0.910	0.340	0.715	0.269	0.794	0.385	0.597	0.263	0.121	1.195	0.408
0.576	0.160	0.609	0.596	0.609	0.092	0.124	0.080	0.217	1.262	0.412	0.665
0.352	0.759	1.005	0.159	0.278	0.058	0.056	0.047	0.061	0.316	0.770	0.311
0.656	0.180	1.594	1.564	0.385	0.144	0.204	0.447	0.150	0.866	1.711	1.493
0.843	1.121	0.875	0.831	0.148	0.312	0.059	0.071	0.299	0.988	1.317	0.902
1.614	1.061	1.496	0.614	0.148	0.400	0.151	0.168	0.468	0.213	1.195	0.547
0.680	1.758	1.546	1.002	0.571	0.110	0.106	0.109	0.449	0.455	0.547	1.724
0.988	0.409	0.299	0.076	0.148	0.553	0.174	0.100	0.137	1.064	1.063	1.417
2.064	0.599	1.138	0.220	0.189	0.079	0.201	0.290	0.199	1.555	1.020	0.644
0.177	1.152	1.058	0.394	0.139	0.061	0.724	0.035	0.076	0.322	0.938	0.503
1.047	1.227	1.431	0.516	0.171	0.040	0.035	0.047	0.030	0.266	0.914	0.201
0.450	0.517	1.023	0.205	0.062	0.067	0.100	0.198	0.281	0.597	0.171	0.432
1.029	0.488	0.077	0.083	0.544	0.208	0.189	0.529	0.299	0.408	0.651	0.361
0.313	0.386	0.251	0.321	0.248	0.101	0.287	0.047	1.228	0.680	1.326	0.603
0.988	1.149	0.038	0.177	0.435	0.214	0.059	0.044	0.137	0.644	0.226	0.385
0.263	0.265	0.461	0.043	0.015	0.159	0.071	0.760	0.856	0.092	1.142	0.541
0.503	0.154	1.946	0.950	0.393	0.107	0.417	0.018	0.273	0.021	0.272	0.479
2.170	0.544	0.086	0.196	0.142	0.140	0.027	0.718	0.498	0.272	1.195	0.367
0.340	0.350	0.225	0.110	0.032	0.043	0.047	0.032	0.046	0.488	0.718	0.943
0.387	1.110	0.234	0.211	0.154	0.037	0.180	0.263	0.617	0.382	1.097	0.453
0.872	0.681	0.698	0.489	0.565	0.083	0.032	0.245	0.046	0.056	1.714	0.435
0.470	0.419	0.364	0.174	0.062	0.058	0.041	0.260	0.132	0.893	0.627	0.822
0.573	0.508	0.071	0.336	0.293	0.351	0.080	0.068	0.516	0.148	0.498	0.281
0.550	0.841	0.420	0.104	0.618	0.098	0.142	0.899	0.498	1.233	0.794	0.603
0.680	0.301	1.005	0.125	0.171	0.024	0.166	0.189	0.021	0.044	0.113	0.804
1.177	1.520	0.331	0.306	0.121	0.810	0.831	2.853	0.889	0.517	0.330	0.843
0.489	0.745	0.499	0.100	0.091	0.029	0.119	0.578	0.333	0.328	0.392	0.297
0.200	0.632	0.588	0.461	0.268	0.552	0.540	0.455	0.354	0.490	0.187	0.774
0.573	0.296	0.095	0.264	0.053	0.062	0.185	0.007	0.303	0.040	1.023	0.753
0.682	0.633	0.496	0.190	0.167	0.058	0.184	0.108	0.152	0.574	0.438	0.375
1.018	0.619	0.146	0.426	0.200	0.061	0.251	0.203	0.145	0.560	0.519	0.547
0.783	0.259	0.481	0.495	0.413	0.056	0.102	0.501	0.665	0.206	0.781	0.722
0.155	0.127	1.502	0.637	0.088	0.247	0.207	0.685	0.414	0.158	1.454	0.690
0.529	0.417	0.954	0.502	0.080	0.236	0.042	0.141	0.090	0.083	0.200	0.727

Table A16.10 *Continued*

0.258	0.446	0.650	0.135	0.103	0.066	0.317	0.124	0.214	0.437	0.491	0.699
0.494	0.801	0.269	0.964	0.373	0.193	0.037	0.570	0.321	0.701	0.582	1.175
0.603	0.470	0.267	0.285	0.717	0.103	0.165	0.526	0.266	0.771	0.584	0.548
0.345	0.494	0.475	0.446	0.389	0.132	0.429	0.232	1.028	0.441	0.587	0.465
0.862	0.347	0.436	0.330	0.996	0.451	0.097	0.109	0.263	0.076	0.937	1.096
1.496	1.116	0.699	0.808	0.094	0.056	0.146	0.203	0.029	0.301	0.637	0.582
0.886	0.316	0.670	0.423	0.282	0.470	0.230	0.520	0.087	0.097	0.535	0.183
1.351	1.284	0.586	0.433	0.234	0.272	0.151	0.146	0.066	0.051	0.160	0.347
0.388	0.161	0.123	0.153	0.413	0.079	0.199	0.220	0.075	0.289	0.098	0.588
0.453	0.633	0.519	0.167	0.094	0.061	0.088	0.082	0.267	0.473	1.048	0.755
0.787	0.432	0.519	0.464	0.464	0.176	0.059	0.118	0.211	0.155	0.193	0.285
0.332	0.344	0.845	0.380	0.290	0.116	0.043	0.040	0.531	1.488	0.355	0.618
1.190	1.416	0.491	0.418	0.290	0.550	0.038	0.055	0.061	0.167	0.463	0.698
0.889	1.240	0.713	0.559	0.510	0.049	0.064	0.318	0.130	0.087	0.296	1.600
1.038	0.816	1.830	0.726	0.621	0.142	0.036	0.139	0.075	0.359	0.787	0.753
0.539	0.938	1.055	0.250	0.068	0.567	0.245	0.250	0.116	0.470	0.906	0.855

Table A16.11 *Variables measured on clay samples from a site in Northumberland, UK*
Dataset:
Grid index number, Easting, Northing, Cohesion (10 kPa), Depth (m), Slope (Deg)
Note that the maximum measurable depth was 1.2 m , therefore some of the 1.2 m values represent depths greater than this, these are indicated with stars

1	247691	722795	4.1	0.85	22.33
2	247668	722800	5.4	0.55	23.67
3	247646	722805	5.7	1.05	25
4	247623	722810	4.1	0.8	24.17
5	247601	722815	5.5	0.7	26.83
6	247578	722820	4.1	0.85	24.33
7	247556	722825	3.3	1.2	30.17
8	247533	722830	3.8	0.8	25.5
9	247511	722835	5.1	1.2*	23.33
10	247488	722840	5.5	1.2*	17.33
11	247694	722805	3.8	0.9	24
12	247672	722810	4.8	0.45	22.67
13	247649	722815	6.4	0.5	24.5
14	247627	722820	4.9	0.7	23.5
15	247604	722825	4.5	0.4	25
16	247582	722830	4.6	0.5	25
17	247559	722835	4.8	0.45	29.83
18	247537	722840	4.2	0.85	26.83
19	247514	722845	4.4	1.2*	23.83
20	247492	722850	6	1.2*	18
21	247698	722814	4.9	0.75	23.67

continued

Table A16.11 *Continued*

22	247675	722819	4.5	0.55	24
23	247653	722824	5.8	0.95	24
24	247630	722829	4.2	0.75	24.5
25	247608	722834	4	0.9	25.17
26	247585	722839	5.1	0.6	24.5
27	247563	722844	5.8	0.85	28.67
28	247540	722849	3.8	0.85	26.5
29	247518	722854	5.8	0.8	23
30	247495	722859	5	1.2*	18.83
31	247701	722824	7.9	0.6	27.5
32	247679	722829	5.7	0.6	28.17
33	247656	722834	5.3	0.8	23.83
34	247634	722839	3.9	0.75	21.83
35	247611	722844	4.5	1.1	24.5
36	247589	722849	4.6	0.75	24.17
37	247566	722854	3.5	1.05	28
38	247544	722859	4.1	0.8	25.83
39	247521	722864	5.7	0.95	23.5
40	247499	722869	4.1	1.2*	19.17
41	247705	722833	3.8	0.75	26.83
42	247682	722838	4.7	0.7	25
43	247660	722843	5.3	0.55	22.17
44	247637	722848	4.7	0.9	23.17
45	247615	722853	3.2	0.35	25.5
46	247592	722858	7	0.45	29.17
47	247570	722863	5.8	0.8	25
48	247547	722868	5.1	0.35	25.17
49	247525	722873	2.6	0.85	21.17
50	247502	722878	3.7	0.6	16
51	247708	722843	5.5	0.35	27
52	247686	722848	5.3	0.9	26
53	247663	722853	6	0.75	22.5
54	247641	722858	3.7	0.95	22
55	247618	722863	2.8	0.5	26.33
56	247596	722868	4.2	0.75	28
57	247573	722873	6.3	0.45	26.17
58	247551	722878	4.1	0.4	23.67
59	247528	722883	2.8	1.2*	22.83
60	247506	722888	3.2	0.9	13.17
61	247712	722852	3.8	0.5	26.67
62	247689	722857	7.5	0.55	25.5
63	247667	722862	7.1	0.6	19.67
64	247644	722867	4.2	1.1	21.5
65	247622	722872	5	0.7	27.67
66	247599	722877	4.9	0.8	28.83
67	247577	722882	7.1	1.2	27.83
68	247554	722887	5	0.65	22
69	247532	722892	2.3	0.65	23.67
70	247509	722897	2.8	0.85	12.5
71	247715	722862	6.2	0.65	27.5
72	247693	722867	3.7	0.7	25.17
73	247670	722872	3.9	0.75	20.5
74	247648	722877	5.6	0.75	22.17
75	247625	722882	6.4	0.4	28.83
76	247603	722887	4	0.55	26.17
77	247580	722892	3.1	0.65	27.67
78	247558	722897	6.9	0.5	24
79	247535	722902	6.6	0.5	23.33

Table A16.11 *Continued*

80	247513	722907	6.7	0.55	9.5
81	247719	722871	6.7	0.55	27.67
82	247696	722876	4	0.6	26
83	247674	722881	4.3	1.05	22.83
84	247651	722886	3	0.8	22.33
85	247629	722891	3.8	0.3	33.67
86	247606	722896	5.2	0.7	22.83
87	247584	722901	4.8	0.55	26.67
88	247561	722906	3.1	0.55	24.5
89	247539	722911	5	0.65	25
90	247516	722916	8.1	0.4	15.33
91	247722	722881	6.7	0.6	26.83
92	247700	722886	4.5	0.6	24.17
93	247677	722891	5.6	0.7	22.5
94	247655	722896	3.7	0.8	24
95	247632	722901	4.5	0.5	32
96	247610	722906	6.8	0.35	25.67
97	247587	722911	6	0.45	26.83
98	247565	722916	4.7	0.75	25.17
99	247542	722921	4.1	0.8	24.5
100	247520	722926	5.2	0.6	16.5
101	247726	722890	5.3	0.6	29.83
102	247703	722895	5.4	0.6	22.5
103	247681	722900	4.4	0.6	22
104	247658	722905	3.9	0.65	27.17
105	247636	722910	4.3	0.65	29.83
106	247613	722915	4.9	0.4	28
107	247591	722920	3.3	0.6	27.83
108	247568	722925	6.9	0.4	27.5
109	247546	722930	3.9	0.75	24.5
110	247523	722935	2.9	0.85	15.5
111	247729	722900	5.5	0.8	30
112	247707	722905	3.6	0.65	23
113	247684	722910	2.5	0.85	22.83
114	247662	722915	2.8	1.1	26
115	247639	722920	3.6	0.65	29.5
116	247617	722925	5.9	0.25	28.5
117	247594	722930	3.9	0.4	28.67
118	247572	722935	3.7	0.5	26.83
119	247549	722940	1.7	1.2*	22.17
120	247527	722945	4.7	0.85	15.67
121	247733	722909	6.1	0.7	29.67
122	247710	722914	4.7	0.45	23.5
123	247688	722919	2.9	0.7	20.83
124	247665	722924	5.6	0.5	26
125	247643	722929	7.4	0.7	31.17
126	247620	722934	4.8	0.35	31.67
127	247598	722939	6	0.4	27
128	247575	722944	6.2	0.45	25.67
129	247553	722949	4.9	1	22.67
130	247530	722954	7.1	0.7	15.83
131	247736	722919	5.3	0.6	30.17
132	247714	722924	4.6	0.6	23.17
133	247691	722929	3.3	0.55	21.17
134	247669	722934	5.9	0.5	26.33
135	247646	722939	5.4	0.65	25.17

continued

Table A16.11   *Continued*

136	247624	722944	5.2	0.35	31
137	247601	722949	4.3	0.35	25.33
138	247579	722954	4	0.35	25.17
139	247556	722959	5.3	0.55	21.83
140	247534	722964	6.8	0.55	14.5
141	247740	722928	6.7	0.45	29.67
142	247717	722933	6.1	0.5	21
143	247695	722938	3.7	0.85	20.33
144	247672	722943	3.4	0.65	19.5
145	247650	722948	6.9	0.35	28.5
146	247627	722953	4	0.65	32.67
147	247605	722958	5.6	0.65	24.5
148	247582	722963	4	0.5	22.67
149	247560	722968	4.2	0.65	18.33
150	247537	722973	2.9	1.1	15.5
151	247743	722938	4.6	0.35	28.17
152	247721	722943	4.9	0.55	21.67
153	247698	722948	5.2	0.75	20.83
154	247676	722953	3.6	0.8	21.5
155	247653	722958	5.5	0.65	27.83
156	247631	722963	6.2	0.4	31
157	247608	722968	5.6	0.7	24.33
158	247586	722973	3	0.85	22.17
159	247563	722978	4.9	0.7	19
160	247541	722983	4.1	1.05	17.33
161	247747	722947	8.3	0.4	27.67
162	247724	722952	7.9	0.75	21
163	247702	722957	8	0.75	21
164	247679	722962	4.6	0.5	24.17
165	247657	722967	6.1	0.45	23.5
166	247634	722972	4.9	0.65	31.17
167	247612	722977	5.4	0.55	26.83
168	247589	722982	1.9	0.75	22.67
169	247567	722987	3	0.75	19.17
170	247544	722992	3.6	0.55	23
171	247750	722957	5.4	0.25	27.5
172	247728	722962	4.6	0.75	21.17
173	247705	722967	3.6	1	24
174	247683	722972	7.2	0.45	22
175	247660	722977	3.2	0.55	23.83
176	247638	722982	5.7	0.45	28.67
177	247615	722987	6	0.5	26.5
178	247593	722992	5	0.7	24.17
179	247570	722997	6.2	0.65	22.67
180	247548	723002	5.2	0.95	20.83
181	247754	722966	4.5	0.65	24.5
182	247731	722971	3.2	0.55	20.83
183	247709	722976	4.8	1.05	24.5
184	247686	722981	2.8	0.7	19.83
185	247664	722986	4.8	0.85	21.17
186	247641	722991	6.9	0.75	25.83
187	247619	722996	6.6	0.45	27.5
188	247596	723001	5.5	0.4	26.33
189	247574	723006	4.4	0.55	25.83
190	247551	723011	3.6	0.75	23.17
191	247757	722976	3.4	0.5	30.83
192	247735	722981	2.8	0.85	22
193	247712	722986	3.5	0.55	24.33

Table A16.11 *Continued*

194	247690	722991	3	1	19.5
195	247667	722996	4	1.05	19
196	247645	723001	3.9	1.2*	25.5
197	247622	723006	4	0.5	28.17
198	247600	723011	5.6	0.55	28
199	247577	723016	5.7	0.65	24.67
200	247555	723021	5	0.9	23.33
201	247761	722985	4.6	0.3	32.5
202	247738	722990	2.7	0.6	22.67
203	247716	722995	3.8	0.9	21.5
204	247693	723000	3	1.15	19
205	247671	723005	4.5	0.75	18.83
206	247648	723010	5	0.65	25
207	247626	723015	4.9	0.2	28.83
208	247603	723020	5.9	0.6	28.33
209	247581	723025	5.9	0.7	23.67
210	247558	723030	4.7	0.8	23.83
211	247764	722995	4.4	0.3	26.83
212	247742	723000	6.7	0.55	21.17
213	247719	723005	4.2	0.4	22.5
214	247697	723010	4.8	0.9	19.83
215	247674	723015	5.4	0.85	19.17
216	247652	723020	1.9	1.2*	23.33
217	247629	723025	7.5	0.8	25.33
218	247607	723030	4.3	0.4	28.17
219	247584	723035	6.2	0.65	25.17
220	247562	723040	6.9	0.95	23.83
221	247768	723004	7.2	0.65	24
222	247745	723009	9.4	0.65	21.33
223	247723	723014	7.2	0.35	24
224	247700	723019	7.6	1.05	22.33
225	247678	723024	5.2	0.9	21.5
226	247655	723029	2.5	1.2*	22.17
227	247633	723034	5.6	0.55	24.5
228	247610	723039	4.9	0.6	29
229	247588	723044	7.5	0.65	27
230	247565	723049	2.9	1.05	24.17
231	247771	723014	6.4	0.9	25
232	247749	723019	6.5	0.7	19.5
233	247726	723024	6.4	0.35	23.83
234	247704	723029	5.7	1.2*	23
235	247681	723034	9.2	0.4	19
236	247659	723039	5.7	0.7	23.17
237	247636	723044	4.8	0.45	32
238	247614	723049	5.6	0.35	28.67
239	247591	723054	7.6	0.55	25.5
240	247569	723059	3.2	1.2*	22.33
241	247775	723023	5.8	1.2	26.33
242	247752	723028	5.8	0.65	18.83
243	247730	723033	5.5	0.3	25
244	247707	723038	5	0.95	23.17
245	247685	723043	6.5	0.65	18.17
246	247662	723048	8.6	0.65	22.5
247	247640	723053	6.7	0.4	32.33
248	247617	723058	6.6	0.4	28.5
249	247595	723063	7.7	0.5	25.33
250	247572	723068	4.4	0.95	22

Table A16.12   *Wave heights (in millimetres relative to still water level) from a wave tank. Sampled at 0.1 s intervals over 39.7 s*

367	−439	−258	116	326	76	−41	−34	253	235
407	−121	−400	438	−71	−29	−184	260	511	172
−255	367	−162	235	−306	113	−331	178	30	−161
−515	478	−196	−120	−150	64	76	164	−249	−186
−500	4	29	−209	17	32	276	322	−262	10
−342	−315	721	36	−112	296	−111	−293	−56	283
−188	−3	118	−132	−296	−201	−171	−522	46	58
77	194	−356	−201	−238	−513	−53	−468	115	−152
494	−45	−340	192	190	−339	226	28	183	−160
737	50	1	310	578	167	4	425	2	−131
375	136	166	−116	268	167	−114	365	15	−52
−221	−42	138	−282	−33	255	171	41	−85	38
−313	−296	93	−172	12	271	354	−114	−234	116
−301	−394	93	−41	−148	122	−133	6	−245	124
−311	−145	−148	204	−132	−500	−345	−117	72	−48
−109	209	−326	34	−163	−416	−218	−134	441	−19
−10	536	−95	−12	−389	129	−307	−173	282	276
150	116	279	276	−81	195	−270	−19	−92	−29
178	136	−10	167	328	9	93	180	−237	−331
47	−167	−99	−328	19	105	574	239	−222	−382
47	−244	96	−260	−32	343	622	−56	−37	−158
767	18	227	16^	268	−136	45	93	90	162
−74	−33	18	−62	380	−522	−162	−94	−20	316
−594	−204	61	−204	−26	−196	−51	−340	−78	375
−541	−20	−125	−107	−245	480	12	−362	−56	119
−133	120	−460	172	−309	159	−312	189	139	−174
148	−89	−337	469	−289	−77	−517	756	530	−120
116	−83	59	50	30	−54	−265	600	−68	−173
169	176	211	−96	125	−107	−33	−120	−437	−41
417	160	73	−62	170	−102	474	−622	−351	−50
295	166	95	−40	141	−14	529	−752	94	−251
−317	94	119	342	−7	324	−97	−526	351	−262
−472	−65	−1	−205	−114	137	−160	115	124	154
−266	−311	−123	−488	−29	−312	127	875	0	391
−12	−430	−56	−363	134	−236	191	607	46	266
314	−398	−26	138	104	−28	−281	−44	48	231
241	199	92	323	−65	239	−315	−349	−366	177
375	659	125	189	−365	176	−20	−478	−429	
59	488	−362	120	−128	−24	24	−321	−78	
−550	22	−324	64	221	29	−120	30	396	

Glossary

Addition rule of probability

The probability of one or the other or both of two events occurring is: the sum of their individual probabilities of occurrence less the probability they both occur.

Aliases (design of experiments)

If there are a large number of factors to be investigated in an experimental programme a full factorial design may not be feasible. If a fraction of the full factorial is run, there will be some ambiguity over the cause of significant results. For example, an increase in strength might be due to the main effect of A or the interaction BDC or both. A and BDC are termed aliases. In practice, high order interactions are often assumed to be negligible.

Aliasing (spectral analysis)

The apparent contribution to the variance of a signal at a certain frequency could be due to higher frequencies, which have any whole number of cycles between sampling intervals. The sampling interval must be shorter than half the wavelength of the highest frequency component in the signal. Spectral analysers remove any very high frequencies, above about 1 MHz, with analogue filters before digitizing (sampling) the signal. This removes the possibility of aliasing.

Analysis of variance (ANOVA)

The sum of squares, about the mean, for some variable is resolved into contributions which can be attributed to different sources. These sources can be different levels of variability (see components of variance), or predictor variables in regression, or treatments in a designed experiment.

Asset management plan (AMP)

A business plan for water companies. It includes estimates of the costs of maintaining, and where necessary improving, the water supply and sewerage systems over the following 20 years. UK companies have to produce an AMP every five years, and submit it to an independent government body.

Asymptotic result

A result which is proved by some limiting process, in which the sample size tends to infinity. Many such results still give reasonable approximations with small sample sizes. The Central Limit Theorem is a well known example.

Auto-covariance function (acvf), auto-correlation function (acf)

The acvf is the covariance between two variables as a function of the time lag which separates them. The acf is the corresponding correlation.

Bayes' Theorem

Bayes' Theorem is used to update our knowledge, expressed in probabilistic terms, as we obtain new data. The posterior distribution is proportional to the product of the likelihood and the prior distribution.

Bias

A systematic difference between the estimator and the parameter being estimated, i.e. the difference will persist when averaged over imaginary replicates of the sampling procedure. Formally, the difference between the mean of the sampling distribution and the parameter being estimated. If the bias is small by comparison with the standard deviation of the sampling distribution the estimator may still be useful. For example, s^2 is unbiased for σ^2 but s is slightly biased for σ.

Binomial distribution

The distribution of the number of successes in a set number of trials, with a constant probability of success.

Class intervals

Before drawing a histogram the data are grouped into classes which correspond to convenient divisions of the variable range. Each division is defined by its lower and upper limits, and the difference between them is the length of the class interval.

Coefficient of variation

The ratio of the standard deviation to the mean.

Conditional probability

The probability of an event conditional on other events having occurred or an assumption they will occur. (All probabilities are conditional on the general context of the problem.)

Confidence interval

A 95%, or whatever, (frequentist) confidence interval for some parameter is an interval constructed in such a way that, on average, if you imagine millions of random samples of the same size, 95% of them will include the parameter. From a practical point of view we think there is a 95% chance the interval contains the parameter, and Bayesian confidence intervals are properly interpreted in this way.

Correlation

A dimensionless measure of linear association between two variables, that lies between -1 and 1. Zero represents no association and negative values correspond to one variable increasing as the other decreases.

Covariance

A measure of linear association between two variables, which equals the average value of the mean corrected products.

Cumulative distribution function (cdf)

A function that gives the probability a continuous variable is less than any value. It is the population analogue of the cumulative frequency polygon. Its derivative is the pdf.

Cumulative frequency polygon

A plot of the proportion, often expressed as a percentage, of data less than or equal to any value.

Degrees of freedom

The degrees of freedom are the number of data used to calculate some statistic, less the number of physically independent parameters estimated from the data and used in the calculation. Each such estimate imposes a constraint. For example, if we estimate a population variance from a sample of n data we first estimate the population mean. Deviations from the sample mean are constrained to equal zero. Therefore we have $(n-1)$ degrees of freedom for the sample variance. In multiple regression the number of constraints equals the number of coefficients to be estimated, which is one more than the number of explanatory variables (k). So, the estimate of the variance of the errors, based on the residuals, has $(n-k-1)$ degrees of freedom.

Discrete white noise (DWN)

A sequence of independent random variables.

Ensemble

An imaginary population of all possible time series that might be generated by the underlying random process.

Ergodic

A random process is ergodic in the mean, for example, if the time average from a single time series tends to the ensemble average (expected value) as the length of the time series increases.

Expected value

An average value in the population.

Explanatory variable

In a multiple regression the dependent variable, usually denoted by y, is expressed as a linear combination of the explanatory variables. In designed experiments it is helpful to subdivide the explanatory variables into control variables, whose values are chosen by the experimenter, and concomitant variables which can be monitored but not preset. Alternative names for the explanatory variables include predictor variables. The dependent variable is often called the response, especially in an experimental context.

Exponential distribution

A continuous distribution of the times between occurrences, when occurrences are random and independent.

Extreme value distributions

Distributions that provide plausible models for the maximum value from repeated samples of the same notional size. For example, annual maximum wind speeds.

Factorial experiment

An experiment designed to examine the effects of two or more factors. Each factor is applied at two, or more, levels and all combinations of these factor levels are tried in a full factorial. This allows interactions to be investigated, and is also an efficient means of estimating the main effects. If the number of runs for the full factorial is prohibitive, fractional factorial designs, which retain the benefits, can be used.

Fitted value

The predicted value of y corresponding to the x values for a datum used in fitting the model. The residual is the difference between the observed value of y and the fitted value of y.

Frequency

The number of times an event occurs. Also, when talking about vibration, the number of cycles per second (Herz) or radians per second.

Gamma distibution

The distribution of the time until the cth occurrence in a Poisson process has a gamma distribution. It generalizes to non-integer c, and is often used in applications which are unrelated to Poisson processes because it gives a good empirical fit to data.

Gamma function

A generalization of the factorial function to values other than positive integers.

Gaussian

A Gaussian distribution is an alternative name for the normal distribution.

Histogram

A chart consisting of rectangles drawn above class intervals with areas equal to the proportion of data in that interval. It follows that the heights of the rectangles equal the relative frequency density, and the total area equals one. If all the class intervals are of the same length the heights are proportional to the frequencies.

Imaginary infinite population

The population we are sampling from is often imaginary and arbitrarily large. A sample from a production line is thought of as a sample from the population of all items that will be produced if the process continues on its present settings. An estimate is thought of as a single value from the imaginary distribution of all possible estimates, so that we can give its precision.

Independent

Two events are independent if the probability one occurs does not depend on whether the other does.

Interaction

Two explanatory variables interact if the effect of one depends on the value of the other. Their product is then included as an explanatory variable in the regression. If this interaction depends on the value of some third variable a three-variable interaction exists, and so on.

Lag

The time between two variables.

Likelihood

The probability of obtaining the observed data, considered as a function of the unknown parameters.

Marginal distribution

The marginal distribution of a variable is the distribution of that variable. The 'marginal' refers to the fact that multivariate data are available, or being modelled, but information on the other variables has been ignored.

Markov property

'The future is independent of the past, given the present' (Guttorp, 1995).

Maximum likelihood estimation

A method of estimating parameters as those which maximize the likelihood function.

Meal

A mixture of powders used as raw material for a chemical process.

Mean

The sum of several quantities divided by their number. Also used as an alternative to 'average' in 'average value of ... '.

Mean corrected

Data are mean corrected if their mean is subtracted from them. Mean corrected data therefore have an average value of 0.

Median

The middle value if data put into ascending order.

Method of moments estimation

A method of estimating parameters by equating sample moments to the corresponding population moments which are expressed in terms of the unknown parameters.

Mode

The most commonly occurring value. Also the value of the variable at which the pdf has its maximum.

Multiple regression

Some variable, called the dependent variable or the response, is expressed as a linear combination of predictor variables plus random error. The coefficients of the variables in this combination are the unknown parameters of the model and are estimated from the data. The predictor variables can be non-linear functions of each other, for example x_1, x_2, $x_1 \times x_2$, $x_2 \times x_2$, and $x_1 \times x_2$ represents a quadratic surface. Other names for predictor variables include 'explanatory' variables.

Multiplicative rule

The probability of two events both occurring is the product of the probability that one occurs with the probability the other occurs conditional on the first occurring.

Mutually exclusive

Two events are mutually exclusive if they cannot occur together. A set of events is mutually exclusive and exhaustive if exactly one must occur.

Normal distibution

A bell-shaped pdf which is a plausible model for random variation if it can be thought of as the sum of a large number of smaller components. It is also important as a sampling distribution, especially of the sample mean.

Null hypothesis

A supposition that is set up to be tested. If there is no evidence against the null hypothesis we will act as though it is true, i.e. accept the hypothesis. If there is evidence against the hypothesis, quantified by a small P-value, we claim evidence to reject the null hypothesis at the $100 \times P$-value % level.

or

In probability, 'A or B' is conventionally taken to include both.

Order statistic

The ith order statistic is the ith largest if the sample of n is sorted into ascending order.

Orthogonal

In a designed experiment the values of the control variables are usually chosen to be uncorrelated, when possible, or nearly so. If they are uncorrelated they are said to be orthogonal. (If the values are put in columns and thought of as 'vectors', the vectors are orthogonal.)

P-value

The probability of an event as extreme, or more extreme, than that observed conditional on the null hypothesis being true.

Parameter

A constant which is a salient feature of a population, its value is usually unknown.

Paver

A paving block. Modern ones are made from concrete in a variety of shapes and colours. Also called paviors.

Percentage point

The upper α percentage point of a pdf is the value beyond which a proportion α of the area under the pdf lies, a lower point is defined similarly.

Percentiles

The lower x percentile is the value of the variable below which $x\%$ of the data lie.

Point process

A random system consisting of point events occurring in space or time. These can be developed into rainfall models, for example, by superimposing pulses at the points.

Poisson distribution

The number of occurrences in some length of continuum if occurrences are independent.

Poisson process

Occurrences in some continuum, often time, form a Poisson process if they are random, independent, and occur at some constant average rate (in the standard case at least). The time from now until the next occurrence is independent of past occurrences.

Probability density function (pdf)

A curve such that the area under it between any two values represents the probability that a continuous variable will be between these values. The population analogue of a histogram.

Probability function

A formula that gives the probabilities that a discrete variable takes any of its possible values. The population analogue of a line diagram.

Precision

The precision of an estimator is a measure of how close replicate estimates are to each other. Formally, the reciprocal of the variance of the sampling distribution.

Priority controlled junction

A road junction which is controlled by 'Give Way' signs and road markings, rather than by traffic lights.

Probability

A measure of how likely some event is to occur on a scale ranging from 0, representing impossibility, to 1, representing certainty. It can be thought of as the long run proportion of times the event would occur if the scenario were to be repeated.

Pseudo-random numbers

A sequence of numbers, generated by a deterministic algorithm in which the numbers appear to be random; the numbers are indistinguishable from genuine random numbers by empirical tests. Computer generated random numbers are actually pseudo-random.

Quartiles

The upper (lower) quartile (UQ, LQ) is the datum above (below) which one quarter of the data lie.

Random digits

A sequence in which each one of the digits from 0 up to 9 is equally likely to occur next.

Random numbers

A sequence of numbers from a specified distribution, such that the next is independent of the existing sequence.

Random sample

A sample which has been selected so that every member of the population has a known, non-zero, probability of appearing.

Range

Difference between the largest datum and the smallest datum when the data are put into ascending order.

Relative frequency

The ratio of the frequency of some event occurring to the number of scenarios in which it could potentially have occurred. That is, the proportion of times on which it occurred.

Relative frequency density

Relative frequency divided by the length of the class interval.

Reliability function

The reliability function, of t say, is the probability a lifetime exceeds t. It is the complement of the cdf.

Residual

The observed value minus the fitted value in a regression analysis.

Return period

The reciprocal of the probability of exceedance in one year. It is the mean time between exceedances.

R-squared

The proportion of the variability in a set of data that is accounted for by the model which is being fitted. This is usually defined as the ratio of the: (corrected sum of squares with no model – sum of residuals squared)/ (corrected sum of squares with no model). It is also known as the coefficient of multiple determination, and the squared multiple correlation coefficient. In a linear regression with one predictor variable it is the square of the correlation coefficient.

Sample space

A list of all possible outcomes of some operation which involves chance.

Sampling distribution

An estimate is thought of as a single value from the imaginary distribution of all possible estimates, known as the sampling distribution. The idea of a sampling distribution is necessary to measure the precision of an estimator. The term 'sampling distribution' refers to the context in which the distribution arises, rather than the form of the distribution. For example, the sampling distribution of the mean is approximately normal for large samples. The t, chi-square and F distributions are usually introduced in the context of sampling.

Simple random sample

A sample chosen so that every possible choice of n from N has the same chance of occurring as the final sample. It implies all members of the population have the same probability of selection, but many other sampling schemes also have this property (see stratified sampling).

Skewness

A measure of asymmetry of a distribution. Positive values correspond to a tail to the right.

Slump

The drop of the apex of a cone of fresh concrete under standard conditions. It is used to check the relative water content of nominally identical mixes.

Spectral analysis

The variance of a signal is attributed to components which have frequencies ranging over some continuous scale.

Spectrum

The spectrum gives the average frequency composition of a signal.

Standard deviation

Square root of the variance.

Standard error

The standard deviation of some estimator. Commonly used, without qualification, for the standard deviation of the sample mean.

Standard normal distribution

The normal distribution scaled to have a mean of 0 and a standard deviation of 1. Its cdf, and percentage points are tabled.

Standardized residual

The residual divided by the estimated standard deviation of the errors in the regression.

State variable

A variable which changes over time is often referred to as the state variable to distinguish it from the time variable.

Stationarity

A random process is stationary in the mean, if the mean does not change over time. That is, there are no trends or seasonal variations in the mean. It is second-order stationary if neither the mean nor the autocovariance structure change over time. In particular, this requires that the standard deviation does not change seasonally. It is strictly stationary if all moments are time invariant.

Statistic

A summary number calculated from the sample, usually to estimate the corresponding population parameter.

Stochastic

A synonym for random.

Student's *t*-distribution

The sampling distribution of many statistics is normal, or at least approximately so, and can be scaled to standard normal. If the standard deviation of the statistic, which is used in the scaling, is replaced by its sample estimate, with v degrees of freedom, the normal distribution becomes a t-distribution with v degrees of freedom. If v exceeds about 30 there is little practical difference.

Studentized residual

A residual divided by its estimated standard deviation, which is not quite the same as the standard deviation of the errors. Hence, studentized residuals are close to, but not identical with, standardized residuals.

Survivor function

An alternative name for the reliability function.

t-ratio

The ratio of an estimate to its standard deviation. If the absolute value of the *t*-ratio is less than one, a 66% confidence interval for the parameter will include 0. If the *t*-ratio exceeds about 1.7 the 90% confidence interval will exclude 0, and we can at least claim to be confident about the sign of the coefficient.

Test statistic

A statistic which has a sampling distribution that is specified by the null hypothesis and is sensitive to any departures from the null hypothesis. Then, unlikely values of the statistic, conditional on the null hypothesis being true, are evidence against the null hypothesis.

Time series

A sequence of values of a variable which changes over time, usually thought of as a realization of some underlying random process.

Transfer function for a linear system

If a linear system is forced at some frequency, it responds at that frequency. However, there is a change in amplitude and a phase shift. Both of these depend on the frequency. The transfer function gives the relative amplitude and phase shift as a function of the frequency.

Unbiased estimator (estimate)

An estimator is unbiased, for some parameter, if the mean of its sampling distribution is equal to that parameter. That is, averaged over imagined replicates of the sampling procedure we would obtain the parameter value. An unbiased estimate is a particular value of the estimator.

Uniform distribution

A variable has a uniform distribution between two limits, if the probability it lies within some interval, between those limits, is proportional to the length of the interval. Limits of 0 and 1 are by far the most usual, and arise in the context of random number generation.

Variable

A quantity that varies from one member of the population to the next. It can be measured on some continuous scale, be restricted to integer values (discrete), or be restricted to descriptive categories.

Variance

Average of the squared deviations from the mean. It has units equal to the square of the units of the variable. Variances of independent variables are additive.

Weighted mean

An average in which the data are multiplied by numbers called weights, added, and then divided by the sum of the weights. If the weights are all the same this is the usual mean.

References

Abramowitz, M. and Stegun, I.A. (1965): *Handbook of Mathematical Functions*. New York: Dover.

Adamson, P.T. (1989): *Robust and Explanatory Data Analysis in Arid and Semi-arid Hydrology. Department of Water Affairs Technical Report TR138*. Pretoria, Department of Water Affairs, Republic of South Africa.

Adamson, P.T. (1994): *Flood Risk Analysis in Extremely Large Catchments – Studies in the La Plata Basin*. Burderop Park, Swindon, Sir William Halcrow & Partners Ltd.

Adamson, P.T., Metcalfe, A.V., Parmentier, B. (1997): *An Application of the Gibbs Sampler to a Joint Distribution of Flood and Volume*. Burderop Park, Swindon, Sir William Halcrow & Partners.

Al Janahai, A.A.M. (1995): An evaluation of the implications of imposing speed limits on major roads. PhD Thesis, Derpartment of Civil Engineering Transport Division, University of Newcastle upon Tyne, UK.

Anderson, C.W. and Nadarajah, S. (1993): Environmental factors affecting reservoir safety. In V. Barnett and K.F. Turkman (eds), *Statistics for the Environment*, Wiley, Chichester, 163–182.

Anderson, V.L. and McLean, R.A., (1974): *Design of Experiments: A Realistic Approach*. Marcel Dekker, New York.

Andrews, J.D. and Moss, T.R. (1993): *Reliability and Risk Assessment*. Longman, Harlow, Essex.

Ang, A.H-S. and Tang, W.H. (1984): *Probability Concepts in Engineering Planning and Design. Volume II, Decision, Risk and Reliability*. Wiley, New York.

ANSI/ASCE7 (1988): *Minimum Design Loads for Buildings and Other Structures*. American Society of Civil Engineers, New York.

Appleby, J.C. (1994): *BUSTLE Program and Teaching Note*. Department of Engineering Mathematics, University of Newcastle upon Tyne.

Atan, I.B. and Metcalfe, A.V. (1994): Estimation of seasonal flood risk using a two stage transformation. *Water Resources Research*, **30**, 2197–2206.

Atkinson, A.C. and Donev, A.N. (1992): *Optimum Experimental Designs*. Oxford Scientific Publications, Oxford.

Azimi-Zonooz, A., Krajewski, W.F., Seo, D.J. and Bowles, D.S. (1989): Spatial rainfall estimation by linear and non-linear co-kriging of radar-rainfall and raingauge data. *Stochastic Hydrology and Hydraulics*, **3**(1), 51–67.

Bathurst, J.C. Wicks, J.M. and O'Connell, P.E. (1993): The SHE/SHESED basin scale water flow and sediment transport modelling system. In V.P. Singh (ed.) *Computer Models of Watershed Hydrology*. Water Resources Publications, Fort Collins, Colorado.

Beran, J. (1994): *Statistics for Long Memory Processes*. Chapman & Hall, New York.

Billah, K.Y. and Scnalan, R.H. (1991): Resonance, Tacoma Narrows bridge failure, and undergraduate physics textbooks. *American Journal of Physics*, **59**(2), 118–124.

Box, G.E.P. and Cox. D.R. (1964): An analysis of transformations. *Journal of the Royal Statistical Society Series B*, **26**, 211–243 discussion 244–252.

British Maritime Technology Ltd (1986): *Global Wave Statistics*. Unwin, Old Woking.

BS 6841 (1987): Measurement and evaluation of human exposure to whole-body mechanical vibration and repeated shock. British Standards Institution.

BS 812: Part 114: (1989): Testing aggregates Part 114. Method for determination of the polished-stone value. British Standards Institution.

BS 5497: Part 1 (1987): Precision of test methods Part 1. Guide for the determination of repeatability and reproducibility for a standard test method by inter-laboratory tests. British Standards Institution.

Bs 5700 (1984): Guide to process control using quality control chart methods and cusum techniques. British Standards Institution.

BS 812: Part 102 (1989): Testing aggregates Part 102. Methods for sampling. British Standards Institution.

Bull, J.W. (1988): The design of footway paving flags. *Highways*, **56**(1936), 44–45.

Bunday, B.D. (1996): *Queueing Theory*. Arnold, London.

Burton, A., Arkell, T.J. and Bathurst, J.C. (1997): Field variability of landslide model parameters. Accepted by *Environmental Geology*.

Burton, P.W. and Makropoulos, K.C. (1985): Seismic risk of circum-Pacific earth-quakes II. Extreme values using Gumbel's third distribution and the relationship with strain energy release. *Pure and Applied Geophysics*, **123**, 849–866.

Butcher, W.S. 1971: Stochastic dynamic programming for optimum reservoir operation. *Water Resources Bulletin*, **7**(1), 115–123.

Carmichael, D.G. (1987): *Engineering Queues in Construction and Mining*. Ellis Horwood, Chichester, Sussex.

Casella, G. and George, E.I. (1992): Explaining the Gibbs sampler. *The American Statistician*, **6**(3), 167–174.

Caulcutt, R. (1995): The rights and wrongs of control charts. *Applied Statistics*, **44**(3), 279–288.

Cavanié, A. (1985): Joint occurrence of extreme wave heights and wind gusts during severe storms on the Frigg Field. Exploration and Production Forum Workshop on Applications of Joint Probability of Metocean Phenomena in the Oil Industry's Structural Design Work, London, UK.

Chang, T.J., Kavvas, M.L. and Delleur, J.W. (1984): Daily precipitation modelling by discrete autoregressive moving average processes. *Water Resources Research*, **20**(5), 565–580.

Changery, M.J. (1982): *Historical Extreme Winds for the United States-Atlantic and Gulf of Mexico Coastlines*. US Nuclear Regulatory Commission, NUREG/CR-2639.

Chatfield, C. (1989): *The Analysis of Time Series* (4th Edition). Chapman & Hall, London.

Chatfield, C. and Collins, A.J. (1980): *Introduction to Multivariate Analysis*. Chapman & Hall, London.

Chouinard, L.E. (1995): Estimation of burial depths for pipelines in Arctic Regions. *Journal of Cold Regions Engineering ASCE*, **9**(4), 167–182.

Chouinard, L.E., Bennett, D.W. and Nadia, F. (1995): Statistical analysis of monitoring data for concrete arch dams. *Journal of Performance of Constructed Facilities ASCE*, **9**(4), 286–301.

Clark, I. (1979): *Practical Geostatistics*. Applied Science Publishers Ltd, London.

Coles, S.G. and Tawn, J.A. (1991): Modelling extreme multivariate events. *Journal of the Royal Statistical Society B*, **53**, 377–392.

Coles, S.G. and Tawn, J.A. (1994): Statistical methods for multivariate extremes: an application to structural design (with discussion). *Applied Statistics*, **43**(1), 1–48.

Cooke, D., Craven, A.H. and Clarke, G.M. (1990): *Basic Statistical Computing* (second edition). Edward Arnold, London.

Cowpertwait, P.S.P. (1991): Further developments of the Neyman-Scott clustered point process for modelling rainfall. *Water Resources Research*, **27**(7), 1431–1438.

Cowpertwait, P.S.P. (1994): A generalised point process for rainfall. *Proceedings of the Royal Society of London A*, **447**, 23–27.

Cowpertwait, P.S.P. (1995): A generalised spatial-temporal model of rainfall based on a clustered point process. *Proceedings of the Royal Society of London A*, **450**, 163–175.

Cowpertwait, P.S.P. and O'Connell, P.E. (1993): A Neyman-Scott shot noise model for the generation of daily streamflow time series. In J.P.J. O'Kane (ed.) *Advances in Theoretical Hydrology – A Tribute to James Dooge*, Elsevier, Amsterdam, 75–94.

Cowpertwait, P.S.P., O'Connell, P.E., Metcalfe, A.V. and Mawdsley, J.A. (1996): Neyman-Scott modelling of rainfall time series: 1. Fitting procedures for hourly and daily data. 2. Regionalisation and disaggregation procedures. *Journal of Hydrology*, **175**(1–4), 17–46, 47–65.

Cox, D.R. (1972): Regression models and life tables (with discussion). *Journal of the Royal Statistical Society B*, **34**, 187–202.

Cox, D.R. and Isham, V. (1980): *Point Processes*. Chapman & Hall, London.

Crowder, M.J. (1991): A statistical approach to a deterioration process in reinforced concrete. *Applied Statistics*, **40**(1), 95–103.

Crowder, M.J., Kimber, A.C., Smith, R.L. and Sweeting, T.J. (1991): *Statistical Analysis of Reliability Data*. Chapman & Hall, London.

Cunnane, C. (1988): Methods and merits of regional flood frequency analysis. *Journal of Hydrology*, **100**, 269–290.

Dales, M.Y. and Reed, D.W. (1989): Regional flood and storm hazard assessment. IH Report No 102, Wallingford, UK, Institute of Hydrology.

Dalzell, J.F. (1974): Cross-bispectra analysis: application to ship resistance in waves. *Journal of Ship Research*, **18**(1), 62–72.

Daubechies, I. (1988): Orthonormal bases of compactly supported wavelets. *Communications on Pure and Applied Mathematics*, **XLI**, 909–996.

Davison, A.C. and Smith, R.L. (1990): Models for exceedances over high thresholds. *Journal of the Royal Statistical Society Series B*, **52**(3), 393–442.

Day, K.W. (1995): *Concrete Mix Design, Quality Control and Specification*. E & FN Spon, London.

De Haan, L. and Resnik, S.I. (1977): Limit theory for multivariate samples extremes. *Z. Wahrsch Theor*, **40**, 317–337.

Dempster, A.P., Laird, N.M. and Rubin, D.B. (1977): Maximum likelihood from incomplete data via the EM algorithm (with discussion). *Journal of the Royal Statistical Society Series B*, **39**, 1–38.

Deutsch, C.V. and Journel, A.G. (1992): *GSLIB Geostatistical Software Library and User's Guide*. Oxford University Press, Oxford.

Dietrich, C.R. (1995): A simple and efficient space domain implementation of the turning bands method. *Water Resources Research*, **31**(1), 147–156.

Dietrich, C.R. and Newsam, G.N. (1993): A fast and exact method for multi-dimensional Gaussian stochastic simulations. *Water Resources Research*, **29**(8), 2861–2869.

Doole, S.H. and Hogan, S.J. (1996): A piecewise linear suspension bridge model: nonlinear dynamics and orbit continuation. *Dynamics and Stability of Systems*, **11**(1), 19–47.

Efron, B. and Gong, G. (1983): A leisurely look at the bootstrap, the jacknife, and cross-validation. *American Statistician*, **37**, 36–48.

Evans, M., Hastings, N. and Peacock, B. (1993): *Statistical Distribution* (2nd edition). Wiley, New York.

Everitt, B.S. (1987): *Introduction to Optimisation Methods and their Application in Statistics*. Chapman & Hall, London.

Firth, J.M. (1992): *Discrete Transforms*. Chapman & Hall, London.

Fisher, N.I. (1993): *Statistical Analysis of Circular Data*. Cambridge University Press, Cambridge.

Fisher, N.I., Lewis, T. and Embleton, B.J.J. (1987): *Statistical Analysis of Spherical Data*. Cambridge University Press, Cambridge.

Futter, M., Mawdsley, J.A. and Metcalfe, A.V. (1990): Risk of flooding dependent on prevailing catchment conditions. In W.R. White (ed.) *Proceedings of the International Conference on River Flood Hydraulics*, 17–20 September 1990, Wallingford, UK, organized by Hydraulics Research Ltd. Wiley, Chichester.

Futter, M.R., Mawdsley, J.A. and Metcalfe, A.V. (1991): Short-term flood risk prediction: a comparison of the Cox regression model and a conditional distribution model. *Water Resources Research*, **27**(7), 1649–1656.

Gardner, G., Harvey, A.C., Philips, G.D.A. (1980): An algorithm for exact maximum likelihood estimation of augoregressive-moving average models by means of Kalman filtering. *AS154 Applied Statistics*, **29**(3), 311–322.

Gelman, A., Carlin, J.B., Stern, H.S. and Rubin, D.B. (1995): *Bayesian Data Analysis*. Chapman & Hall, London.

Goupillaud, P., Grossman, A. and Morlet, J. (1984): Cycle-Octave and related transforms in seismic signal analysis. *Geoexploration*, **23**, 85–102.

Graham, G. and Martin, F.R. (1946): Heathrow. The construction of high-grade quality concrete paving for modern transport aircraft. *Journal of the Institution of Civil Engineers*, **26**(6), 117–190.

Greenfield, A.A. and Savas, D. (1992): DEX: A program for the design and analysis of experiments. Tony Greenfield, Middle Cottage, Little Hucklow, Derbyshire, U.K.

Greenwood, J.A., Landwehr, J.M., Matalas, N.C. and Wallis, J.R. (1979): Probability weighted moments: definition and relation to parameters of several distributions expressible in inverse form. *Water Resources Research*, **15**, 1049–1054.

Gumbel, E.J. (1958): *Statistics of Extremes*. Colombia University Press, New York.

Gurney, T. (1992): *Fatigue of Steel Bridge Decks*. HMSO, London.

Guttorp, P. (1995): *Stochastic Modeling of Scientific Data*. Chapman & Hall, London, UK.

Hashash, Y.M.A. and Whittle, A.J. (1996): Ground movement prediction for deep excavations in soft clay. *Journal of Geotechnical Engineering ASCE*, **122**(6), 474–486.

Hay, J. (1992): *Responses of Bridges to Wind*. HMSO, London.

Hearn, G.E. and Metcalfe, A.V. (1995): *Spectral Analysis in Engineering*. Arnold, London.

Hogg, R.V. and Craig, A.T. (1978): *Introduction to Mathematical Statistics* (4th edition). Macmillan, New York.

Hosking, J.R.M. (1984): Modelling persistence in hydrological time series using fractional differencing. *Water Resources Research*, **20**(12), 1898–1908.

Hosking, J.R.M. (1990): *L*-moments: analysis and estimation of distributions using linear combinations of order statistics. *Journal of the Royal Statistical Society Series B*, **52**(1), 105–124.

Hosking, J.R.M. and Wallis, J.R. (1987): Parameter and quantile estimation for the Generalised Pareto distribution. *Technometrics*, **29**(3), 339–349.

Hosking, J.R.M. and Wallis, J.R. (1995): A comparison of unbiased and plotting-position estimators of *L* moments. *Water Resources Research*, **31**(8), 2019–2025.

Hosking, J.R.M., Wallis, J.R. and Wood, E.F. (1985): Estimation of the generalised extreme value distribution by the method of probability-weighted moments. *Technometrics*, **27**(3), 251–261.

Hurst, H.E. (1951): Long-term storage capacity of reservoirs. *Transactions of the American Society of Civil Engineers*, **116**, 770–799.

Hutchinson, P. (1996): Waiting for buses: size-weighted means. *Teaching Statistics*, **18**(1), 9.

Ibrahim, K.B. and Metcalfe, A.V., (1993): Bayesian overview for evaluation of mini-roundabouts as a road safety measure. *The Statistician (Journal of the Royal Statistical Society Series D)*, **42**, 525–540.

Isaaks, E.H. and Srivastava, R.M. (1989): *Applied Geostatistics*. Oxford University Press, New York.

Israel, R.B. (1990): Gibbs distributions II. In S. Kotz, (ed.) *Encyclopaedia of Statistical Science*. Wiley, New York.

Jancauskas, E.D., Mahendran, M. and Walker, G.R. (1994): Computer-simulation of the fatigue behaviour of roof cladding during the passage of a tropical cyclone. *Journal of Wind Engineering and Industrial Aerodynamics*, **51**(2), 215–227.

Jenkins, G.M. and Watts, D.G. (1968): *Spectral Analysis and its Applications*. Holden Day, San Franscisco.

Johnson, M.E. (1987): *Multivariate Statistical Simulation*. John Wiley, New York.

Johnson, N.L. (1981): A simple theoretical approach to cumulative sum charts. *Journal of the American Statistical Association*, **56**, 835–40.

Johnson, N.L. and Kotz, S. (1972): *Distributions in Statistics: Continuous Multivariate Distributions*. John Wiley, New York.

Johnson, N.L., Kotz, S. and Balakrishnan, N. (1995): *Continuous Univariate Distributions Volume 2* (2nd edition). John Wiley, New York.

Jones, R.H. (1980): Maximum likelihood fitting of ARMA models to time series with missing observations. *Technometrics*, **22**(3), 389–395.

Jones, R.T., Pretlove, A.J. and Eyre, R. (1981): Two case studies in the use of tuned vibration absorbers on footbridges. *The Structural Engineer*, **59B**, 27.

Kalbfleisch, J. and Prentice, R.L. (1980): *The Statistical Analysis of Failure Time Data*. John Wiley, New York.

Kalman, R.E. and Bucy, R.S. (1961): New results in linear filtering and prediction theory. *Journal of Basic Engineering. Transactions of ASME Series D*, **83**, 95–108.

Kendall, M.G. and Ord, J.K. (1990): *Time Series* (3rd Edition). Edward Arnold, Sevenoaks, UK.

Kinnison, R.R. (1985): *Applied Extreme Value Statistics*. Battelle Press, Columbus, Richland.

Klemeš, V., Drikanthan, R., and McMahon, T.A. (1981): Long memory flow models in reservoir analysis: what is their practical value? *Water Resources Research*, **17**(3), 737–751.

Kreyszig, E. (1993): *Advanced Engineering Mathematics* (7th Edition). John Wiley, New York.

Krige, D.G. (1951): A statistical approach to some basic mine valuation problems on the Witwatersrand. *Journal of the Chemical Metallurgy and Mineral Society of South Africa*, **52**(6), 119–139.

Kwakernaak, H. and Sivan, R. (1972): *Linear Optimal Control Systems*. Wiley, New York.

Landwehr, J.M., Matalas, N.C. and Wallis, J.R. (1979): Estimation of parameters and quantiles of Wakeby distributions. *Water Resources Research*, **15**, 1361–1379 correction 1672.

Leadbetter, M.R., Lindgren, G. and Rootzén, H. (1982): *Extremes and Related Properties of Random Sequences and Processes*. Springer-Verlag, New York.

Lee, P.A. (1989): *Bayesian Statistics: an Introduction*. Edward Arnold, London.

Lin, G.-F. and Lee, F.-C. (1994): Assessment of aggregated hydrologic time series modelling implementations in two and three season flows of Tanshui River. *Journal of Hydrology*, **156**, 447–458.

Liu, H. (1991): *Wind Engineering - A Handbook for Structural Engineers.* Prentice Hall, Englewood Cliffs, New Jersey.

Mackay, R. and Cooper, T.A. (1996): Contaminant transport in heterogeneous porous media: a case study. 1. Site characterisation and deterministic modelling. *Journal of Hydrology*, **175**, 383–406.

Mackay, R., Cooper, T. and Metcalfe, A.V. (1996): Contaminant transport in heterogeneous porous media: a case study. 2. Stochastic modelling. *Journal of Hydrology*, **175**, 429–452.

Mallat, S. (1989): A theory for multiresolution signal decomposition: the wavelet representation. *IEEE Transactions on Pattern Analysis and Machine Intelligence*, **11**, 674–693.

Mantoglou, A. and Wilson, J.L. (1982): The turning bands method for simulating random fields using line generation by a spectral method. *Water Resources Research*, **18**, 1379–1394.

Mardia, K.V. (1972): *Statistics of directional data.* Academic Press, London.

Marquardt, D.W. (1963): An algorithm for least squares estimation of nonlinear parameters. *Journal of the Society for Industrial and Applied Mathematics*, **11**, 431–441.

McCleod, A.I. (1985): A remark on AS183. An efficient and portable pseudorandom number generator. *Applied Statistics*, **34**, 198–200.

Mellor, D. (1996): The modified turning bands (MTB) model for space-time rainfall. I. Model definition and properties. *Journal of Hydrology*, **175**, 113–127.

Mellor, D. and O'Connell, P.E. (1996): The modified turning bands (MTB) model for space time rainfall. II. Estimation of raincell parameters. *Journal of Hydrology*, **175**, 129–159.

Mellor, D. and Metcalfe, A.V. (1996): The modified turning bands (MTB) model for space time rainfall. III. Estimation of the storm/rainband profile and a discussion of future model prospects. *Journal of Hydrology*, **175**, 161–180.

Metcalfe, A.V. (1991): Probabilistic modelling in the water industry. *Journal of the Institution of Water and Environmental Management*, **5**(4), 439–449.

Metcalfe, A.V. (1994): *Statistics in Engineering.* Chapman & Hall, London.

Miaou, S.-P. (1994): The relationship between truck accidents and geometric design of road sections: Poisson versus negative binomial regressions. *Accident Analysis and Prevention*, **26**(4), 471–482.

Michie, D., Spiegelhalter, D.J. and Taylor, C.C. (1994): *Machine Learning, Neural and Statistical Classification.* Ellis Horwood, New York.

Miller, R.B., Bell, W., Ferreiro, O. and Wang, R.Y-Y. (1981): Modelling daily river flows with precipitation input. *Water Resources Research*, **17**(1), 209–215.

Montgomery, D.C. (1991): *Design and Analysis of Experiments* (3rd edition). Wiley, New York.

Montgomery, D.C. and Peck, E.A. (1992): *Introduction to Linear Regression Analysis.* Wiley, New York.

Moran, P.A.P. (1954): Theory of dams and storage systems. *Australian Journal of Applied Science*, **5**.

Morison, J.R., O'Brien, M.P., Johnson, J.W. and Shaaf, S.A. (1950): The force exerted by surface waves on piles. *AIME Petroleum Transactions*, **189**, 149–154.

Myers, R.H. (1986): *Classical and Modern Regression with Applications*. Duxbury Press, Boston, MA, USA.

Najafian, G., Burrows, R. and Tickell, R.G. (1995): A review of the probabilistic description of Morison wave loading and response of fixed offshore structures. *Journal of Fluids and Structures*, **9**, 585–616.

Natural Environment Research Council (1975): *Flood Studies Report*. HMSO, London.

Newland, D.E. (1993): *Random Vibrations, Spectral and Wavelet Analysis* (3rd edition). Longman, Harlow, Essex.

Nordquist, J.M. (1945): Theory of largest values, applied to earthquake magnitudes. *Transactions of the American Geophysical Union*, **26**, 29–31.

North, M. (1980): Time-dependent stochastic models of floods. *Journal of the Hydraulics Division of the American Society of Civil Engineers*, **106**, 649–655.

O'Connell, P.E., Mellor, D., Sheffield, J. and Metcalfe, A.V. (1996): The development of a stochastic space–time rainfall forecasting system for real-time runoff forecasting. Results of the hydrological radar experiment. A joint meeting of the British Hydrological Society, Natural Environment Research Council and the Royal Meteorological Society at the Institution of Civil Engineering, Great George Street, London on 6 November 1996.

O'Hagan, A. (1994): *Kendall's Advanced Theory of Statistics - Volume 2B Bayesian Inference*. Edward Arnold, London.

O'Hagan, A., Glennie, E.B. and Beardsall, R.E. (1992): Subjective modelling and Bayes' linear estimation in the UK water industry. *Journal of the Royal Statistical Society, Series C*, **41**, 563–577.

Önöz, B. and Bayazit, M. (1995): Best-fit distributions of largest available flood samples. *Journal of Hydrology*, **167**, 195–208.

Paris, P. and Erdogan, F. (1963): A critical analysis of crack propagation laws. *ASME Journal of Basic Engineering*, **85**, 528–534.

Paskalov, T.A., Petrovski, J.T. and Jurukovski, D.V. (1981): Full scale forced vibration studies and mathematical model formulation of arch concrete dams. In *Dams and earthquake – Proceedings of a conference held at the Institution of Civil Engineering*, London, 1–2 October 1980. Thomas Telford Ltd, London.

Pearson, E.S. and Hartley, H.O. (eds) (1976): *Biometrika Tables for Statisticians*. 3rd edition. Biometrika Trust, London.

Pearson, K. (1901): On lines and planes of closest fit to systems of points in space. *Philosophical Magazine*, **6**(2), 559–572.

Peel, D. (1991): *Risk of Unplanned Interruptions of Supply to the Customer*. Yorkshire Water Services, Leeds.

Pegram, G.G.S. (1971): A note on the use of Markov chains in hydrology. *Journal of Hydrology*, **13**, 216–230.

Phillips, D.L., Dolph, J. and Marks, D. (1992): A comparison of geostatistical procedures for spatial analysis of precipitation in mountainous terrain. *Agricultural and Forest Meteorology*, **58**, 119–141.

Pierson, W.J. and Holmes, P. (1965): Irregular wave forces on a pile. *ASCE Journal of the Waterways and Harbours Division 91*, **WW4**, 1–10.

Plackett, R.L. (1950): Some theorems in least squares. *Biometrika*, **37**, 149–157.

Pole, A., West, M. and Harrison, J. (1994): *Applied Bayesian Forecasting and Time Series Analysis*. Chapman & Hall, New York.

Press, W.H., Teukolsky, S.A., Vettering, W.T. and Flannery, B.P. (1992): *Numerical Recipes in Fortran* (2nd Edition). Cambridge University Press, Cambridge, UK.

Rathi, A.K. and Santiago, A.J. (1990): Urban network traffic simulation: TRAF-NETSIM program. *Journal of Transportation Engineering ASCE*, **116**, 734–743.

Rebeiz, K.S., Rosett, J.W. and Craft, A.P. (1996): Strength properties of polyester mortar using PET and fly ash wastes. *Journal of Energy Engineering ASCE*, **122**(1), 10–19.

Reinhorn, A.M., Soong, T.T., Riley, M.A., Lin, R.C., Aizawa, S. and Higashino, M. (1993): Full-scale implementation of active control II; installation and performance. *Journal of Structural Engineering ASCE*, **19**, 1935–1960.

Reitsma, R., Zigurs, I., Lewis, C., Wilson, V. and Sloane, A. (1996): Experiment with simulation-models in water resources negotiations. *Journal of Water Resources Planning and Management-ASCE*, **122**(1), 64–70.

Resnick, S.I. (1987): *Extreme Values, Regular Variation and Point Processes*. Springer, New York.

Robins, A.J. (1982): The extended Kalman filter. *Aerospace Dynamics*, **8**, 16–24.

Rodriguez-Iturbe, I., Cox, P.R. and Isham, V. (1987): Some models for rainfall based on stochastic point processes. *Proceedings of the Royal Society of London A*, **410**, 269–288.

Roeder, C.W. and Foutch, D.A. (1996): Experimental results for seismic resistant steel moment frame connections. *Journal of Structural Engineering ASCE*, **122**, 581–588.

Rooke, D. (1996): How do I explain that we have now had two 100-year events in four years? *Proceedings of the Institution of Civil Engineering – The Municipal Engineer*, **115**, 79–85.

Rossi, F., Fiorentino, M. and Versace, P. (1984): Two-component extreme value distribution for flood frequency analysis. *Water Resources Research*, **20**(7), 847–856.

Royle, A.G. (1977): Global estimates of ore reserves. *Transactions of the Institution of Mining and Metallurgy*, **86**, A9–17.

Sahabandu, K.L.S. (1988): The analytical and experimental relationships between the size, loading and interlock of paving flags. M.Sc. dissertation, Department of Civil Engineering, University of Newcastle upon Tyne.

Scanlan, R.H. and Gade, R.H. (1977): Motion of suspended bridge spans under gusty wind. *Proceedings of the ASCE Journal of the Structural Division*, **103**(ST9), 1867–1883.

Senevirante, P.N. (1990): Analysis of on-time performance of bus services using simulation. *Journal of Transportation Engineering ASCE*, **116**, 517–531.

Sharp, J.I. (1988): The climate of Newcastle upon Tyne: Seminar Paper Number 64. Department of Geography, University of Newcastle upon Tyne.

Shaw, R. (1982): *Wave Energy: A Design Challenge*. Ellis Horwood, Chichester.

Silverman, B.W. (1982): AS176 Kernel density estimation using the fast Fourier transform. *Applied Statistics*, **31**(1), 93–99.

Silverman, B.W. (1986): *Density Estimation for Statistics and Data Analysis*. Chapman & Hall, London.

Sinclair, A.J. (1974): A geostatistical study of the Eagle copper vein, Northern British Columbia. *Canadian Institution of Mining and Metallurgy*, **67**(746), 131–142.

Sinclair, A.J. (1978): Geostatistical investigation of the Kutcho Creek deposit, Northern British Columbia. *Mathematical Geology*, **10**, 273–288.

Singhal, A. and Kiremidjian, A.S. (1996): Method for probabilistic evaluation of seismic structural damage. *Journal of Structural Engineering ASCE*, **122**, 1459–1467.

Skipp, B.O. and Renner, L. (1963): The improvement of the mechanical properties of sand. In *Grouts and Drilling Muds in Engineering Practice* (Symposium organized by the British National Society of the International Society of Soil Mechanics and Foundation Engineering at the Institution of Civil Engineers, May 1963) Butterworths, London.

Smith, D.K. (1991): *Dynamic Programming – A Practical Introduction*. Ellis Horwood, New York.

Smith, J.A. and Carr, A.F. (1986): Flood frequency analysis using the Cox regression model. *Water Resources Research*, **22**, 890–896.

Smith, R.L. 1986: Extreme value theory based on the *r* largest annual events. *Journal of Hydrology*, **86**, 27–43.

Spiegel, M.R. (1963): *Schaum's Outline Series: Advanced Calculus*. McGraw-Hill, New York.

Stahl, F.L. and Gagnon, C.P. (1995): *Cable Corrosion in Bridges and Other Structures*. ASCE Press, New York.

Suyanto, A., Metcalfe, A.V. and O'Connell, P.E. (1995): The influence of storm characteristics and catchment conditions on extreme flood response: a case study based on the Brue River basin, U.K. *Surveys in Geophysics*, **16**(2), 201–225.

Taffe, J. and Garnham, N. (1996): Resampling, the bootstrap and Minitab. *Teaching Statistics*, **18**(1), 24–25.

Tawn, J.A. (1988): An extreme-value theory model for dependent observations. *Journal of Hydrology*, **101**, 227–250.

Thompson, J.M.T. and Stewart, H.B. (1986): *Nonlinear Dynamics and Chaos*. Wiley, Chichester.

Thomson, W.T. (1993): *Theory of Vibration with Applications* (4th edition). Chapman & Hall, London.

Tickell, R.G. (1977): Continuous random wave loading on structural members. *The Structural Engineer*, **55**(5), 209–222.

Tompson, A.F.B., Ababou, P. and Gelhar, L.W. (1989): Implementation of the three dimensional turning bands random field generator. *Water Resources Research*, **25**(1), 2227–2243.

Tong, H. (1975): Determination of the order of a chain by Akaike's information criterion. *Journal of Applied Probability*, **12**, 488–497.

Tong, H. (1990): *Non-linear Time Series*. Oxford University Press, Oxford.

Tong, H. and Lim, K.S. (1980): Threshold autoregression, limit cycles and cyclical data (with discussion). *Journal of the Royal Statistical Society B*, **42**, 245–292.

Trenberth, K.E. (1984): Some effects of finite sample size and persistence on meteorological statistics, Part 1: autocorrelations. *Monthly Weather Review*, **112**, 2359–2368.

Troutman, B.M. and Karlinger, M.R. (1992): Gibbs' distribution on drainage networks. *Water Resources Research*, **28**(2), 563–577.

Troutman, B.M. and Karlinger, M.R. (1994): Inference for a generalised Gibbsian distribution on channel networks. *Water Resources Research*, **30**, 2325–2338.

Tsang, W.-W. (1991): Estimation of short term flood risk. M.Sc. dissertation, Department of Civil Engineering, University of Newcastle upon Tyne.

Vogel, R.M. and Fennessey, N.M. (1993): *L*-moment diagrams should replace product moment diagrams. *Water Resources Research*, **29**, 1745–1752.

Vogel, R.M., McMahon, T.A. and Chiew, F.H.S. (1993): Flood frequency model selection in Australia. *Journal of Hydrology*, **146**, 421–447.

Wadhams, P. 1983: The prediction of extreme keel depths from sea ice profiles. *Cold Regions Science and Technology*, **6**, 257–266.

Wald, A. (1943): *Sequential Analysis of Statistical Data: Theory*. Statistical Research Groups, Columbia University.

Walden, A.T. (1994): Interpretation of geophysical borehole data via interpolation of fractionally differenced white noise. *Applied Statistics*, **43**, 335–345.

Wallis, J.R. and Wood, E.F. (1985): Relative accuracy of log Pearson III procedures. *Journal of Hydraulic Engineering ASCE* **111**(7), 1043–1056.

Waymire, E., Gupta, V.K. and Rodriguez-Iturbe, I. (1984): A spectral theory of rainfall intensity at the meso-β scale. *Water Resources Research*, **20**(10), 1453–1465.

Weibull, W. (1939): The phenomenon of rupture in solids. *Ingenior Vetenskaps Akademiens Handlingar*, **153**(2).

Weissman, I. (1978): Estimation of parameters and large quantiles based on the *k* largest observations. *Journal of the American Statistical Association*, **73**, 812–815.

Wetherill, G.B. (1975): *Sequential Methods in Statistics*. Methuen, London.

Wheater, H.S., Isham, V.S., Cox, D.R., Chandler, R.E., Kakou, A., Northrop, P.J., Oh, L. and Onof, C. (1997): Spatial-temporal rainfall fields: modelling and statistical aspects. Draft from Department of Civil Engineering, Imperial College, London.

Whyte, I.D. (1995): *Climatic Change and Human Society*. Arnold, London.

Wichmann, B.A. and Hill, I.D. (1982): An efficient and portable pseudo-random number generator. *Applied Statistics*, **31**, 188–190.

Wolfgram, C., Rothe, D., Wilson, P. and Sozen, M. (1985): Earthquake simulation tests of three one-tenth scale models. In J.K. Wright (ed.) *Earthquake Effects on Reinforced Concrete Structures*. American Concrete Institute, Detroit.

Wood, J. (1981): *Car travel from Scotland to England on the A74 road in October 1977: Supplementary Report 643*. Transport and Road Research Laboratory, Berkshire.

Woolhiser, D.A. and Osborn, H.B. (1985): A stochastic model of dimensionless thunderstorm rainfall. *Water Resources Research*, **21**(4), 511–522.

Wray, J. and Green, G.G.R. (1995): Calculation of the Volterra kernels of nonlinear dynamic systems using an artificial neural network. *Biological Cybernetics*, **71**(3), 187–195.

Wright, C.C., Abbess, C.R. and Jarrett, D.F. (1988): Estimating the regression-to-mean effect associated with road accident blackspot treatment: towards a more realistic approach. *Accident Analysis and Prevention*, **20**(3), 199–214.

Wu, W.F. (1993): Computer-simulation and reliability analysis of fatigue crack propagation under random loading. *Engineering Fracture Mechanics*, **45**(5), 697–712.

Young, P. (1974): Recursive approaches to time series analysis. *Bulletin of the Institute of Mathematics and its Applications*, **10**(5/6), 209–224.

Author Index

Ababou, P. 408
Abbess, C.R. 164, 409
Abramowitz, M. 29, 399
Adamson, P.T. 19, 107, 110, 399
Aizawa, S. 5, 407
Al Janahai, A.A.M. 148, 399
Anderson, C.W. 105, 111, 399
Anderson, V.L. 270, 399
Andrews, J.D. 56, 399
Ang, A.H-S. 3, 399
ANSI/ASCE7 81, 399
Appleby, J.C. 8, 399
Arkell, T.J. 221, 400
Atan, I.B. 204, 399
Atkinson, A.C. 286, 399
Azimi-Zonooz, A. 220, 400

Balakrishnan, N. 84, 88, 91, 94, 139, 404
Bathurst, J.C. 212, 221, 400
Bayazit, M. 26, 406
Beardsall, R.E. 166, 167, 406
Bell, W. 205, 206, 405
Bennett, D.W. 14, 81, 401
Beran, J. 208, 400
Billah, K.Y. 5, 247, 400
Bowles, D.S. 220, 400
Box, G.E.P. 20, 400
British Maritime Technology Ltd 32, 400
BS-812 43, 44, 47, 48, 50, 400
BS-5497 43, 44, 400
BS-5700 51, 400
BS-6841 244, 400
Bucy, R.S. 255, 404
Bull, J.W. 277, 400
Bunday, B.D. 67, 400
Burrows, R. 260, 406
Burton, A. 221, 400
Burton, P.W. 81, 400

Carlin, J.B. 173, 402
Carmichael, D.G. 68, 400
Carr, A.F. 73, 74, 408
Casella, G. 109, 400
Caulcutt, R. 51, 400
Cavani, A. 105, 400
Chandler, R.E. 215, 409
Chang, T.J. 182, 400
Changery, M.J. 401
Chatfield, C. 136, 199, 401
Chiew, F.H.S. 90, 409
Chouinard, L.E. 3, 14, 81, 401
Clark, I. 217, 219, 401
Clarke, G.M. 8, 401
Coles, S.G. 105, 106, 110, 401
Collins, A.J. 136, 401
Cooke, D. 8, 401
Cooper, T. 7, 139, 217, 219, 405
Cooper, T.A. 223, 405
Cowpertwait, P.S.P. 7, 9, 213, 214, 215, 216, 217, 348, 401
Cox, D.R. 20, 73, 74, 213, 215, 345, 346, 400, 401, 409
Cox, P.R. 9, 213, 346, 347, 407
Craft, A.P. 127, 130, 407
Craig, A.T. 308, 321, 403
Craven, A.H. 8, 401
Crowder, M.J. 63, 69, 73, 401
Cunnane, C. 401

Dales, M.Y. 104, 115, 401
Dalzell, J.F. 258, 401
Daubechies, I. 353, 402
Davison, A.C. 99, 402
Day, K.W. 2, 14, 50, 55, 117, 124, 125, 402
De Haan, L. 111, 402
Delleur, J.W. 182, 400
Dempster, A.P. 17, 402

Deutsch, C.V. 221, 402
Dietrich, C.R. 226, 402
Dolph, J. 221, 406
Donev, A.N. 286, 399
Doole, S.H. 237, 402
Drikanthan, R. 208, 404

Efron, B. 326, 402
Embleton, B.J.J. 31, 402
Erdogan, F. 26, 406
Evans, M. 17, 402
Everitt, B.S. 15, 16, 29, 64, 327, 402
Eyre, R. 247, 404

Fennessey, N.M. 90, 91, 409
Ferreiro, O. 205, 206, 405
Fiorentino, M. 100, 407
Firth, J.M. 240, 402
Fisher, N.I. 31, 402
Flannery, B.P. 285, 329, 407
Foutch, D.A. 117, 407
Futter, M. 73, 74, 402

Gade, R.H. 250, 407
Gagnon, C.P. 63, 408
Gardner, G. 200, 402
Garnham, N. 326, 408
Gelhar, L.W. 408
Gelman, A. 173, 402
George, E.I. 109, 400
Glennie, E.B. 166, 167, 406
Gong, G. 326, 402
Goupillaud, P. 262, 402
Graham, G. 39, 40, 403
Green, G.G.R. 258, 260, 409
Greenfield, A.A. 283, 403
Greenwood, J.A. 88, 403
Grossman, A. 262, 402
Gumbel, E.J. 100, 403
Gupta, V.K. 215, 409
Gurney, T. 60, 403
Guttorp, P. 403

Harrison, J. 4, 138, 174, 175, 176, 407
Hartley, H.O. 135, 318, 406
Harvey, A.C. 200, 402
Hashash, Y.M.A. 117, 403
Hastings, N. 17, 402
Hay, J. 237, 247, 250, 403
Hearn, G.E. 258, 349, 403
Higashino, M. 5, 407
Hill, I.D. 319, 349, 409
Hogan, S.J. 237, 402
Hogg, R.V. 308, 321, 403
Holmes, P. 261, 406
Hosking, J.R.M. 86, 87, 89, 90, 93, 94, 96,
    97, 105, 208, 209, 210, 343, 344,
    403
Hurst, H.E. 207, 403
Hutchinson, P. 292, 403

Ibrahim, K.B. 151, 163, 172, 403
Isaaks, E.H. 220, 403
Isham, V. 9, 213, 345, 346, 347, 401, 407
Isham, V.S. 215, 409
Israel, R.B. 231, 403

Jancauskas, E.D. 1, 403
Jarrett, D.F. 164, 409
Jenkins, G.M. 242, 404
Johnson, J.W. 261, 405
Johnson, M.E. 320, 404
Johnson, N.L. 84, 88, 91, 94, 106, 139, 334,
    404
Jones, R.H. 200, 404
Jones, R.T. 247, 404
Journel, A.G. 221, 402
Jurukovski, D.V. 251, 252, 406

Kakou, A. 215, 409
Kalbfleisch, J. 73, 404
Kalman, R.E. 255, 404
Karlinger, M.R. 230, 409
Kavvas, M.L. 182, 400
Kendall, M.G. 195, 404
Kimber, A.C. 63, 401
Kinnison, R.R. 100, 404
Kiremidjian, A.S. 3, 408
Klemes, V. 208, 404
Kotz, S. 84, 88, 91, 94, 139, 404
Krajewski, W.F. 220, 400
Kreyszig, E. 365, 404
Krige, D.G. 217, 404
Kwakernaak, H. 255, 404

Laird and Rubin, D.B. 17, 402
Landwehr, J.M. 88, 102, 103, 403, 404
Leadbetter, M.R. 335, 404
Lee, F.-C. 204, 404
Lee, P.A. 166, 404
Lewis, C. 7, 407
Lewis, T. 31, 402
Lim, K.S. 210, 409
Lin, G.-F. 204, 404
Lin, R.C. 5, 407
Lindgren, G. 335, 404
Liu, H. 23, 81, 247, 404

McCleod, A.I. 320, 405
Mackay, R. 7, 139, 217, 219, 223, 405
McLean, R.A. 270, 399
McMahon, T.A. 90, 208, 404, 409
Mahendran, M. 1, 403
Makropoulos, K.C. 81, 400
Mallat, S. 355, 405
Mantoglou, A. 225, 227, 405
Mardia, K.V. 31, 405
Marks, D. 221, 406
Marquardt, D.W. 329, 405
Martin, F.R. 39, 40, 403
Matalas, N.C. 88, 102, 103, 403, 404

Mawdsley, J.A. 7, 73, 74, 213, 214, 401, 402
Mellor, D. 212, 215, 405, 406
Metcalfe, A.V. 7, 73, 74, 110, 139, 151, 163, 166, 168, 172, 204, 212, 215, 217, 219, 258, 349, 399, 402, 403, 405, 406, 408
Metcalfe, P.E. 7, 213, 214, 401
Miaou, S.-P. 17, 405
Michie, D. 260, 405
Miller, R.B. 205, 206, 405
Montgomery, D.C. 133, 283, 405
Moran, P.A.P. 183, 405
Morison, J.R. 261, 405
Morlet, J. 262, 402
Moss, T.R. 56, 399
Myers, R.H. 144, 406

Nadarajah, S. 105, 111, 399
Nadia, F. 14, 81, 401
Najafian, G. 260, 406
Natural Environment Research Council 97, 406
Newland, D.E. 263, 353, 354, 406
Newsam, G.N. 226, 402
Nordquist, J.M. 406
North, M. 99, 406
Northrop, P.J. 215, 409

O'Brien, M.P. 261, 405
O'Connell, P.E. 7, 212, 213, 214, 215, 216, 217, 348, 400, 401, 405, 406, 408
Oh, L. 215, 409
O'Hagan, A. 166, 167, 341, 406
Onof, C. 215, 409
Önöz, B. 26, 406
Ord, J.K. 195, 404
Osborn, H.B. 32, 33, 409

Paris, P. 26, 406
Parmentier, B. 110, 399
Paskalov, T.A. 251, 252, 406
Peacock, B. 17, 402
Pearson, E.S. 135, 318, 406
Pearson, K. 406
Peck, E.A. 133, 405
Peel, D. 56, 57, 406
Pegram, G.G.S. 184, 406
Petrovski, J.T. 251, 252, 406
Philips, G.D.A. 200, 402
Phillips, D.L. 221, 406
Pierson, W.J. 261, 406
Plackett, R.L. 137, 406
Pole, A. 4, 138, 174, 175, 176, 407
Prentice, R.L. 73, 404
Press, W.H. 285, 329, 407
Pretlove, A.J. 247, 404

Rathi, A.K. 7, 407
Rebeiz, K.S. 127, 130, 407
Reed, D.W. 104, 115, 401
Reinhorn, A.M. 5, 407

Reitsma, R. 7, 407
Renner, L. 122, 408
Resnick, S.I. 111, 407
Resnik, S.I. 111, 402
Riley, M.A. 5, 407
Robins, A.J. 258, 407
Rodriguez-Iturbe, I. 9, 213, 215, 346, 347, 407, 409
Roeder, C.W. 117, 407
Rooke, D. 3, 407
Rootzn, H. 335, 404
Rosett, J.W. 127, 130, 407
Rossi, F. 100, 407
Rothe, D. 263, 409
Royle, A.G. 217, 407
Rubin, D.B. 173, 402

Sahabandu, K.L.S. 277, 407
Santiago, A.J. 7, 407
Savas, D. 283, 403
Scanlan, R.H. 5, 247, 250, 400, 407
Senevirante, P.N. 8, 27, 407
Seo, D.J. 220, 400
Shaaf, S.A. 261, 405
Sharp, J.I. 25, 407
Shaw, R. 250, 408
Sheffield, D. 212, 406
Silverman, B.W. 72, 408
Sinclair, A.J. 217, 408
Singhal, A. 3, 408
Sivan, R. 255, 404
Slipp, B.O. 122, 408
Sloane, A. 7, 407
Smith, D.K. 264, 408
Smith, J.A. 73, 74, 408
Smith, R.L. 63, 99, 105, 401, 402, 408
Soong, T.T. 5, 407
Sozen, M. 263, 409
Spiegel, M.R. 408
Spiegelhalter, D.J. 260, 405
Srivastava, R.M. 220, 403
Stahl, F.L. 63, 408
Stegun, I.A. 29, 399
Stern, H.S. 173, 402
Stewart, H.B. 210, 226, 408
Suyanto, A. 212, 408
Sweeting, T.J. 63, 401

Taffe, J. 326, 408
Tang, W.H. 3, 399
Tawn, J.A. 104, 105, 106, 110, 401, 408
Taylor, C.C. 260, 405
Teukolsky, S.A. 285, 329, 407
Thompson, J.M.T. 210, 226, 408
Thomson, W.T. 251, 408
Tickell, R.G. 260, 262, 406, 408
Tompson, A.F.B. 408
Tong, H. 182, 210, 211, 408, 409
Trenberth, K.E. 213, 409
Troutman, B.M. 230, 409

Tsang, W.-W. 182, 409

Versace, P. 100, 407
Vettering, W.T. 285, 329, 407
Vogel, R.M. 90, 91, 409

Wadhams, P. 3, 81, 409
Wald, A. 330, 409
Walden, A.T. 210, 409
Walker, G.R. 1, 403
Wallis, J.R. 88, 90, 93, 94, 96, 97, 102, 103,
    104, 403, 404, 409
Wang, R.Y.-Y. 205, 206, 405
Watts, D.G. 242, 404
Waymire, E. 215, 409
Weibull, W. 58, 409
Weissman, I. 104, 409
West, M. 4, 138, 174, 175, 176, 407
Wetherill, G.B. 332, 409

Wheater, H.S. 215, 409
Whittle, A.J. 117, 403
Whyte, I.D. 409
Wichmann, B.A. 319, 349, 409
Wicks, J.M. 212, 400
Wilson, J.L. 225, 227, 405
Wilson, P. 263, 409
Wilson, V. 7, 407
Wolfgram, C. 263, 409
Wood, E.F. 93, 94, 104, 403, 409
Wood, J. 15, 409
Woolhiser, D.A. 32, 33, 409
Wray, J. 258, 260, 409
Wright, C.C. 164, 409
Wu, W.F. 7, 409

Young, P. 137, 409

Zigurs, I. 7, 407

Subject Index

Acceptance–rejection (AR) algorithm 320
Acceptance–rejection (AR) methods 13–14
Addition rule 289
Akaike's Information Criterion (AIC) 182
Analysis of variance (ANOVA) 271, 273, 276
Anderson–Darling statistic 20, 317, 318
Annual flood series 371–6
Annual maximum daily mean discharges 367
Annual maximum flows 370
Annual maximum peak discharges 377
Annual maximum sea levels 368
AR(1) process, spectrum 242
AR(2) process, spectrum 242
ARIMA models 203, 208–9, 226
ARIMA processes 343–4
ARMA models 197–201, 208–11
  integrated 203–4
  periodic 204
  seasonal 203–4
Arrangements and choices (permutations and
  combinations) 291
Artificial neural network (ANN) 258–60
Asset management plans (AMPs) 166
Assumptions, checking reasonableness 130–3
Autocorrelation function (acf) 189
  estimator 195–7
Autocovariance function (acvf) 189, 241, 243,
  249
  estimator 195–7
Autoregressive processes 198–201
  second-order 242

Bandwidth 72, 240
Baseflow 73
Bayes' linear estimator (BLE) 166–72, 341–2
Bayes' theorem 163, 166, 176, 290
Bayesian analysis 4, 173–4
Bayesian Analysis of Time Series (BATS) 174
Bayesian methods 163–77
  introduction 163–6
  overview 172–4

Bernoulli trials 18
Bessel function 31
Bias 305
Biased estimator 305
Bilinear models, first-order 211
Binomial distribution 293
Binomial response, GLIM 151–6
Bivariate distribution simulation 109
Bivariate Gumbel distributions
  Type A 106–7
  Type B 107–9
Bivariate normal distribution 302–3
Blocking 269
Bootstrap methods 1–2, 22, 94
  for confidence intervals 325–6
Bootstrap sample 325
Boundary function 105
Box–Cox transform 20
Box–Muller method 13
Box plot 19
Bus scheduling 8
BUSTLE program 8, 9

Canonical form 185–6
Categorical variables 124
Censored data 60
Central Limit Theorem 13, 296, 298–300, 306
Centred moving average 190–1
Chapman–Kolmogorov equation 181
Chi-square distribution 19, 27, 102, 306–7,
  311, 361
Chi-square goodness of fit statistic 17
Chi-square goodness of fit test 25
Claremont Tower 26
Clay sample variables 379–83
Climatic change predictions 3
Coefficient of determination 125
Coefficient of variation (CV) 292
Co-kriging 220
Common cause variation 50
Complement 289

Completely randomized design (CRD)
  with one factor 269–74
  with two factors 276–81
  with two factors and RBD design 281–2
Composite design 282
Compound Poisson process 15
Compressive strength and relative density 122
Computer programs 6
Computer simulation 1–2, 7, 56, 109, 201–2
Concentration parameter 31
Concomitant variables 269
Concrete
  contractually defective 50
  reducing variability 39–42
  samples 2
  structurally defective 50
  test cubes 39–42
Condition of separation 102
Conditional distribution 302
Conditional intensity function 345
Conditional probability 289
Confidence interval 1, 22, 48, 123, 306, 314
  bootstrap methods 325–6
  difference in proportions 315
  proportions 315
Conjugate gradients method 328
Consulting engineers 6
Continuous distributions 19–34
Continuous time random process 242
Continuous variable 180
Contractually defective concrete 50
Control charts 50–5, 365
Control theory 5
Control variables 269
Controlled experiments 5
Convolution integral 247
Convolution theorem 247–9
Correlation 301
Correlogram 189
Covariance 33, 300
Covariance matrix 120, 121, 139, 144, 207, 303–4
Covariance structure 217
Cross-covariance function 248–9
Crossing analysis 261–2
Cross-spectrum 248–9
Cumulative distribution function (cdf) 11, 23
  inverse 296
Cumulative hazard function 58
CUSUM charts 53–5, 330–4
  Johnson's approximate approach 332–4
CUSUM technique 2

Dams 183–5
Daniell window 240
Decision trees 289–90
Defence structures 2
Dependent variables 117
Descriptive statistics 292–3
Deseasonalized data 191

Design matrix 284
Design of experiments 5, 269–87
Difference operator 203
Differences in means 314–15
Differencing 203
Di-gamma function 29
Dirichlet distribution 32–4
Disaggregation 32–4
Discrete distributions 14–19
Discrete Fourier transform 241, 349
Discrete time changes 8
Discrete time model, linear dynamic
  system 254–5
Discrete variable 180
Discrete wavelet transform (DWT) 262–3, 355
Discrete white noise 195
Distribution of peaks 261–2
Distribution of sample mean 306
Division modulo 6075 8, 9
D-optimum designs 283–6
Dynamic linear models (DLM),
  definition 174–5
Dynamic systems 1, 5

Eigenvalues 139
Eigenvectors 139
EM algorithm 17
Empirical distribution function (edf)
  definition 70
  tests 317–18
Ergodicity 66, 189
Erlang distributions 27
Errors, variance 121–3
Estimated vibration dose value (eVDV) 244
Expected monetary value (EMV) 289
Expected value 300
Explanatory variables 206
Exponential distribution 27, 295
Exponential model 218
Exponentially weighted past averages 138
Extreme value distribution 2–3, 81–115
  mixtures 100–1
  two component 99–100
  Type 1 81–5, 93, 94, 100, 102, 335–6
  Type 2 92–3, 110
  Type 3 92–3
Extreme wind speeds 370

F-distribution 311–13, 362–3
F-test 274
Factorial design
  2^k 282–3
  fractional 283
Failure rate function 58
Finite Fourier series 238, 349–50
Finite population correction 306
Fisher distribution 31
Flood frequency analysis 99
  precision 103–5

Flood water levels 2–3
Floodpeaks 20
 inflows 19
Forecasting 201–2
Fortran 6, 9
Fourier line spectrum 238
Fourier series 349–52
Fourier transforms 243, 248, 249, 349–52
Fractional differencing 207–10
 of time series 209–10
Fractional factorial design 283
Fractionally differenced processes 210
Friction test readings (FTR) 44–6
Functions of variables 304–5
Fundamental matrix 186

Gamma distribution 26–9, 34, 306
Gamma function 18, 291, 365
Gaussian distribution 204–5
Gaussian model 218
Gaussian process 204–5
Generalized extreme value distribution 26
Generalized least squares 138–40
Generalized linear regression model 139
Generalized extreme value (GEV)
 distribution 90, 92–5
Generalized linear model 149
Generalized Pareto distribution 90, 94–8
 definition 95
 estimation of parameters 95
Genetic algorithms 327
Gibbs' distributions 230–1
Gibbs sampler 109
GLIM 6
 binomial response 151–6
 Poisson response 148–51
Goodness-of-fit tests 317–18
Gradient methods
 using first derivatives 328–9
 using second derivatives 329
Greatest known value 105
Gumbel distribution 81–5
Gumbel distributions, *see also* Bivariate
 Gumbel distributions

Haar wavelets 263, 264, 357
Hardap Dam 19, 22, 27, 29
Harmonic cycles 192
Hazard function 58, 73
Hessian matrix 329
Hierarchical design 40
Hierarchical model 172
Hurst exponent 208
Hydraulic flow models 7
Hyperparameters 172
Hypothesis testing 316–17

Ill conditioning 126
Impulse response 247
Independent samples 314

Indicator variables 124
Intensity 345
Interference 69–71
Inter-laboratory trial 43–50
Internet 5
Intervention 176
Intrinsically linear models 140–3
Inverse cumulative distribution function 296
Inverse discrete wavelet transform
 (IDWT) 355
Inverse form 88
Inverse function 22
Inverse transformation 243
Isotropic medium 217

Jacobian 310
Joint posterior distribution 173–4

Kalman filter 255–8
 background 255
 equations 257
 extended 258
 principle of 256
Kernel 71–2
 density estimation 71–2
Kolmogorov-Smirnov statistic 60, 317
Kriging 217, 219–23
Kurtosis 23, 27, 84, 85, 87, 95, 261, 292

L-kurtosis 87, 90
L-moment diagram 90, 91
L-moments
 definition 86–8
 estimation 89–90
L-skewness 87, 90
L-statistics 85
Lagrange multiplier 135–6, 219–20
Laplace distribution 132
Large sample approximation 308–9
Least significant differences (LSD) 274, 276
Least squares estimators 119, 120, 272, 275
Life distributions 56–8
Likelihood 16
Likelihood function 14, 28, 64, 73, 146, 321
Likelihood principle 14
Likelihood ratio test 102, 323–4
Limits of prediction 123–4, 201
Linear dynamic systems 247–58
 discrete time model 254–5
 matrix solution 251–4
Linear function 304
Linear model 119
Load–strength interference 69
Log-gamma 90
Log-likelihood 15, 16, 24, 28, 29, 64, 73, 102
Log-normal distribution 19–24, 27, 29
Log-Pearson III (LP3) 27, 90

Mallat's pyramid algorithm 355–7

Maple 6, 31
Marginal distribution 300
Markov analysis 65–9
Markov chains 179–88
  absorbing 185–8
  aperiodic 185
  irreducible 185
Markov processes 212
Markov property 65, 180, 295
Matlab 6
Matrix differential equations 67
Matrix exponential 253
Matrix inversion lemma 138
Matrix solution of linear dynamic
    system 251–4
Maximal model 151
Maximum likelihood estimator (MLE) 14–17,
    24–6, 28, 60, 73, 82–3, 132, 146, 321–4
  asymptotic efficiency 322–3
Maximum Likelihood Program (MLP) 6, 65
Mean charts 51–3
Mean squared error (MSE) 135, 305
Mean time before failure (MTBF) 56
Mean time to repair (MTTR) 56
Means 18, 21, 23, 27, 30, 33, 95, 188, 195,
    306
  differences in 314–15
Median 23
Memoryless process 65
Method of moments (MOM) 15, 18, 24, 27,
    28, 33, 82, 84, 197
Method of moments estimator (MOME) 16,
    29, 109
Michaelis–Menten model 142
Minimum chi-square estimators 317
Minitab 6, 60, 128, 277
Mixed congruential generator 8
MOD function 9
Modified Bayes' linear estimator 167
Moment generating function (mgf) 27, 296–7
Monte Carlo simulation 7–14, 56
Monthly effective inflows 378–9
Monthly maximum wind speeds 368
Morison's equation 261
Moving average processes 197
Multicollinearity 134–7
Multilayer perceptron (MLP) 259
Multiple comparisons 274
Multiple regression analysis 118–33
Multiple regression model 118–19
Multiplicative rule 289
Multi-site models 206–7
Multivariate extreme value
    distributions 105–12
  point process theory of 110–12
Multivariate normal distribution 303
Multivariate probability distributions 300–4

Negative binomial distribution 17–19
Newton-Raphson method 329

Neyman-Scott rectangular pulses (NSRP)
    rainfall model 9, 212–17
Neyman-Scott streamflow model 215–17,
    345–8
Non-Gaussian process 204–5
Non-linear equations 4
Non-linear functions 143, 304–5
Non-linear model 143
Non-linear regression 143–4
Non-linear systems 258–60
Non-linear time series 210–11
Normal distribution 21, 22, 24, 27, 64,
    295–6
  derivation 296
  improved estimate of upper percentile 313
Normal equations 120
Normal mixture 64
Normal probability 21
Normalizing factor 31, 32
Nugget effect 218
Null hypothesis 316
Nyquist frequency 243–7

Observability 255
Office sprinkler system 55
Offshore structures 7
  wave loading on 260–2
One-sided alternative hypothesis 316
One-step ahead forecast 175
One-step transition probabilities 181
Optimization algorithms 327–9
Order statistics 85–92
  definition and distribution 85–8
  plotting against expected values 90–2
Ordinary kriging system 220, 223
Overdispersion 17, 151

Paired comparisons 314
Parameters 118
Parametric bootstrap 326
Pareto distribution 93
PARMA models 204
Parseval's Theorem 239, 350–1
Partial correlation function (pacf) 201
Pascal distribution 18
Pascal variable 17
Peak flows, predicting 142
Peaks over threshold (POT) analysis 94–8
Pearson type III distribution 26
Periodogram 239
Permutations and combinations (arrangements
    and choices) 291
Pierson Holmes distribution 261
Pivotal methods 326
Plotting-position estimates 90
Point forecast 201
Point process theory of multivariate extreme
    distributions 110–12
Point processes, stochastic models based
    on 212–17

Poisson distribution 9, 11, 15–18, 148, 212, 294–5
Poisson model with rectangular pulses 346–7
Poisson probabilities 15
Poisson process 9, 11, 17, 27, 104, 212, 294–5
Poisson random variable 9
Poisson response, GLIM 148–51
Poisson variable 11, 15, 348
Polished stone value (PSV) 43–50
Polynomial relationships 127–30
Posterior distribution 163, 176
Prediction error sum of squares (PRESS) residual 131–2, 221
Predictions 123–4
Predictor variables 117, 130, 206
  measurement errors 145–8
  number of 133
Principal component (PC) regression 135–7
Principle of least squares 119
Prior distribution 163
Probability 289–91
Probability chains 182
Probability density function (pdf) 22, 23, 27, 28, 31, 292
Probability distributions 2, 7
Probability mass function 11
Probability theory 1
Probability vector 181
Probability weighted moments (PWM) 88–9, 93, 103, 104
Proportional hazards modelling 72–5
Proportions 315
Pseudo-angular coordinates 110
Pseudo-radial coordinates 110
Pseudo-random number generator 319–20
Pseudo-random numbers 8–9
Pythagoras theorem 282

Quality assurance 2, 39–55
Quantiles, estimation 96–8
Queues in construction and mining 7

r largest events method 104–5
Radio-nucliides 7
Rainfall 7
Rainfall field models 215
Random digits 364
Random fields, simulating 223–30
Random number generation 319–20
Random numbers 8, 27, 33
  from continuous distributions 12–14
  from discrete distributions 9
Random process 179, 188
  models for 180
  moments 189
  spectrum 240–4
Random variables 8, 22, 27
Randomization 4
Randomized block design (RBD) 274–6

and CRD design with two factors 281–2
Range charts 51–3
Rao–Cramér inequality 321–2
Rao–Cramér lower bound (RCLB) 321, 323
Rate matrix 66
Rayleigh distribution 23–6, 59, 262
Recursive calculations 137–8
Regional flood frequency analysis (RFFA) 103–4
Regionalization methods 3
Regression 302–3
Regression coeficients 118
Regression equation 121
Regular chains 179–82
Relationships between variables 3–4, 117–62
Relative density and compressive strength 122
Reliability 2, 55–75
Repairable processes 67
Repeatability 43, 47–8
Reproducibility 43, 47–8
Residual sum of squares (RSS) 138, 339–40
Residuals 131
Response variable 269
Ridge regression 135
Ridge trace 135
River Browney 74, 75
River Coquet 97–8, 369
River Khan 101
River Swat 30
River Thames 82, 84
Robust regression 133

Sample circular variance 31
Sample covariance and correlation 302
Sample mean, distribution of 306
Sample size 316
Sample spectrum 237–40
Sample variance, sampling distribution of 307–9
Sampling 291–2
Sampling distribution 306
  of sample variance 307–9
Sampling inspection 2
SARIMA model 203, 204
Satellite global positioning systems 5
Scale factor 151
Scale models 2
Scaled deviance 149–50
Screening effect 221
Seasonality 99, 190–5
Second-order stationary 189
Sequential probability ratio test (SPRT) 330–4
Shewhart chart 51–3
Signal processing 5
Simple random sample (SRS) 291
Simplex algorithm 328
Simplex method 327–8
Simpson's rule 32
Simulated annealing 327

Simulation. *See* Computer simulation
Singularity 64
Size weighted means 292
Skewness 23, 26, 27, 84, 85, 87, 90, 95, 102–3, 292
Smoothing parameter 72
Spatial cross-covariances 221
Spatial processes 217–31
Spectral analysis 5, 237–68
Spectral density of derived process 260
Spectrum 237–47
   and transfer function 249–51
   AR(1) process 242
   AR(2) process 242
   random process 240–4
Spherical model 218
Standard deviation 18, 21, 23, 30, 31, 41
Standard normal distribution 13, 21, 295
Standard normal variable 359
Standardization method 192
Star design 282
State space 67
State transition matrix 254
State variables 179
Stationarity 189
Stationary transition mechanism 181
Statistical methods, introduction 289–318
Stochastic dynamic programming 264–6
Stochastic models based on point processes 212–17
Stochastic process. *See* Random process
Stochastic simulation 7–14
Storage systems 183–5
Stratified sampling 291–2
Streeter-Phelps equations 258
Structurally defective concrete 50
Structure variable 105
Studentized residuals 131
Student's *t*-distribution 306, 310–11, 313, 360
Student's *t*-tests 274
Sum of squares, partitioning 272–3
Survivor function 57, 72
Swat River, Pakistan 34
System reliability 55–6
Système Hydrologique Européen (SHE) 212

Tacoma Narrows Bridge 5, 247
Taylor expansion 93, 143, 258, 294
Threshold models 210–11
Time covariates 101–2
Time lag 189
Time series 4–5, 188–211, 237
   definition 188
   fractional differences of 209–10
   methods 142
   non-linear 210–11
Traffic accidents 17
Training set 259
Transfer function and spectra 249–51
Transformation of variables 310

Transition matrix 185–6
Transition models 218
Transition probabilities 65
Trips per occupied dwelling unit day 377
Turning Bands Method 225
Two component extreme value distribution (TCEV) 99–100
Two-sided alternative hypothesis 316
Two-way classification fixed effects model 275

Unbiased estimators 305
Uncertainty 1
Uniform distribution 295
Unit Fréchet (EV2) marginal distributions 110
Urban Pollution Management Research Programme 212
Urban traffic networks 7

Variability modelling in time and space 179–235
Variable width kernel 72
Variables
   functions of 304–5
   relationships between 3–4, 117–62
   transformation of 310
Variance 18, 23, 27, 95
   of errors 121–3
Variance components 39–42
Variance-covariance matrix. *See* Covariance matrix
Variogram 217–19, 226
Variogram models 218
Vector variables 31–2
Volterra kernels 258
Volterra series 258
von Mises distribution 31–2

Wakeby distribution 89, 90, 102–3, 337–8
Washington Bridge 72
Washlands 3
Water resources 4, 7
Wave heights 384
Wave loading on offshore structures 260–2
Wave tanks 2
Wavelets 262–3
   analysis 353–7
   transforms 353–7
Weakest link hypothesis 63
Weibull distribution 25, 58–63, 65, 202
Weibull plot 59, 60
Weighted mean 291–2
White noise 8
Wind tunnels 2
Window width 72
Windspeeds 26

Yule–Walker equations 199

Zero flows 101